International Conference on Theoretical Physics: Dubna-Nano 2012

Journal of Physics: Conference Series Volume 393

Moscow, Russian Federation
9-14 July 2012

Editors:

Vladimir Osipov
Yury Shukrinov
Valentin Nesterenko

ISBN: 978-1-62276-857-8
ISSN: 1742-6588

Printed from e-media with permission by:

Curran Associates, Inc.
57 Morehouse Lane
Red Hook, NY 12571

Some format issues inherent in the e-media version may also appear in this print version.

Printed by Curran Associates, Inc. (2013)

For permission requests, please contact the Institute of Physics
at the address below.

Institute of Physics
Dirac House, Temple Back
Bristol BS1 6BE UK

Phone: 44 1 17 929 7481
Fax: 44 1 17 920 0979

techtracking@iop.org

Additional copies of this publication are available from:

Curran Associates, Inc.
57 Morehouse Lane
Red Hook, NY 12571 USA
Phone: 845-758-0400
Fax: 845-758-2634
Email: curran@proceedings.com
Web: www.proceedings.com

TABLE OF CONTENTS

ATOMIC CLUSTERS, QUANTUM DOTS AND QUANTUM TRANSPORT

JOSEPHSON JUNCTIONS, SUPERCONDUCTING STRUCTURES, TOPOLOGICAL INSULATORS AND SUPERCONDUCTORS

GRAPHENE AND OTHER CARBON NANOSTRUCTURES

MODELLING AND APPLICATIONS

PREFACE

The International Conference 'Dubna-Nano2012' was held on 9–14 July 2012 at the Bogoliubov Laboratory of Theoretical Physics, Joint Institute for Nuclear Research, Dubna, Moscow region, Russia. The conference was the third one in the series started in 2008.

'Dubna-Nano2012' provided an opportunity for presentations and discussions about theoretical and experimental advances in the rapidly growing area of nanophysics. The multidisciplinary character of the conference allowed an effective exchange of ideas between different areas of nanophysics. The following topics were covered: graphene and other carbon nanostructures, topological insulators, quantum transport, quantum dots, atomic clusters, Josephson junctions and applications of nanosystems.

About 100 scientists from 22 countries participated in the conference. The program included 38 oral talks and 39 posters. This volume contains 35 contributions.

We would like to express our gratitude to all participants for their presentations and discussions. We are deeply indebted to the members of the International Advisory Committee Professors K S Novoselov, T Ando, T Chakraborty, J Fabian, V M Galitski, F Guinea, M Z Hasan, P Hawrylak, K Kadowaki, R Kleiner, T Koyama, Yu I Latyshev, Yu E Lozovik, M Machida, B K Nikolic, N F Pedersen, P-G. Reinhard, J M Rost and A Ya Vul.

Financial support from BLTP JINR, Russian Foundation for Basic Research, Heisenberg-Landau Program and Bogoliubov-Infeld Program was of a great importance. Further information about 'Dubna-Nano2012' is available on the homepage http://theor.jinr.ru/~nano12.

Vladimir Osipov, Valentin Nesterenko and Yury Shukrinov
Editors

Peer review statement

All papers published in this volume of *Journal of Physics: Conference Series* have been peer reviewed through processes administered by the proceedings Editors. Reviews were conducted by expert referees to the professional and scientific standards expected of a proceedings journal published by IOP Publishing.

Dubna-Nano 2012 IOP Publishing
Journal of Physics: Conference Series **393** (2012) 012001 doi:10.1088/1742-6596/393/1/012001

Laser excitation of clusters: observables from electron emission

P Wopperer[1,2], P M Dinh[1], P G Reinhard[2] and E Suraud[1]

[1] Laboratoire de Physique Théorique, Université P. Sabatier, 118 Rte de Narbonne, F-31062 Toulouse cedex, France

[2] Institut für Theoretische Physik II, Universität Erlangen-Nürnberg, Staudstr. 7, D-91058 Erlangen, Germany

E-mail: paul-gerhard.reinhard@physik.uni-erlangen.de

Abstract. We give a brief review of the theoretical description of photo-electron spectra (PES) and photo-angular distributions (PAD) and discuss a few selected, typical results. The description is based on time-dependent density-functional theory at the level of the local-density approximation augmented by a self-interaction correction which is crucial for a quantitative assessment of emission processes. Coordinate-space grids are used together with absorbing boundary conditions. We discuss the basic features and trends of PES and PAD for two typical test cases, the clusters Na_8 and C_{60}.

1. Introduction

Photo-induced reactions are key tools to explore the properties of electronic systems as atoms, molecules, and clusters. The basic step are optical absorption measurements which provide information on both, structure and dynamics of clusters, for an overview see [1, 2, 3, 4]. More information can be gathered by measuring the properties of electrons emitted through an excitation by electromagnetic fields, particularly by short laser pulses. A first useful indicator is the mere net ionization yield. More specifically, photo-electron spectroscopy (PES) which measures the distribution of the kinetic energy of emitted electrons is a tool of choice as it delivers information, e.g., on the energies of occupied single electron states, for early applications see [8, 9]. A further, more involved step is to analyze the photo-electron angular distribution (PAD). Such measurements came up more recently in clusters and are usually combining PAD with PES, then providing double differential cross-sections (energy- and angular-resolved). First measurements dealt with cluster anions : W_N^-, $N = 4 - 11$ [10, 11], Hg_N^-, $N = 3 - 20$ [12], and C_N^-, $N = 10 - 22$ [13, 14, 15]. More recent results were published for the C_{60} cluster [16] and medium sized Na_N^- clusters [17, 18]. The latter results nicely demonstrate a dependence of the photo-emission on the electronic ground state wave functions.

The theoretical description of PAD has a long history in atomic physics [19, 20, 21]. PAD have also been investigated theoretically in some molecular systems [24, 25, 26]. Such early investigations of PAD, and also those for PES on atoms and molecules [22], employ (multi-photon) perturbation theory [23]. These perturbative methods require a good knowledge of the continuum states for the outgoing electrons and become extremely inconvenient for systems with low or without any symmetry as it is typically the case for clusters. Particularly for clusters, time-dependent density-functional theory at the level of the local-density approximation (TDLDA)

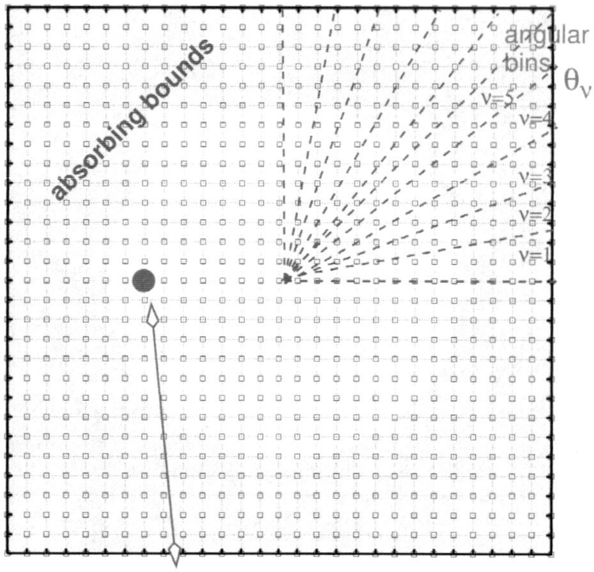

Figure 1. Schematic view of the computation of observables from electron emission. Basis is a Cartesian coordinate space grid. The area where the absorbing boundary conditions are active is indicated by yellow shading. The photo-angular distribution (PAD) of emitted electrons is collected in angular bins as indicated. The photo-electron spectra (PES), i.e. the kinetic energy spectra of emerging electrons, are deduced from collecting phase information at measuring points, here represented by one $\mathbf{r}_{\text{meas}}$, placed short before the absorbing bounds.

propagated directly in the time domain has proven to be a robust, reliable, and efficient tool, see e.g. [28, 29, 27]. This approach gives a convenient framework for analyzing electronic emission properties when augmented by a self-interaction correction (SIC) [30] and when using absorbing boundary conditions [31, 32]. Reliable absorption allows a comparatively inexpensive recording of PES during TDLDA propagation [33]. With developing a careful book-keeping of the directions of the absorbed (=emitted) electrons, one can develop a direct TDLDA-SIC description of PAD [37]. Angular distributions from free Na clusters with fixed orientation have been studied extensively in [37] and from Na clusters deposited on MgO(001) or Ar(001) substrates in [38]. A fixed cluster orientation is naturally given when depositing the cluster on a substrate. But free clusters usually come along as an isotropic ensemble of cluster orientations. Recently, we have developed efficient schemes for orientation averaging of PAD from TDLDA [39, 40, 41]. This contribution presents and discusses typical results on PES and PAD for small Na clusters and for C_{60}. The aim is to review briefly the most important features and trends.

The paper is outlined as follows: Section 2 summarizes the formal framework, discussing in particular the evaluation of PES and PAD. In section 3, we discuss basic properties and trends on the example of Na clusters. Section 4 shows first results for the C_{60} cluster.

2. Formal framework

Cluster dynamics is described with standard coordinate-space techniques as outlined in detail in [32, 46]. Basis of the description is time-dependent density functional theory at the level of the time-dependent local-density approximation (TDLDA) using the exchange-correlation functional of [47] augmented with an averaged self-interaction correction (SIC) and absorbing boundary conditions. The coupling to the ions is mediated by soft local pseudo-potentials for Na [48] and non-local pseudo-potentials for C [49]. Alternatively, we also consider a soft jellium model for Na_8. The ionic background is frozen. This is a legitimate approximation as we consider short laser pulses and direct electron emission which proceeds rather fast. Laser excitation is described by an external classical laser field with a finite temporal envelope $\propto \sin(\pi t/T_{\text{pulse}})^2$.

Observables from electron emission will be explained in somewhat more detail. The description of electron emission requires absorbing boundary conditions. These are indicated in figure 1 which provides a schematic view of gridding and associated computation of observables. The absorption is performed after each time step as follows: First, we perform a standard time step from t to $t+\delta t$ $\phi(\mathbf{r},t) \longrightarrow \tilde{\phi}(\mathbf{r})$, and then we apply a mask function to the preliminary wavefunction $\tilde{\phi}$ to obtain $\phi(\mathbf{r},t+\delta t) = \mathcal{M}(\mathbf{r})\tilde{\phi}(\mathbf{r})$ which removes gradually any flow towards the bounds. We use here a spherically symmetric mask profile

$$\mathcal{M} = \cos\left(\frac{|\mathbf{r}| - R_{\text{in}}}{R_{\text{out}} - R_{\text{in}}}\frac{\pi}{2}\right)^{1/8} \tag{1}$$

which is active in an absorbing margin $R_{\text{in}} < |\mathbf{r}| < R_{\text{out}}$. The spherical profile is needed to minimize gridding artifacts when computing angular distributions [37]. A detailed discussion of absorbing bounds and optimal choice of absorption profiles is found in [50].

The absorbing bounds reduce gradually the norm of the wavefunctions. This mimics the dynamical ionization of the system. The net ionization, i.e the number of escaping electrons can be computed simply from the single-particle norms as $N_{\text{esc}} = N_{\text{el}}(t=0) - \sum_\alpha(\phi_\alpha|\phi_\alpha)$. The PES are evaluated for each state separately from the phase oscillations of the single-electron wavefunction at an analyzing point $\mathbf{r}_{\text{meas}}$ near the absorbing bounds, see figure 1 and ref. [33]. The protocol $\phi_\alpha(\mathbf{r}_{\text{meas}},t)$ is Fourier transformed to $\tilde{\phi}_\alpha(\mathbf{r}_{\text{meas}},\omega)$ and the spectrum $\mathcal{P}(\omega) \propto |\tilde{\phi}_\alpha(\mathbf{r}_{\text{meas}},\omega)|^2$ is translated to a kinetic-energy spectrum of emitted electrons by identifying $\hbar\omega = \varepsilon_{\text{kin}}$. This simple identification is possible by virtue of the absorbing bounds which manage to leave only outgoing waves in its vicinity (thus establishing a unique relation between frequency and momentum). The case of high laser intensity (not employed here) requires slight modifications of this recipe [51].

The PAD are evaluated in angular bins as indicated in figure 1. We collect all probability which was removed by the absorption step (1) and accumulate it in the bin to which $\mathbf{r}$ belongs. That is done for each wavefunction separately. At the end, we dispose of the PAD for emission from each state and, of course, of the total angular distribution as well.

This procedure yields the PAD for a fixed cluster orientation. Free clusters usually come along in isotropic ensembles of orientations (each direction has equal weight). This requires orientation averaging as outlined in [39, 40]. There are two strategies for that. The first is a brute force numerical sampling of orientations. It was found that one can achieve converged results with about 18-32 appropriately chosen reference orientations. This method, although cumbersome, is applicable in all dynamical regimes. The second strategy employs formally perturbation theory to develop a compact evaluation requiring only six reference orientations. This method is applicable only to one-photon emission. Whatever averaging scheme is used, we end up with a PAD which depends only on one angle, the emission angle ϑ relative to the laser polarization axis. The PAD for emission from s.p. state i can thus be expanded as

$$\frac{d\overline{\sigma}_i}{d\Omega} \propto 1 + \beta_2^{(i)}P_2(\cos\vartheta) + \beta_4^{(i)}P_4(\cos\vartheta) + \beta_6^{(i)}P_6(\cos\vartheta) + ... \tag{2}$$

where P_l is a Legendre polynomial of order l. The parameter of the leading term, $\beta_2^{(i)}$, is called the anisotropy parameter. The same expansion applies to the total PAD $d\overline{\sigma}/d\Omega$ with the expansion coefficients β_2..., i.e. without index (i). One-photon processes have only the P_2 term and vanishing β_{2n} for $n \geq 2$. In that case, the values of the anisotropy $\beta_2^{(i)}$ and β_2 range between -1 and 2: $\beta_2^{(i)} = 2$ corresponds to a $\cos^2\vartheta$-shape which is aligned with the laser polarization thus having its maxima at $\vartheta = 0$ and 180°, $\beta_2^{(i)} = -1$ yields a $\sin^2\vartheta$-shape perpendicular to the laser polarization with a maximum at $\vartheta = 90^o$, and $\beta_2^{(i)} = 0$ stands for a strictly isotropic distribution. The expansion becomes richer if more photons are involved. One can have finite values for any β_{2n} and $\beta_2 > 2$ becomes possible.

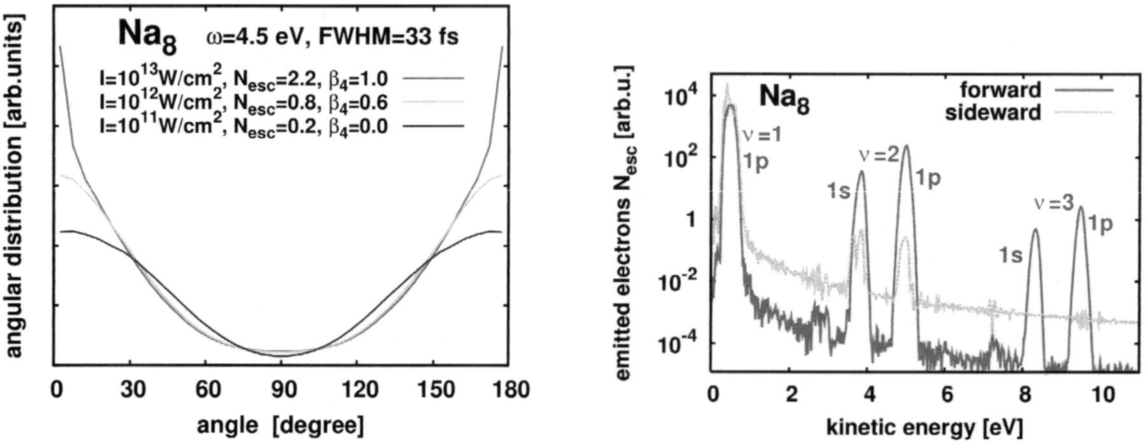

Figure 2. Left panel: PAD for Na_8 with spherical jellium background excited with a laser of frequency $\omega_{\rm las} = 4.5$ eV, pulse length of FWHM=33 fs, and varying intensity as indicated. The anisotropy parameter was approximately $\beta_2 = 2$ in all cases. The next higher anisotropy parameter β_4 varies as indicated. Right panel: PES for the same test case and weak laser intensity $I = 10^{10}\mathrm{W/cm}^2$. Compared are the PES along laser polarization axis (red) with PES orthogonal to it (green). The peaks correspond to different photon numbers ν and states (1s or 1p) as indicated.

3. Basic properties

The basic features of PES for clusters have been collected in a series of papers [33, 34, 35, 36] and of PAD in [37, 38, 39, 40, 41]. We summarize them briefly here. PES for one-photon processes provide a direct map of the ground state s.p. spectra as has been used since long in experiments [8, 9]. This requires a laser frequency which is larger than the span of the s.p. spectra to be measured and sufficiently gentle excitation to avoid perturbation of the ground state under investigation. Larger intensities blur the signal from PES. Lower frequencies lead quickly into the regime of multi-photon ionization (MPI) where one can spot the multiple copies of PES peaks each one standing for a certain photon number ν. The envelope of this multiple peak structure shows exponential decrease with kinetic energy $E_{\rm kin}$ of the outgoing electrons. Using larger intensities in the MPI regime again smoothes the signal to pure exponential decrease which can easily be misinterpreted as thermal emission [36].

PAD show also different pattern in the different dynamical regimes. We illustrate that here for Na_8 as typical test case. The left panel of figure 2 shows the PAD for varying laser intensity. (An orientation averaging is not necessary for jellium background.) The lowest intensity (blue line) shows nearly perfectly the typical pattern of emission in the one-photon regime where the anisotropy β_2 alone characterizes the profile and where all higher β_{2n} vanish. In fact, the $\beta_2 \approx 2$ demonstrates a case of maximum alignment with the laser polarization. Increasing intensity leads to increasing forward/backward emission and enhances the contribution from β_4 and higher moments. It is interesting to relate that to the total ionization yield. A value of $N_{\rm esc} \approx 1$ is found to define the transition region between the frequency dominated perturbative regime below and the field dominated non-linear regime above [33].

The right panel of figure 2 shows the photo-electron spectrum (PES) for the same test case. The calculated s.p. energies are: $\epsilon_{1s} = -5.44\,\mathrm{eV}$ (twofold degenerated) $\epsilon_{1p} = -4.08\,\mathrm{eV}$ (sixfold degenerated). The observed peaks match perfectly with the well-known relation $\epsilon_{\rm out} = \epsilon_i + \nu\omega_{\rm las}$ for a ν-photon process (here $\nu = 1$, 2, and 3). The laser frequency is just too small to move all

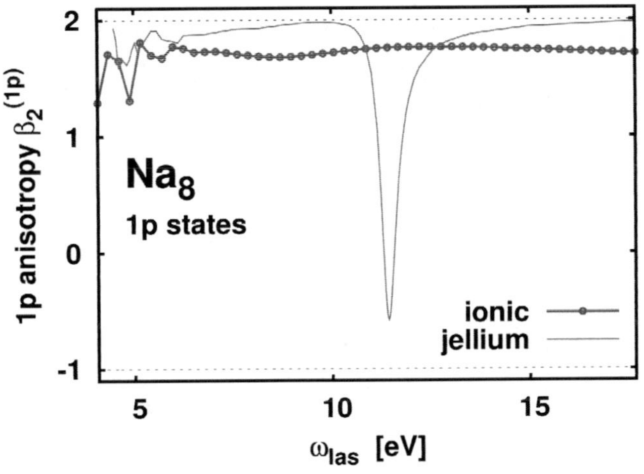

Figure 3. Angular anisotropy $\beta_2^{(1p)}$ for emission from the 1p states of Na_8 with full ionic (red) or spherical jellium (green) background as function of laser frequency $\omega_{\rm las}$. The laser pulse length was FWHM= 60 fs. The intensity was varied with frequency to stay in a low emission regime with $N_{\rm esc} = 10^{-4} - 10^{-1}$. The limiting values $\beta_2^{(1p)} = -1$ and 2 are indicated by faint dashed lines.

occupied states above threshold. Thus we see from the one-photon process only the map of the $1p$ state. On the other hand, the choice of parameters allows to display also the multi-photon processes. The figure shows the PES parallel and perpendicular to the laser polarization. Yields in both directions are about the same for $\nu = 1$. However the yield in perpendicular direction drops dramatically with increasing photon number. This demonstrates, here in the case of fixed laser field strength, that emission becomes the more forward directed the more photons are involved.

The impact of the ionic background structure is demonstrated in figure 3 for the case of emission from the $1p$ states in Na_8. We compare results for the anisotropy parameter $\beta_2^{(1p)}$ as function of laser frequency $\omega_{\rm las}$ produced using detailed ionic background with those produced using soft spherical jellium. For the jellium model we used a Wigner-Seitz radius $r_s = 3.65\,a_0$ and surface thickness $\sigma = 1\,a_0$). Both models deliver about the same IP (jellium: $\mathrm{IP}_{\rm jell} = 4.22\,\mathrm{eV}$; ionic background: $\mathrm{IP}_{\rm ion} = 4.28\,\mathrm{eV}$) in accordance with the empirical value. The general pattern from both models differ very much. For the jellium models, the asymmetry parameter $\beta_2^{(1p)}$ explores almost the full range of possible values between -1 and 2. It stays mostly near the maximum possible $\beta_2 = 2$ and dives occasionally down to nearly -1 in a steep and sharp minimum. It is to be noted that these deep minima are correlated with minima in the total yield $N_{\rm esc}(\omega_{\rm las})$ as function of $\omega_{\rm las}$. The ionic background, in contrary, produces a very smooth trend which stays around $\beta_2^{(1p)}$ =1.7 over the whole range of frequencies shown. Particularly, the sharp minima with negative values observed in the jellium models disappear as soon as ionic structure is taken into account. For most frequencies, however, the jellium models overestimate the asymmetry parameter which was already found earlier [40]. The ionic perturbations rescatter the electronic waves and so produce a sizable isotropic component which reduces the forward/backward orientation in cases where $\beta_2^{(1p)}$ was nearly 2 for jellium. The steep minima in the jellium case are due to particular interference effects which are only possible in the highly symmetric situation of a spherical jellium. (The same interference effect is also responsible for the corresponding minima in $N_{\rm esc}(\omega_{\rm las})$.) The ionic structure destroys these interferences very efficiently such that the trends with frequency become very smooth. It is to be remarked that already a small deformation of the jellium background suffices to suppress the steep minima in $N_{\rm esc}(\omega_{\rm las})$ which, again, is due to reduced wavefunction interferences.

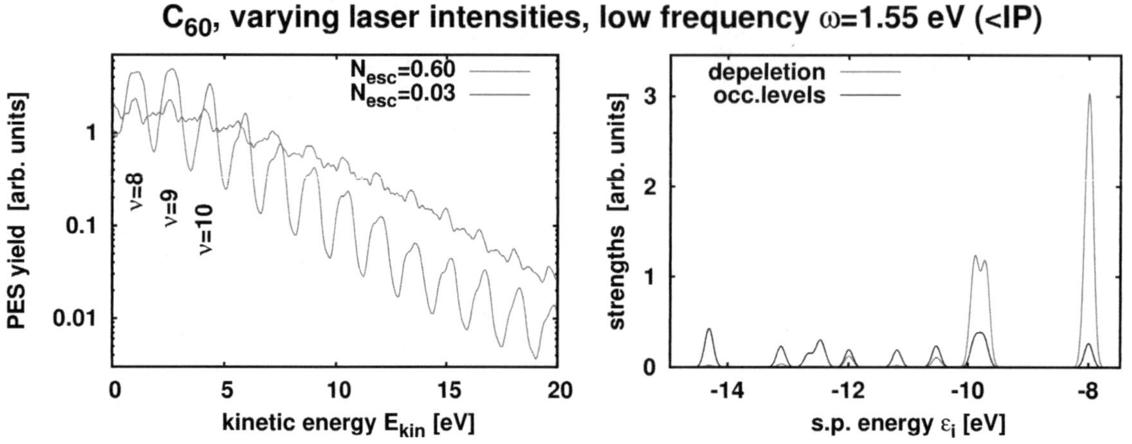

Figure 4. Left panel: PES for C_{60} produced with a laser pulse of low frequency $\omega_{\text{las}} = 1.55$ eV, pulse length FWHM=33 fs and intensity near of order of 10^{10}W/cm^2 fine tuned to deliver the indicated ionization N_{esc}. The various peaks correspond to different photon numbers ν. Some of them are indicated. Right panel: The distribution of occupied s.p. states (black) and the actual depletion of these states (green).

4. C_{60}

The C_{60} cluster consists of 60 C atoms arranged in a highly symmetric configuration of what is called a fullerene [42]. The electrons from the 1s shell of C are considered as being core electrons kept frozen by means of a pseudo-potential during TDLDA calculations. The four electrons from the 1p shell of the C atom constitute the 240 dynamically active electrons of C_{60}. The typically strong binding of C material yields a high ionization potential (IP) of 7.58 eV [43]. The binding is partially metallic with the consequence that one observes a pronounced plasmon resonance [44] which resides around 20 eV, i.e. far above the continuum threshold and which is thus strongly fragmented due to the high density of 1ph states. The spectral span of the occupied valence states is also rather large ranging over about 20 eV from -27 eV up to the IP [16]. To resolve the s.p. spectrum by a one-photon process, one takes a photon with frequency just above the lowest bound state. We have tested that with TDLDA and find that a laser pulse with $\omega_{\text{las}} = 33$ eV reproduces the s.p. spectrum nicely including the occupancy weight. A particularity of C_{60} is that the HOMO is well separated from the rest of the occupied states. This gives the next bound state, called HOMO-1, particular importance. It lies 1.8 eV below the HOMO.

The left panel of figure 4 shows PES produced with a low frequency laser with $\omega_{\text{las}} = 1.55$ eV being far below the ionization threshold. It requires at least $\nu = 6$ photons to move an electron from the HOMO to the continuum. The intense excitation (red line) shows a basically smooth decrease. One is not surprised as one would expect that the broad span of s.p. energies overlays completely the narrow energy difference of 1.55 eV between the multi-photon steps. However, the lower intensity (green line) resolves very clearly various photon orders as indicated. One wonders how that could come about. The explanation is found in the right panel of the figure. Here we show the upper part of the s.p. spectrum, once as it is (black line) and once weighted with the actual depletion, that is $N_{\text{esc},i}$ taken state-wise, from the laser pulse (green line) with the same low intensity as used for the green line in the left panel. It is obvious that the low-frequency pulse removes electrons preferably from the Fermi surface and that deeper lying states are not touched at all. It is a general feature of all molecular systems that low frequency pulses ionize preferably from the Fermi surface while high frequency pulses remove all states with about equal weight [45]. For the particular case here, it has the consequence that the PES emission pattern become rather transparent allowing to resolve the multi-photon steps seen in the left panel (green line). The present case is particularly favorable because the energy difference between HOMO and HOMO-1 is close to the laser frequency such that signals from these two emitting states overlap strongly in the PES (to the extend that the $\nu + 1$ photon signal from the HOMO-1 approximately coincides with the ν photon signal from the HOMO). The feature that the signal is blurred for higher intensity (red line) stems from the gradual Coulomb down-shift of the spectra during ionization, the same effect as was already figured out for the much simpler Na clusters [33].

5. Conclusions

We have reviewed the theoretical description of photo-electron spectra (PES) and photo-angular distributions (PAD). The numerical evaluation relies on a Cartesian coordinate-space grid with absorbing boundary conditions cutting a dynamically active sphere into the 3D grid. PAD are simply computed from gathering the lost particles in angular segments. PES are obtained from recording the time evolution of each wavefunction at a few selected measuring points near the absorbing bounds. Both techniques add only a small overhead to the underlying TDLDA propagation thus allowing large scale studies. Free clusters require orientation averaging for the PAD which amounts to run several TDLDA calculations in parallel, six different computation in the one-photon regime up to 32 for violent excitations.

We have discussed the basic features of electron emission for two typical test cases, the clusters Na_8 and C_{60}. The outcome for PES and PAD depends very much on the dynamical conditions set mainly by laser frequency and intensity. The regime of one-photon processes (low intensity, frequency larger than the lowest bound state) is particularly clean thus allowing a detailed analysis of the single particle states in the cluster ground state. Higher intensities and/or lower frequencies lead into the regime of multi-photon ionization. At the lower end of intensities, one stays still in the frequency dominated regime where one can count photons and sees them acting, e.g., in the repeated peaks of PES or in the increasing forward/backward emission with increasing photon number. Higher intensities lead to the field dominated regime where the electrons are soaked off strongly forward/backward along the laser polarization rather independent of laser frequency and cluster structure. The PES become smooth exponential curves with little information content. The results for C_{60} are particularly interesting as several measurements on this large cluster are to be expected in near future. Accordingly, more calculations for C_{60} are underway.

Acknowledgment: This work was supported by the Deutsche Forschungsgemeinschaft, project RE 322/10-1, Institut Universitaire de France, and Agence Nationale de la Recherche.

References

[1] de Heer W A (1993) *Rev. Mod. Phys.* **65** 611
[2] Brack M (1993) *Rev. Mod. Phys.* **65** 677
[3] Kreibig U and Vollmer M 1993 *Optical Properties of Metal Clusters* (Springer Series in Materials Science 25) (Berlin: Springer)
[4] Haberland H (ed.) 1994 *Clusters of Atoms and Molecules 1- Theory, Experiment, and Clusters of Atoms* (Springer Series in Chemical Physics 52) (Berlin: Springer)
[5] Richardson H H, Hickman Z N, Govorov A O, Thomas A C, Zhang W and Kordesch M E 2006 *Nano Lett.* **6** 783
[6] Khlebtsov B, Zharov V, Melnikov A, Tuchin V, Khlebtsov N 2006 *Nanotechnology* **17** 5167
[7] Haberland H (ed.) 1994 *Clusters of Atoms and Molecules Experiment, and Clusters of Atoms 2- Solvation and Chemistry of Free Clusters, and Embedded, Supported and Compressed Clusters* (Springer Series in Chemical Physics 56) (Berlin: Springer)

[8] McHugh K M , Eaton J G, Lee G H , Sarkas H W , Kidder L H , Snodgrass J T , Manaa M R and Bowen K H 1989 *J. Chem. Phys.* **91** 3792
[9] Lichtenberger D L, Nebesny K W, Ray C D, Huffman D R and Lamb L D 1991 *Chem. Phys. Lett.* **176** 203
[10] Pinaré J, Baguenard B, Bordas C and Broyer M 1999 *Eur. Phys. J.* D **9** 21
[11] Baguenard B, Pinaré J C, Bordas C and Broyer M 2001 *Phys. Rev.* A **63** 023204
[12] Verlet J R R, Bragg A E, Kammrath A, Cheshnovsky O C and Neumark D M 2004 *J. Chem. Phys.* **121** 10015
[13] Wills J, Pagliarulo F, Baguenard B, Lépine F and Bordas C 2004 *Chem. Phys. Lett.* **390** 145
[14] Bordas C, Baguenard B, Climen B, Lépine F, Pagliarulo F, Lebeault M A and Wills J 2005 *Int. J. Mod. Phys.* B **19** 2899
[15] Pagliarulo F, Climen B, Baguenard B, Lépine F, Lebeault M A, Ollagnier A, Wills J and Bordas C 2006 *Int. J. Mass Spectr.* **252** 100
[16] Kjellberg M, Johansson O, Jonsson F, Bulgakov A V, Bordas C, Campbell E E B, and Hansen K 2010 *Phys. Rev.* A **81** 023202
[17] Kostko O, Bartels C, Schwöbel J, Hock C and v Issendorff B 2007 *J. Phys. : Conf. Ser.* **88** 012034
[18] Bartels C, Hock C, Huwer J, Kuhnen R, Schwöbel J and von Issendorff B 2009 *Science* **323** 132
[19] Bethe H 1933 *Handbuch der Physik, Band 24/1* (Berlin: Springer)
[20] Cooper J and Zare R N 1968 *J.Chem. Phys.* **48** 942
[21] Cooper J and Zare R N 1968 *J.Chem. Phys.* **49** 4252
[22] Ghosh P 1983 *Introduction to photoelectron spectroscopy* (New York: Wiley)
[23] Faisal F H M 1987 *Theory of multiphoton Processes* (New York: Plenum)
[24] Seideman T 1997 *J. Chem. Phys.* **107** 7859
[25] Seideman T 2002 *Ann. Rev. Phys. Chem.* **53** 41
[26] Suzuki Y and Seideman T 2005 *J. Chem. Phys.* **122** 234302
[27] Calvayrac F, Reinhard P G and Suraud E 1997 *Ann. Phys. (NY)* **255** 125
[28] Calvayrac F, Reinhard P G and Suraud E 1995 *Phys. Rev.* B **52** R17056
[29] Yabana K and Bertsch G F 1996 *Phys. Rev.* B **54** 4484
[30] Perdew J P and Zunger A 1981 *Phys. Rev.* B **23** 5048
[31] Ullrich C A 2000 *J. Mol. Struct. (THEOCHEM)* **501-502** 315
[32] Calvayrac F, Reinhard P G, Suraud E and Ullrich C A 2000 *Phys. Rep.* **337** 493
[33] Pohl A, Reinhard P G, and Suraud E 2000 *Phys. Rev. Lett.* **84** 5090
[34] Pohl A, Reinhard P G and Suraud E 2001 *J. Phys.* B **34** 4969
[35] Pohl A, Reinhard P G and Suraud E 2003 *Phys. Rev.* A **68** 053202
[36] Pohl A, Reinhard P G and Suraud E 2004 *J. Phys.* B **37** 3301
[37] Pohl A, Reinhard P G and Suraud E 2004 *Phys. Rev.* A **70** 023202
[38] Bär M, Dinh P M, Moskaleva L V, Reinhard P G, Rösch N and Suraud E 2009 *Phys. Rev* B **80** 195404
[39] Wopperer P, Faber B, Dinh P M, Reinhard P G and Suraud E 2010 *Phys. Lett.* A **375** 39
[40] Wopperer P, Faber B, Dinh P M, Reinhard P G and Suraud E 2010 *Phys. Rev.* A **82** 063416
[41] Wopperer P, Dinh P M, Suraud E and Reinhard P G 2012 *Phys. Rev.* A **85** 015402
[42] David W I F, Ibberson R M, Matthewman J C, Prassides K, Dennis T J S, Hare J P, Kroto H W, Taylor R and Walton D R M 1991 *Nature* **353** 147
[43] de Vries J, Steger H, Kamke B, Menzel C, Weisser B, Kamke W and Hertel I V 1992 *Chem. Phys. Lett.* **188** 159
[44] Hertel I V, Steger H, de Vries J, Weisser B, Menzel C, Kamke B and Kamke W 1992 *Phys. Rev. Lett.* **68** 784
[45] Vidal S, Wang Z P, Dinh P M, Reinhard P G and Suraud E 2010 *J. Phys.* B **43** 165102
[46] Reinhard P G and Suraud E 2003 *Introduction to cluster dynamics* (New York:Wiley)
[47] Perdew J P and Wang Y 1992 *Phys. Rev.* B **45** 13244
[48] Kümmel S, Brack M and Reinhard P G 1999 *Euro. Phys. J.* D **9** 149
[49] Goedecker S, Teter M and Hutter J 1996 *Phys. Rev.* B **54** 1703
[50] Reinhard P G, Stevenson P D, Almehed D, Maruhn J A and Strayer M R 2006 *Phys. Rev.* E **73** 036709
[51] Dinh P M, Romaniello P, Reinhard P G and Suraud E 2012 *subm. Phys. Rev.* A; arXiv:1206.3918.

Journal of Physics: Conference Series **393** (2012) 012002 doi:10.1088/1742-6596/393/1/012002

Variations on a QM/MM approach to cluster dynamics

P M Dinh[1], D Berger[2], B Faber[2], P-G Reinhard[2], E Suraud[1]

[1] Laboratoire de Physique Théorique, Université P. Sabatier, 118 Rte de Narbonne, F-31062 Toulouse cedex, France

[2] Institut für Theoretische Physik II, Universität Erlangen-Nürnberg, Staudstr. 7, D-91058 Erlangen, Germany

E-mail: suraud@irsamc.ups-tlse.fr

Abstract.
We discuss two extensions of a hierarchical QM/MM approach developed to describe clusters embedded into a polarizable environment. The first extension consists in considering the presence of a supporting metal. More specifically we consider the structure and dynamics of Na clusters on a surface built from Ar layers grown on a metal support. We study the effect of the Ar substrate and of the metal support. We find a faint binding on the surface and the effect of the dielectric response of the metal (DRM) turns out to be negligible. Deposit of small Na clusters is crucially influenced by the mechanical hardness of the metal support and the number of Ar layers, while the DRM makes little effect. The second extension consists in using the QM/MM approach to study color centers at an MgO surface. The particular efficiency of the QM/MM approach allows to consider large system's sizes and thus test the impact of the (necessarily) finite (surface) sample. We find that our results compare favorably with ab-initio approaches as well as with available experimental results.

1. Introduction

Cluster physics has been a lively research field over the last decades, see e.g. [1, 2]. The case of clusters in contact with a substrate is more involved and covers an extremely rich range of physical and chemical situations. Indeed many experiments can better be performed for non-isolated systems but new phenomena also appear which could possibly lead to practical applications as, for example, chromophore effects to induce well localized heating of a substrate [3, 4]. The field is still much a fast evolving one, see e.g. the collections [5, 6, 7, 8, 9].

Substrates carry an enormous amount of degrees of freedom and a fully detailed quantum mechanical description of such systems is thus beyond today's possibilities. A good way out is attained by considering Quantum Mechanical/Molecular Mechanical (QM/MM) approaches, which rely on a hierarchy of importance from the active zone of interest down to the farther outskirts of the system. These QM/MM approaches couple a quantum description of the active piece to a classical description of the outer environment. These methods were first developed in bio-chemistry [10] and surface chemistry [11]. The QM/MM approach is the method of choice for active electron systems (molecules, clusters, color centers...) in contact with an environment as there is a clear distinction between the electrons and the environment. The applications we plan to consider, however, require a careful modeling of the dynamical response of the substrate.

We have thus built an extension of the usual QM/MM methods which includes the dynamical polarizability of the substrate [12, 13, 14]. This is perfectly adapted for a cluster deposited on an insulating surface. But we will now apply this approach to 2 slightly different scenarios.

We will concentrate first on the particular combination of a Na_N cluster deposited on a few Ar layers themselves deposited on a metal support, following typical experimental setups [15, 16, 17]. This indeed allows gentle deposition of a metal cluster on a hard surface [15, 18]. But the structural and spectral properties of such systems are also very interesting [17], as they combine an optically active metal cluster with an inert and insulating raregas material, itself coupled to bulk metal. We will thus mostly focus here on structural aspects and explore how deposition dynamics proceeds in the presence of the hard underlying metal.

As a second extension of our approach we shall explore the case of color centers by considering an oxygen vacancy near MgO(001) surfaces, as color centers in solids have attracted attention since long [20]. Studies in MgO have also a long history, see the book [21] for a summary of the early stages. Reliable theoretical descriptions of color centers in MgO have come up only recently, because of the high computational effort required [22, 23, 24, 25, 26, 27, 28, 29].

These calculations employ an embedded cluster approach which imposes to treat only a very small fraction of the system, so that there is a demand for simplifications. Again one can simplify the picture by realizing that the most active pieces are the electrons in the vacancy while the MgO surroundings are only slightly perturbed, which again justifies the use a hierarchical approach such as our QM/MM approach. The idea is now to apply our modeling to the electrons confined in a color center of MgO. This is not as detailed as the full ab-initio methods, but it allows to proceed to larger samples in and at the MgO substrate, and also to account for a pertinent description of the long-range tail of the electron cloud and of the optical response. In that respect our simplified QM/MM approach is to say complementary to ab-initio ones.

We concentrate here on a so called F-center, produced from an O vacancy in the MgO substrate. For that we replace an O-site in MgO by 2 electrons, thus ensuring proper neutrality. This configuration produces large optical activity stemming from the two vacancy electrons. We shall only consider such an F-center on the surface, and not in the bulk, although the principle is the same in the latter case.

2. An extended QM/MM approach for dynamics of molecules and clusters in contact with an environment

2.1. Basics of the QM/MM approach

The QM/MM model for metal clusters in contact with a rare gas substrate has been developed in [30, 13]. A detailed description of the model is provided, e.g., in the review [31]. The such developed QM/MM approach provides a good basis for exploring the case of clusters deposited on a rare gas layer itself topping a metal surface. Similarly the case of color centers (obtained by creating a vacancy) can be attacked by omitting the cluster itself and simply replacing it by electrons. These are the two extensions we explore in this paper. Figure 1 provides a schematic picture of both setups: O vacancy and Ar covered metal environments. More details can be found in [32, 33]. We give here only a short summary.

The (cluster or vacancy) electrons are described by the time-dependent local-density approximation (TDLDA) augmented by an averaged self-interaction correction [34]. The cluster ions, when present, are propagated by means of molecular dynamics (MD), while the electron-ion interaction is described by soft, local pseudo-potentials [35]. This forms together the TDLDA-MD approach, an approach which was well validated for linear and non-linear dynamics of free metal clusters [36, 37]. The substrate atoms (Ar or Mg, O) are described as purely classical particles, but (for Ar and O) dressed by a classical time dependent dipole, which in particular allows to account for charge effects [31]. These time dependent dipoles may play a very important role for example in deposition scenarios. They constitute a specific feature of

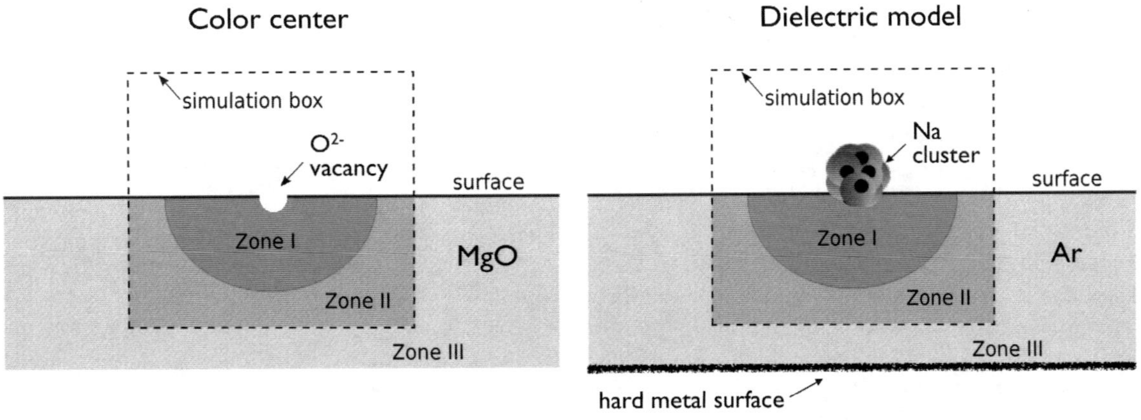

Figure 1. Schematic plot of the different areas in the QM/MM model. The "simulation box" corresponds to the coordinate-space grid for wavefunctions and fields.
Left panel: the "defect" is the vacancy filled by two quantum-mechanically active electrons (neutral F center). The three other zones contain Mg^{2+} and O^{2-} ions described at different levels: "Zone I" maintains full mobility of all ions while in "Zone II" ionic positions are frozen while the electron cloud of O^{2-} is still responsive . "Zone III" consists of all ions (and their electron cloud) frozen at crystalline positions and accounts for the long-range Madelung potential.
Right panel: the separation in 3 zones is similar in Ar as in MgO but is now complemented by the hard metal surface. The vacancy is replaced by a small Na cluster.

the present QM/MM approach.

The numerical solution proceeds with standard methods [37], on the basis of space-grid techniques to solve the (time-dependent) Kohn-Sham equations for electrons. Electronic wave functions are propagated by a time-splitting method [38] and ions by velocity Verlet. We use a Cartesian 3D grid for computing ground state configurations and optical response. For the more demanding dynamics of deposition, we use an axially averaged approximation for the electronic mean-field [39, 40].

2.2. The case of color centers: O vacancy on MgO surface

The substrate is composed of two species: Mg^{2+} cations and O^{2-} anions. While the cations are electrically inert and can thus be treated as charged point particles the anions are easily polarizable, an aspect which is accounted for by the associated dynamical dipole. The combined system is sorted in several stages of decreasing activity as sketched in figure 1. The (quantum) vacancy electrons are represented in a large numerical simulation box. The Mg^{2+} and O^{2-} ions of the substrate are arranged in the fcc crystalline order of bulk MgO. All dynamical degrees-of-freedom for Mg^{2+} and O^{2-}, are taken into account in an active cell of the MgO(001) surface region near the vacancy (zone I in the figure). This fully active area is continued by an outer region of bulk MgO material with fixed MgO and O ionic centers but active O dipoles (zone II in the figure), thus accounting for long range polarization effects. Zone III contains Mg and O ions frozen at crystalline configuration in order to obtain the correct Madelung potential for the whole system [41, 42].

2.3. Extension to account for a dielectric: Na cluster deposited on Ar and metal

Each Ar atom is associated with two classical degrees of freedom: center-of-mass and electrical dipole moment, allowing polarizability effects via polarization potentials [41]. A local pseudo-potential is added for the electron-Ar short-range repulsion following [43]. The Na-Ar Van-der-Waals interaction is absorbed here for simplicity in the Ar-Na coupling parameters, while the atom-atom interactions are described by a standard Lennard-Jones potential. The Ar-Na^+ subsystem is treated by means of known effective potentials [44].

The underlying metal support interacts with the Na-Ar system in two ways. It delivers a strong coupling with the lowest layer of Ar atoms directly in contact with the metal. This is modeled by fixing the atomic position in the bottom layer: in parallel direction to their crystal equilibrium positions and in perpendicular direction at a fixed distance of 3.5 a_0 from the metal surface. Furthermore the dielectric response of the metal (DRM) establishes a long range interaction with all charges in the Na-Ar system. This is described in standard textbook manner by the method of image charges [45], assuming the metal to be an ideal conductor.

3. Some illustrative examples

3.1. Structure of a sodium clusters deposited on Ar coated metal

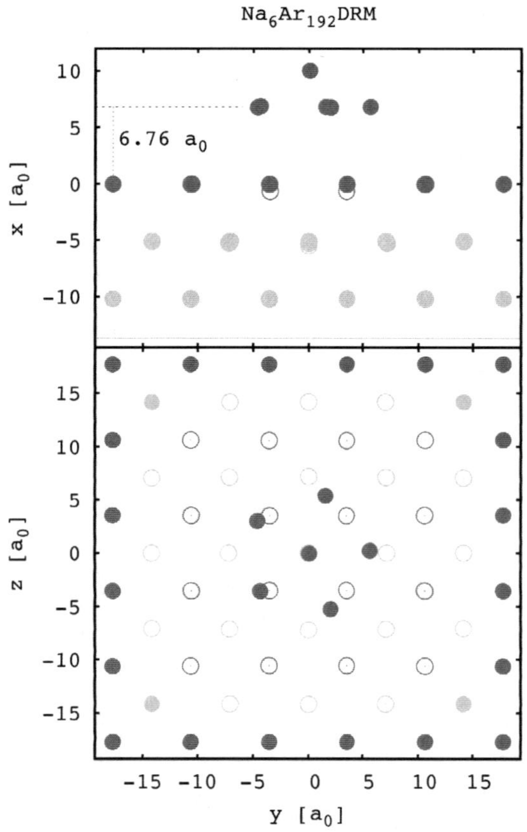

Figure 2. Structure of Na_6Ar_{192} with DRM. The surface is in the y- and z plane. Na ions are represented by red filled circles and Ar atoms in green. Dark green indicates the uppermost layer and open circles indicate the mobile atoms.

As a first example we consider in figure 2 a small sodium cluster (Na_6) deposited on an Ar surface (actually three layers of Ar, here represented by an Ar_{192} system) topping a metal surface. The lower panel visualizes the areas of mobile atoms (open symbols) surrounded by fixed atoms (full symbols). It turns out that the deposited structure (which integrates the DRM) very little differs from the case in which the DRM is omitted. Switching on/off the DRM changes the distance between the cluster and the top Ar layer by 0.15 a_0, i.e. by 2.5%, which is negligible as the binding pocket has a rather flat bottom (whence some uncertainty in the binding distance at negligible energetic cost). Still, the robustness of the ionic/atomic configuration is a bit surprising as the Ar dipoles are modified substantially by the DRM. The equilibrium distance thus emerges from a subtle balance of counter-acting effects (dipole attraction, core repulsion, tail of the electronic wavefunction, metal polarization), the net result being here in stabilizing the positions. The resulting small impact of the metal on the structure has also been noted in other explored cases. This a posteriori validates approaches avoiding the account of the metal, at least what concerns the actual structures. It is then interesting to check how the metal affects the deposition process itself, a situation which corresponds to many experimental cases.

3.2. Dynamics of deposit

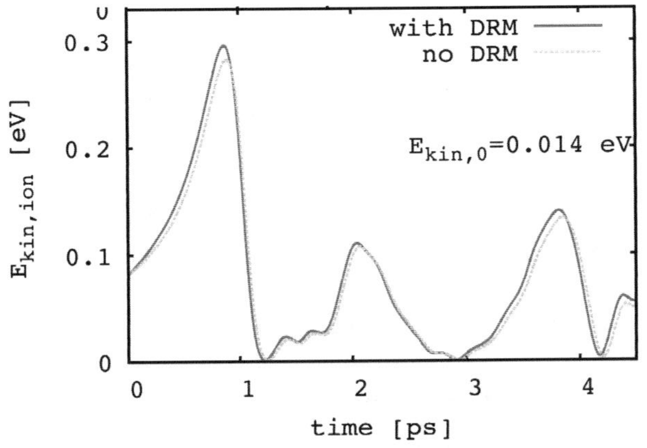

Figure 3. Time evolution if ionic kinetic energy for the deposition dynamics of Na_6 on Ar_{192} at $E_{kin} = 0.014 eV$ impact kinetic energy, with DRM ("with DRM') and without ("no DRM").

We now want to explore the dynamics of deposition for Na_6 impinging on Ar layers grown on a metal support. We handle here the electronic potentials approximately, in axial symmetry (cylindrical averaged pseudo-potential method [39]), which is acceptable as discussed in [14, 46]. We initially place a free Na_N cluster at a large distance from a free surface (typically 17–20 a_0) with the cluster symmetry axis perpendicular to the surface. Then the cluster is boosted instantaneously with the velocity along the symmetry axis towards the substrate. There are different possible choices for the position of the collision axis on the substrate but this plays a minor role for the Na_6 we consider here, for which the cluster symmetry does not match the surface geometry anyway. The collision axis is on a top site, i.e. it hits an Ar atom in the uppermost layer.

In figure 3 we illustrate this deposition process. The most important effect of the metal support, namely the mechanical stability, fixes the lowest Ar layer, thus stops the propagation of absorbed energy. This leads to cluster reflection in the more energetic cases. The DRM had not been considered here but one can check that the differences are extremely small, except maybe for very slow initialization. This holds particularly for the impact phase where the metal dipole attraction produces slightly more initial acceleration. The differences for the dynamics after impact remain smaller and not significant at the level of our modeling. The dominant role of the mechanical hardness makes the number of Ar layers a key ingredient for energy absorption. We have systematically varied the number of Ar layers and found that energy absorption levels off from six layers on, so that an Ar surface with at least six layers behaves like a bulk Ar surface. Energy absorption from the impinging cluster is then almost complete.

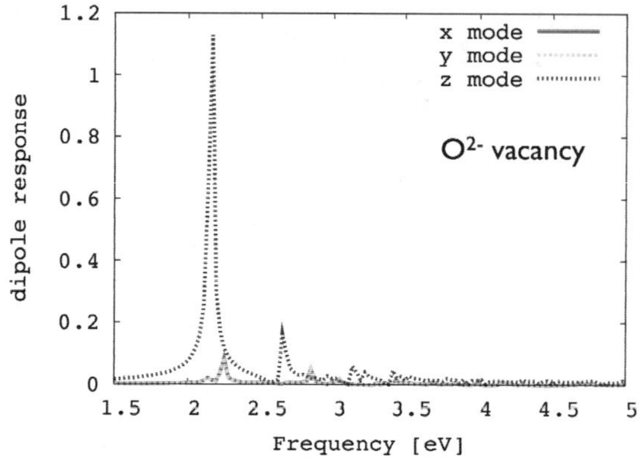

Figure 4. Dipole excitation spectrum for the modes in x, y-, and z-direction for an O vacancy on the surface. The surface is here parallel to the $x - y$ plane.

3.3. Optical response

A usually well accessible and crucial observable in electronic systems is the optical absorption. For systems without symmetry it is most efficient to compute these spectra directly from real time electronic evolution and subsequent spectral analysis. The present QM/MM model is well suited for such a purpose as electrons are propagated with time-dependent LDA (with self-interaction correction) [47, 31]. The optical absorption spectra were thus computed in the usual manner by starting from the stationary ground state, and by initializing dynamics by an instantaneous small dipole boost. The emerging time-evolution of the dipole moment $D(t)$ is finally Fourier transformed to the frequency domain [48, 37] and the imaginary part of the Fourier transform $\Im\{\tilde{D}(\omega)\}$ becomes the optical absorption strength.

Fig. 4 shows an example of dipole strength in all three principal directions for an O vacancy in MgO. There is a strong resonance mode at about 2.2 eV for the z mode which has some resemblance with the collective Mie resonance seen in metal clusters [1]. This resonance is determined by global features and thus insensitive to details. The spots of smaller strength above that dominant peak depend very sensitively on the active area chosen around the vacancy. The modes in x- and y-direction have much less strength as compared to z direction, which is plausible as the motion parallel to the surface is much more restricted. They furthermore exhibit no prominent resonance peak. The numerous small structures can be attributed to single-electron excitations.

It is interesting to compare our results to available experimental ones and to the ones from ab-initio methods. The experimental analysis is difficult, as one cannot easily associate observed peaks to the configurations as calculated here. Still, one can claim that a peak for a surface site has been observed at 2.05–2.3 eV [21]. The ab-initio calculations of [25] yield the "same" peak for optical response of the surface vacancy at 3.24 eV, namely about 30% higher than our results and than the experimental values. Still the matter is debated, especially what concerns the interpretation of experimental results and one should thus not overconclude here. The situation is thus mixed and experiments might very well reside in between the two different approaches.

3.4. Conclusion

We have discussed some extensions of a hierarchical QM/MM approach developed to study the dynamics of clusters in contact with a polarizable environment. In this approach the optically active electrons are treated fully quantum-mechanically at the level of time-dependent local-density approximation (TDLDA) while cluster ions, atoms and dipole moments are described by classical molecular mechanics.

We have first considered the case of a compound consisting out of a Na cluster in contact with the surface of an Ar systems of several layers resting on a metal support. The latter provides a hard mechanical boundary for the motion of the lowest Ar layer and its dielectric response is also accounted for. We have seen that equilibrium structure of deposited Na_6 is fairly insensitive to the dielectric response of the metal support (DRM). We also studied the dynamics of cluster deposition. It turns out that the dominant effect of the metal support is its mechanical hardness which limits energy absorption from the impinging Na cluster by the Ar layers. Above typically 6 Ar layers one reaches full energy absorption and thus cluster stopping.

As a second extension we have presented a study of color centers in MgO, again on the basis of our hierarchical approach, which allows to treat rather large samples as compared to standard ab-initio approaches, still at an acceptable degree of accuracy. A direct comparison between both theoretical approaches and experiments is possible for the optical response of the surface F center. It shows a good qualitative agreement between theory and experiments, similar between the QM/MM and ab-initio methods, what concerns the differences with experiments.

Acknowledgment: This work was supported by the Deutsche Forschungsgemeinschaft, project RE 322/10-1. Institut Universitaire de France and Agence Nationale de la Recherche are also acknowledged for support.

References

[1] Kreibig U, Vollmer M, *Optical Properties of Metal Clusters*, Vol. 25 (Springer Series in Materials Science, 1993)
[2] Haberland H, ed., *Clusters of Atoms and Molecules 1- Theory, Experiment, and Clusters of Atoms*, Vol. 52 (Springer Series in Chemical Physics, Berlin, 1994)
[3] Richardson H H et al, Nano Lett. **6** (2006)
[4] Khlebtsov B et al Nanotechnology **17**, 5167 (2006)
[5] Haberland H, ed., *Clusters of Atoms and Molecules 2- Solvation and Chemistry of Free Clusters, and Embedded, Supported and Compressed Clusters*, Vol. 56 (Springer Series in Chemical Physics, Berlin, 1994)
[6] Meiwes-Broer K-H, ed., *Metal clusters at surfaces* (Springer, Berlin, 2000)
[7] Meiwes-Broer K-H, ed., *Clusters at Surfaces: Electronic Properties and Magnetism*, Vol. 82 (2006), Applied Phys. A, special issue
[8] Meiwes-Broer K-H, Berndt R, eds., *Atomic Clustes at Surfaces and in Thin Films*, Vol. 45 (2007), Eur. Phys. J. D, topical issue
[9] Meiwes-Broer K-H, ed., *Clusters at surfaces*, Phys. Stat. Sol. B **247**, No. 5 (2010)
[10] Warshel A, Levitt M, J. Mol. Biol. **103**, 227 (1976)
[11] Matveev A V et al, Chem. Phys. Lett. **299**, 603 (1999)
[12] Gervais B et al, Phys. Rev. A **71**, 015201 (2005)
[13] Fehrer F et al, Ann. Phys. (Leipzig) **14**, 411 (2005)
[14] Fehrer F et al, Appl. Phys. A **82**, 151 (2006)
[15] Bromann K et al, Surf. Sci **377**, 1051 (1997)
[16] Schaub R et al, Phys. Rev. Lett. **86**, 3590 (2001)
[17] Irawan T et al, Appl. Phys. A **82**, 81 (2005)
[18] Schlipper R et al, Appl. Phys. A **72**, 255 (2001)
[19] Gresh N et al, THEOCHEM **458**, 27 (1999)
[20] Schulman J H, Compton W D, *Color Centers in Solids* (Pergamon, Oxford, 1962)
[21] Henrich V E, Cox P, *The surface science of metal oxides* (Cambridge University Press, Cambridge, 1994)
[22] Pandey R, Vail J M, J.Phys.: Condens. Matter **1**, 2801 (1989)
[23] Ferrari A-M, Pacchoni G, J.Phys.Chem. **99**, 17010 (1995)
[24] Scorza E, Birkenheuer U, Pisani C, J. Chem. Phys. **107**, 22 (1997)
[25] Francesc Illas G-P, Journal of Chemical Physics **108**, 18 (1998)
[26] Sousa C, Pacchioni G, Illas F, Surface Science **429**, 217 (1999)
[27] Sushko P V, Shluger A L, Catlow C R A, Surface Science **450**, 153 (2000)
[28] Suhsko P V, Gavartin J L , Shluger A L, J. Phys. Chem. B (2002)
[29] Tobita M, Ho S, J.Chem.Theory Comput. **4**, 1057 (2008)
[30] Gervais B et al, J. Chem. Phys. **121**, 8466 (2004)

[31] Dinh P M et al, Phys. Rep. **485**, 43 (2009)
[32] Faber B et al, Euro. Phys. J. D **66**, 194 (2012)
[33] Berger D et al, Euro. Phys. J. D **66**, 164 (2012)
[34] Legrand C et al, J. Phys. B **35**, 1115 (2002)
[35] Kümmel S et al, Euro. Phys. J. D **9**, 149 (1999)
[36] Reinhard P-G, Suraud E, *Introduction to Cluster Dynamics* (Wiley, New York, 2003)
[37] Calvayrac F et al, Phys. Rep. **337**, 493 (2000)
[38] Feit M D et al, J. Comp. Phys. **47**, 412 (1982)
[39] Montag B, Reinhard P-G, Phys. Lett. A **193**, 380 (1994)
[40] Montag B, Reinhard P-G, Z. f. Physik D **33**, 265 (1995)
[41] Dick B G, Overhauser A W, Phys. Rev. **112**, 90 (1958)
[42] Nasluzov A et al,J. Chem. Phys. **115**, 17 (2001)
[43] Duplàa F, Spiegelmann F, J. Chem. Phys. **105**, 1492 (1996)
[44] Ahmadi G R et al, Chem. Phys. **199**, 33 (1995)
[45] Jackson J D, *Classical Electrodynamics* (Wiley, New York, 1962)
[46] Dinh P M et al, Euro. Phys. J. D **45**, 415 (2007)
[47] Bär M et al, Euro. Phys. J. D **45**, 507 (2007)
[48] Calvayrac C et al, Ann. Phys. (NY) **255**, 125 (1997)
[49] Blum V et al, J. Comp. Phys **100**, 364 (1992)

Effects of self-assembled gold nanoparticles on $YBa_2Cu_3O_{7-\delta}$ thin films and devices

P. Michalowski[1], **C. Katzer**[1], **F. Schmidl**[1] **and P. Seidel**[1]

[1] Institut für Festkörperphysik, Friedrich-Schiller-Universität Jena, D-07743 Jena, Germany

Peter.Michalowski@uni-jena.de

Abstract. In our work we prepared $YBa_2Cu_3O_{7-\delta}$ (YBCO) thin films with self-assembled gold nanoparticles on $SrTiO_3$ (STO) substrates. We carried out different experiments to determine the effects on the crystallographic properties of the YBCO matrix as well as of the gold nanoparticles. Furthermore, we investigated how the particles influence the superconducting parameters of the film, e.g. the critical temperature T_C and the critical current density j_C. To ascertain j_C we employed magneto-optical Faraday microscopy. In addition, the YBCO film was deposited and structured on STO bi-crystal substrates, thus producing grain boundary Josephson junctions. We studied those junctions with respect to the normal state resistance R_N, and the dependence of the critical current I_C on the temperature T as well as on the magnetic flux Φ. Finally, we prepared direct current superconducting quantum interference device (dc-SQUID) gradiometers and embedded gold nanoparticles at well-defined areas such as only the antenna or the SQUID region. We measured the flux noise in a shielded environment using an ac-bias reversal technique and compared it with that of sensors without gold nanoparticles. Thus, we demonstrate a new preparation method and an innovative application of gold nanoparticles.

1. Introduction

YBCO is one of the best known high temperature superconductors and is used for superconducting thin film devices for more than twenty years. There are different approaches to the realization of dc-SQUIDs and dc-SQUID gradiometers [1, 2]. Noise characteristics have been studied thoroughly and have shown that a small critical current I_C is favorable to minimize the contribution of current noise [3]. On the other hand, especially for applications in planar gradiometer structures, a high critical current density in the antenna areas crossing the grain boundary is necessary [4]. Therefore an independent optimization of the critical current density of YBCO thin films across the grain boundary in the region of the Josephson junctions and in the region of the antennas is very helpful. The scaling law between the critical current density of the Josephson junction j^{JJ}_C and I_CR_N is $I_CR_N \sim (j^{JJ}_C)^n$ with $n \approx 0.5$ [5]. In former investigations on grain boundary engineering a reduction of I_C correlated with a decrease of the I_CR_N product [6]. This is unfavorable, because a high I_CR_N product enables high transfer functions for SQUID applications [7]. In our current work we are following another approach: adjusting the properties of bi-crystal grain boundary Josephson junctions by using gold nanocrystals [8].

During the last years there have been various attempts to manipulate the properties of superconducting thin films by using nanoparticles. We are going to give a short summary of different approaches by other groups.

The superconducting properties of the YBCO films like critical temperature T_C, critical current density j_C and critical magnetic fields H_{C1} and H_{C2} of the type II superconductor depend on the oxygen stoichiometry as well as on the crystal structure and the possible structure of defects. Therefore, it is possible to influence a lot of experimental parameters through the embedding of nanoparticles in the YBCO matrix. There is the chance to reduce the interface resistance [9]. Artificial pinning centers (APCs) can be created by adding non-superconducting material to the target [10], when depositing the film by pulsed laser deposition (PLD). Another approach to generate APCs is the preparation of the substrate with arrays of non-superconducting particles before the deposition of the superconducting film [11]. Mikheenko et al. report an improvement of flux pinning leading to an increase of j_C [12, 13]. The difference to our work is that the particles have either been on the surface of the film [11] or at the interface to the substrate [14]. This leads to 1D or 2D APCs with a strong angular dependence of the $j_C(B)$ characteristics, where B is the magnetic induction [15]. A combination of different APCs and the arising improvement of j_C are presented by Ercolano et al. [16].

In contrast to the widely-used method to embed nanoparticles in thin films we do not use non-superconducting additives to the target. We prepared very thin gold films which cluster during the deposition of the subsequent film of YBCO leading to self-organized crystalline nanoparticles Besides, the technique developed by us enables the systematic modification and optimization of selected regions of the sample.

2. Sample Preparation

We prepared our samples on polished STO (001) single-crystal substrates with sizes of 5 x 10 mm² and 10 x 10 mm². Using PLD we started with an initial gold layer with a thickness of 1 to 5 nm. This process was executed with a KrF excimer laser ($\lambda = 248$ nm, $\tau = 25$ ns) at a repetition rate of 10 Hz and a laser fluence of 1.3 J/cm² under high vacuum conditions ($p_0 < 5{\cdot}10^{-4}$ Pa). With these parameters we achieved homogeneous Au-growth at a rate of approximately 4 nm/min. To get comparable YBCO films with and without gold nanocrystals on one substrate we used a shadow mask to cover one half of the substrate. After the deposition of the gold layer the shadow mask was removed and the substrate was heated to 780°C leading to a clustering of the gold. Using the KrF excimer laser again, a 150 nm thick YBCO layer was deposited in an oxygen atmosphere of $p_{O2} = 50$ Pa. To achieve optimum growth conditions, the repetition rate was set to 5 Hz and the laser fluence to 2.2 J/cm2. Under these conditions the growth rate was 20 nm/min. After deposition the substrate was cooled down to room temperature at a rate of 50 K/min in an atmosphere of pure oxygen ($p_{O2} = 8 \cdot 10^4$ Pa). During this process, gold nanoparticles evolved from the initial layer (see figure 1). If electrical characterization was planned, a 20 nm thick gold layer was deposited by dc-sputtering using a shadow mask to provide a bonding area. Finally, a possible structuring of the sample was performed using photolithography and ion beam etching. For the latter we attached the samples to a liquid-nitrogen-cooled sample holder. Applying an acceleration voltage of 500 V and an ion current density of 1 mA/cm² an etching rate of 20 to 30 nm/min, depending on the angle between ion beam and sample, was achieved. In this process, the superconducting properties of the YBCO films were not affected.

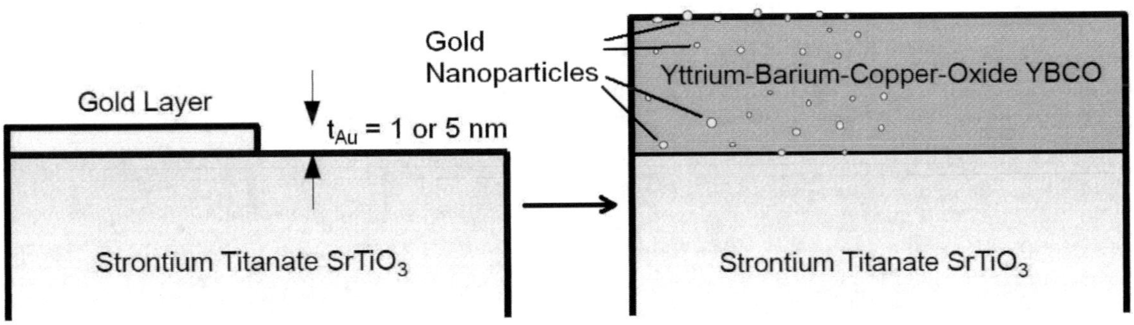

Figure 1. Scheme of the formation of the gold nanoparticles.

For the preparation of Josephson junctions, dc-SQUIDs or dc-SQUID gradiometers we used 10 x 10 mm² STO bi-crystal substrates with grain boundary angles of 24° as well as of 36.8° and carried out the steps mentioned above. In addition to the application of a shadow mask we went for another approach to produce regions with and without gold nanocrystals: To restrict the nanocrystals only to well defined positions, e.g. only the SQUID or only the antenna structures, we used spin coating to apply the photoresist AZ5214E from Microchemicals to the cleaned substrates. After the exposure of our samples to ultraviolet light and development of the structure we went on with the deposition of the gold layer and subsequently stripped off the remaining photoresist.

3. YBCO films with gold nanoparticles

TEM images show a distribution of the gold nanoparticles in the entire YBCO film, e.g. they can be found at the interface to the STO substrate, in the film and at the surface [17, 18]. The particles are found to be crystalline and orientated in (111) or (110) direction with regard to the substrate normal. Also in the case of changes in the preparation conditions, e.g. substrate temperature, we found that the gold nanocrystals influence the growth of the YBCO film. SEM images show a reduced roughness by less voids and no spiral growth as well as fewer outgrowths in the vicinity of the gold nanocrystals (figure 2) [17]. Furthermore, a-axis growth (needle-like structures seen in the left side of figure 2) can be completely prevented in the modified regions. Enhanced pinning of flux vortices leads to a significant increase of the critical current density j_C, which is shown on patterned samples via transport measurements as well as for unpatterned regions via magneto-optical Faraday microscopy [18]. Another positive effect of the gold nanoparticles is the slightly increased critical temperature T_C.

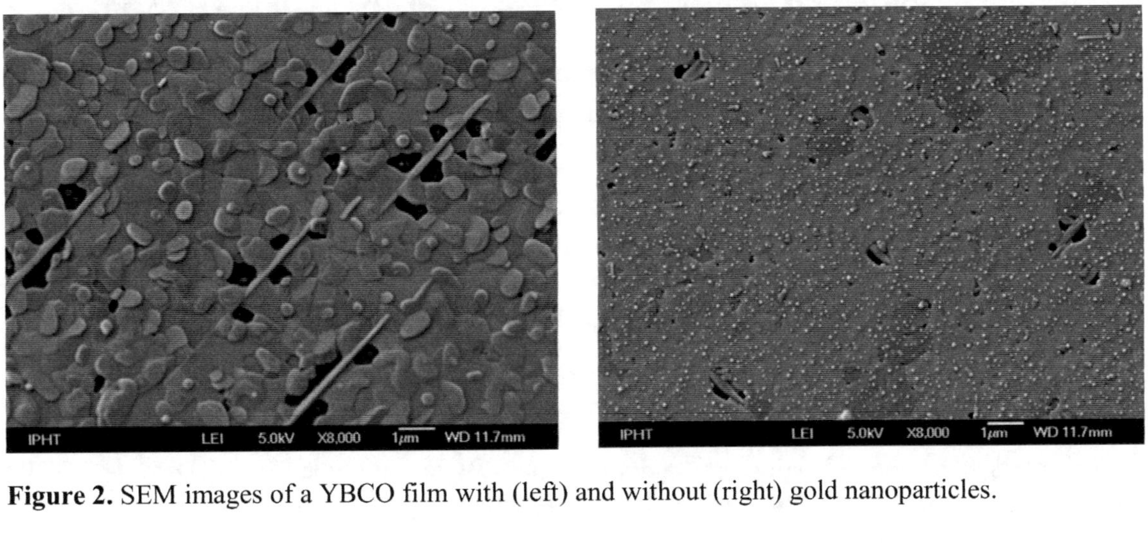

Figure 2. SEM images of a YBCO film with (left) and without (right) gold nanoparticles.

4. Grain boundary Josephson junctions, dc-SQUIDs and dc-SQUID gradiometers

In addition to YBCO films we also prepared bi-crystal grain boundary Josephson junctions and characterized them. We found, that the gold nanoparticles cause a modification of the superconducting parameters of the YBCO film crossing the grain boundary. Measuring the dependence of the critical Josephson current I_C on the magnetic induction B, I_C was more suppressed for Josephson junctions with gold nanocrystals [8]. Thus, we assume a smaller amount of a possible leakage current compared to junctions without gold nanoparticles. Unlike for YBCO films without grain boundary I_C is reduced for gold-modified junctions. We measured the temperature dependence of I_C for the SQUIDs S2 (initial gold layer thickness of 3 nm) and S9 (5 nm). As figure 3 shows I_C of S9 is smaller than that of S2 over the whole temperature range.

Both curves exhibit the typical behavior of grain boundary Josephson junctions in high-T_C superconductors [19]. This indicates that the gold nanocrystals do not change the transport mechanism of Cooper pairs, whereas the value of the critical current is significantly different depending on the thickness of the initial gold layer. Since this decrease in I_C is correlated with an increase of the normal state resistance R_N the I_CR_N product stays nearly constant [8]. Figure 4 illustrates this fact by showing data of Josephson junctions on two different samples, each containing junctions with and without gold nanoparticles. While I_C is reduced by a factor of 1.5 to 2 the I_CR_N product does not change. As the transfer function of a dc-SQUID depends on the I_CR_N product, the decrease of the critical current without a reduction of the I_CR_N product should be in favor of a high transfer function. This opens up new interesting perspectives for the improvement of sensors based on bi-crystal grain boundary Josephson junctions.

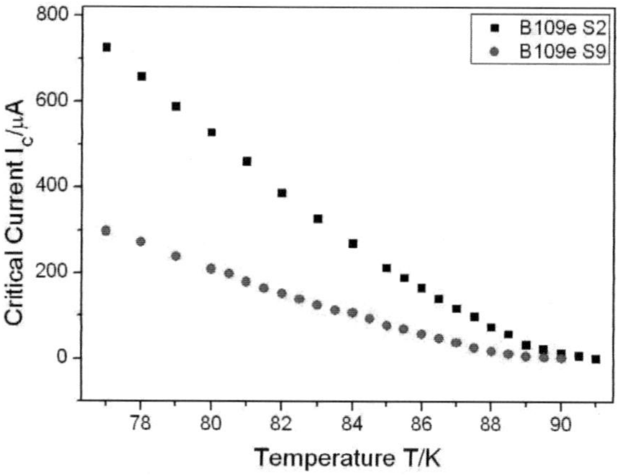

Figure 3. Critical current I_C of SQUIDs B109e S2 (3 nm initial gold layer) and S9 (5 nm initial gold layer) depending on temperature T.

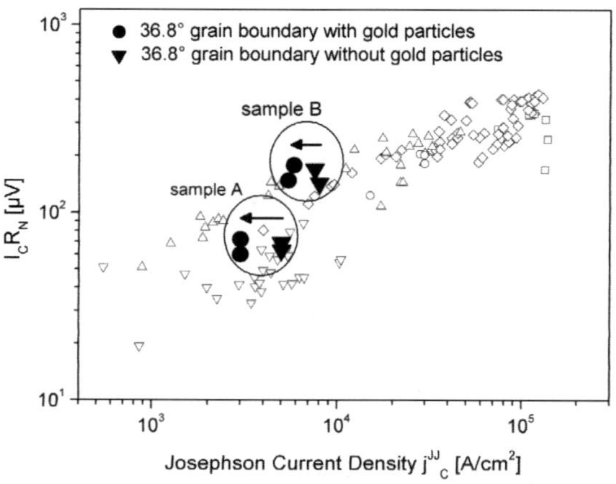

Figure 4. I_CR_N (j^{JJ}_C) characteristics of bi-crystal grain boundary Josephson junctions with and without gold nanoparticles. The open symbols denote data from past measurements and illustrate the general dependence of the two parameters.

To study the modulation depth ΔV or transfer function we also exposed the SQUIDs to the varying magnetic field of a small coil and measured the dependence of the voltage V on the coil current I_{coil} for different bias currents I_B. We found modulations for SQUIDs with different initial gold thicknesses. An example can be seen in figure 5.

The influence of LC-resonances in the I-V-curves of the dc-SQUID is clearly visible as different maxima and minima occur. To compare ΔV of different SQUIDs we carried out modulation measurements for different temperatures and ascertained ΔV for the respective I_B. Data for SQUID B109e S9 is given in figure 6 and shows a higher maximum ΔV for the SQUID with the larger initial gold layer. For all investigated SQUIDs ΔV passes through several maxima and rises with smaller temperatures T. An interesting point is the shift of the maximum of ΔV from the first peak at lower I_B to the second peak with lower temperatures, a behavior we observed for the first time. In contrast to former investigations of dc-SQUID structures with the same geometry and film thickness we found that from around 70 K downwards the second maximum starts to dominate the transfer function.

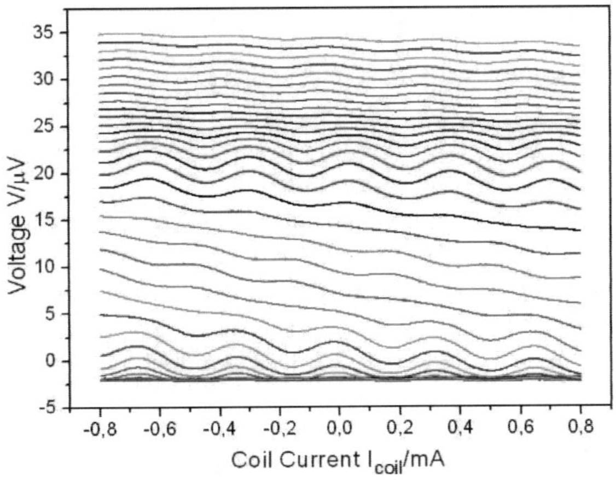

Figure 5. Modulation curves of SQUID B109e S9 for bias currents I_B=190..780 µA at T=69 K.

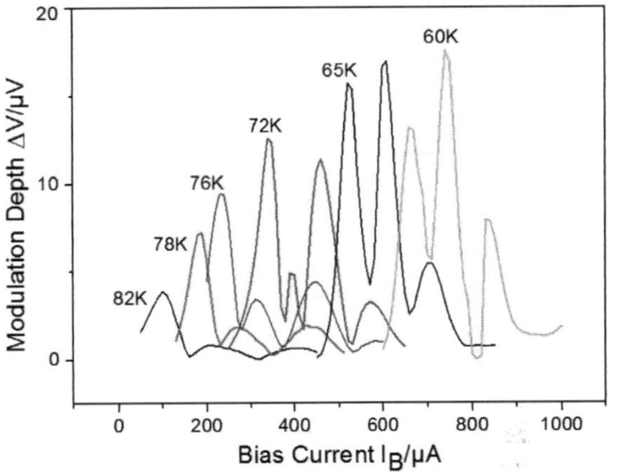

Figure 6. Modulation depth ΔV depending on bias current I_B of B109e S9 measured for different temperatures.

As stated above, additional pinning centers should not only increase the critical current density of the YBCO thin film, but also influence the noise properties of the dc-SQUID gradiometer. First noise measurements indicate improved voltage or flux noise characteristics [18]: We achieved a white noise between 5 $\mu\Phi_0/\sqrt{Hz}$ and 6 $\mu\Phi_0/\sqrt{Hz}$ with our gold-modified gradiometers. This is a very low value compared to conventional SQUID gradiometers based on bi-crystal Josephson junctions, as it is comparable to the best values reported for single dc-SQUIDs without antenna structures. Furthermore, we did also observe a shift of the corner frequency of the 1/f noise from about 10 Hz (as reported for gradiometers used in biomagnetic systems [20]) to a frequency of about 2 Hz. This is a far more interesting point because it strongly influences the sensor performance in the low frequency range e.g. at 1 Hz. Nevertheless, the impact of the thickness of the initial gold layer on the noise properties has not been investigated yet.

5. Summary

In conclusion, we have developed a new application of gold nanoparticles. Self-assembling from an initial gold layer during the growth of the YBCO film, these crystalline particles modify the superconducting properties of the latter in a favorable way. Used in plain YBCO films the nanoparticles lead to an enhancement of the critical current density j_C. At Josephson junctions they decrease the critical Josephson current I_C while increasing the normal state resistance R_N leading to an unchanged I_CR_N product. Using photolithography the nanoparticles can be restricted to certain regions of the sample. Therefore, an adjustment of e.g. the critical current density j_C in determined areas on a single substrate can be achieved. This opens up the possibility to tune the parameters of superconducting devices based on grain boundary Josephson junctions.

Furthermore, it is also possible to expand this technique to other films than YBCO, e.g. STO. In this case a modification of the optical properties could be observed [21].

Acknowledgement
The authors would like to thank J. Albrecht, S. Treiber, C. Stahl, V. Grosse, D. Kuhwald, M Schmidt, U. Hübner and S. Koch for their support. C. Katzer would like to thank the Landesgraduiertenförderung Thüringen for financial support.

References

[1] Peiselt K, Schmidl F, Linzen S, Anton A S, Hübner U and Seidel P 2003 *Supercond. Sci. Tech.* **16** 1408–12
[2] Zakosarenko V, Schmidl F, Schneidewind H, Dörrer L and Seidel P 1994 *Appl. Phys. Lett.* **65**, 779
[3] Il'ichev E, Dörrer L, Schmidl F, Zakosarenko V, Seidel P and Hildebrandt G 1996 *Appl. Phys. Lett.* **68** 708
[4] Steppke A, Becker C, Grosse V, Dörrer L, Schmidl F, Seidel P, Djupmyr M and Albrecht J 2008 *Appl. Phys. Lett.* **92** 122504
[5] Gross R, Chaudhari P, Kawasaki M and Gupta A 1990 *Phys. Rev B, Condensed matter* **42** 10735–37
[6] Wunderlich S, Schmidl F, Dörrer L, Schneidewind H and Seidel P 1999 *IEEE T. Appl. Supercon.* **9** 71
[7] Enpuku K, Shimomura Y and Kisu T 1993 *J. Appl. Phys.* **73** 7929–34
[8] Michalowski P, Schmidt M, Schmidl F, Grosse V, Kuhwald, D, Katzer C, Hübner U and Seidel P 2011 *Phys. Status Solidi RRL* **5** 268.
[9] Caton R, Selim R, Buoncristiani A M and Byvik C E 1990 *J. Appl. Phys.* **67** 7478–82
[10] Pinto R, Apte P R, Rao M S R, Chandra R, D'Souza C P, Pai S P, Gupta L C,Vijayaraghavan R, Gnanasekar K I and Sharon M 1996 *Appl. Phys. Lett.* **68** 1006–8
[11] Lengl G, Ziemann P, Banhart F and Walther P 2003 *Physica C: Superconductivity* **390** 175–84
[12] Mikheenko P, Abell J S, Sarkar A *et al.* 2010 . *J. Phys.: Conference Series* **234** 022022
[13] Mikheenko P, Dang V-S, Kechik M M A, Sarkar A, Paturi P, Huhtinen H, Abell J S and Crisan A 2011 *IEEE T. Appl. Supercon.* **21** 3184–8
[14] Mikheenko P, Tanner J L, Bowen J, Sarkar A, Dang V-S, Abell J S and Crisan A 2010 *Physica C: Superconductivity* **470** 234–6
[15] Matsumoto K and Mele P 2010 *Supercond. Sci. Tech.* **23** 014001
[16] Ercolano G, Bianchetti M, Wimbush S C, Harrington S A, Wang H, Lee J H and MacManus-Driscoll J L 2011 *Supercond. Sci. Tech.* **24** 095012
[17] Grosse V, Engmann, Schmidl F, Undisz A, Rettenmayr M and Seidel P 2010 *Phys. Status Solidi RRL* **4** 97–9
[18] Katzer C, Schmidt M, Michalowski *et al.* 2011. *Europhys. Lett.* **95** 68005
[19] Hilgenkamp H, Mannhart J 2002 *Rev. Mod. Phys.* **74** 485–549
[20] Weidl R, Brabetz S, Klemm F, Dörrer L, Schmidl F and Seidel P 1997 *Cryogenics* **37** 691-3
[21] Christke S, Katzer C, Grosse V, Schmidl F, Schmidl G, Fritzsche W, Petschulat J, Pertsch T and Rettenmayr M 2011 *Optical Materials Express* **1** 890-7

Journal of Physics: Conference Series **393** (2012) 012004 doi:10.1088/1742-6596/393/1/012004

Coherent spin dynamics of nanomolecules and magnetic nanoclusters

V I Yukalov[1] **and E P Yukalova**[2]

[1]Bogolubov Laboratory of Theoretical Physics, Joint Institute for Nuclear Research, Dubna 141980, Russia

[2]Laboratory of Information Technologies, Joint Institute for Nuclear Research, Dubna 141980, Russia

E-mail: yukalov@theor.jinr.ru

Abstract. Spin dynamics of nanomolecules and nanoclusters are analyzed. The nanosizes of these objects make it possible to consider them as single-domain magnets with a large total spin, where the motion of the spins of all atoms, composing a nanocluster, occurs in a coherent way. Another meaning of coherence in spin dynamics is the coherent spin motion of several nanomolecules or nanoclusters. Different approaches for treating spin dynamics are compared and the main mechanisms influencing the spin motion are studied. Spin dynamics of separate magnetic nanomolecules and nanoclusters are investigated, as well as the spin dynamics of the ensembles of these nano-objects.

1. Introduction

Magnetic nanomolecules and nanoclusters enjoy many similar properties because of which the dynamics of their magnetization can be described by the same type of equations. This is why, we consider both these nano-objects together. Of course, there is difference in their structure and parameters which we shall take into account and characterize them by the appropriate models. The detailed description of general physical properties and applications of different magnetic nanoparticles can be found in review articles [1-9]. Here we briefly mention those of the properties and parameters that will be necessary for the following consideration.

It is worth stressing that there exist two types of magnetic nanoparticles. One large class consists of nanoclusters and nanomolecules, whose magnetic moments are formed by electron spins. Another type includes nanomolecules that possess magnetic moments solely due to polarized proton spins. Examples are propanediol $C_3H_8O_2$, butanol C_4H_9OH, and ammonia NH_3. In such nanomolecules, there is no any other magnetic moment except that caused by polarized protons. So, here the proton magnetic moment is not a contribution, but the main object.

The magnetic moment of an atom is composed of electron and proton moments, with the electron magnetic moment $\mu_e = -g_e\mu_B S = \hbar\gamma_e S = -\mu_B$ and the proton magnetic moment $\mu_p = g_p\mu_N S = \hbar\gamma_p S$, where $g_e = 2$ and $g_p = 5.586$ are the electron and proton Landé factors, $\mu_B = |e|\hbar/2m_e$ and $\mu_N = |e|\hbar/2m_p$ are the Bohr and nuclear magnetons, γ_e and γ_p are the electron and proton gyromagnetic ratios. Since the proton mass is larger than that of an electron, $m_p/m_e \sim 10^3$, the proton magnetic moment is essentially smaller, $\mu_p/\mu_e \sim 10^{-3}$. The electron and proton radii are $r_e \sim 10^{-15}$cm and $r_p \sim 10^{-13}$cm, respectively. An atom is called magnetic, when its total magnetic moment is nonzero. The total spin of a magnetic atom can be between $1/2$ and $S \sim 10$, hence, its magnetic moment can be of order $1\mu_B - 10\mu_B$. Atom radii are of order $r_A \sim 10^{-9} - 10^{-8}$cm. Examples of magnetic atoms are Fe (Iron), Co (Cobalt), Ni (Nickel), Gd (Gadolinium), and Cr (Chromium).

Magnetic nanomolecules are composed of many magnetic atoms an, as is clear from their name, are of the nanometer size. An important property of a magnetic nanomolecule is that its total magnetic moment can be treated as being due to an effective total spin. Generally, the molecule spin can be directed either up or down, with an energy barrier between these directions of order $E_A \sim 10 - 100$ K. At high temperatures, above the *blocking temperature* $T_B \sim 1 - 10$ K, a magnetic molecule behaves as a superparamagnetic particle, whose spin randomly oscillates between the up and down positions. While below the blocking temperature the spin is frozen in one of the directions.

Magnetic nanoclusters are also made of magnetic atoms that are assembled together in a random way. This distinguishes them from magnetic molecules, where atoms are strictly connected by chemical bonds. The sizes of nanoclusters can be in the range between 1 nm and 100 nm, containing about $100 - 10^5$ atoms. These values define the *coherence radius* R_{coh}, below which a nanocluster is in a single-domain state and can be treated as a large particle with an effective spin. A cluster, with a size larger than R_{coh}, separates into domains with opposite magnetizations. Similarly to magnetic molecules at low temperature, the magnetic moment of a nanocluster, below the blocking temperature $T_B \sim 10 - 100$ K, is frozen in one of two possible directions. The effective spin of a nanocluster is formed by electron spins and can be as large as $S \sim 100 - 10^5$.

The often considered nanoclusters are made of the magnetic atoms of Fe, Ni, and Co. They can be made of oxides, such as NiO, Fe_2O_3, $NiFe_2O_4$ or alloys, such as $Nd_2Fe_{14}B$, $Pr_2Fe_{14}B$, $Tb_2Fe_{14}B$, $DyFe_{14}B$, $Pr_2Co_{14}B$, $Sm_1Fe_{11}Ti_1$, $Sm_1Fe_{10}V_2$, $Sm_2Fe_{17}N_{23}$, $Sm_2Fe_{17}C_22$, Sm_2Co_{17}, Sm_2Co_5. To protect nanoclusters from oxidation, one coat them with graphene or noble metals, forming the double-component nanoclusters, such as Fe-Au, Co-Au , Co-Ag, Co-Cu, Co-Pt, Co-Pd, Ni-Au, Ni-Ag, Ni-Pd, and Mn^--Au. The coating is done be means of chemical reactions or laser ablation techniques. The nanoclusters are produced by employing thermal decomposition, microemulsion reactions, and thermal spraying.

Magnetic nanoclusters and nanomolecules find numerous applications, among which we can mention magnetic chemistry, biomedical imaging, medical treatment, genetic engineering, waste cleaning, information storage, quantum computing, and creation of radiation devices. Since both nanomolecules and nanoclusters possess many common properties and can be considered as single particles with a large spin, we shall often talk on nanoclusters, implying that similar effects can be realized with both of them, molecules as well as clusters.

The use of these nano-objects requires the existence of two properties that contradict each other. From one side, to be able to keep memory, a cluster has to enjoy a stable state with its spin frozen in one direction. But from another side, in order to be able to manipulate the cluster magnetization, there should exist a way of suppressing the anisotropy. And it is necessary that the spin manipulation could be done sufficiently fast, so that the cluster magnetization could be quickly reversed. Recall that thermal reversal is characterized by the Arrhenius law giving the longitudinal relaxation time $T_1 \sim \exp\{E_A/k_BT\}$, where E_A is the anisotropy energy, so that, at temperatures below the blocking temperature, the magnetization is frozen.

Magnetization reversal can be realized by different methods, by applying transverse constant or alternating magnetic fields and short magnetic field pulses [10]. To achieve fast reversal, one needs to find optimal values for the amplitude, frequency, and duration of such field pulses.

A very efficient method of achieving ultrafast magnetization reversal of magnetic nanoclusters has been suggested [11] by employing the acceleration effect caused by a resonator feedback field. The efficiency of this method is due to self-optimization of the spin motion producing the resonator field acting back on the spins. Historically, this effect was described by Purcell [12] and considered by Bloembergen and Pound [13] using classical phenomenological equations. Such equations are not sufficient for describing different regimes of spin motion. Microscopic theory of spin dynamics has been developed being applied to polarized proton spins of such molecules as propanediol $C_3H_8O_2$, butanol C_4H_9OH, and ammonia NH_3 (see review articles [4,14]) and to magnetic molecules [15-19].

The aim of the present paper is threefold. First, we concentrate on the spin dynamics of nanoclusters, comparing the peculiarity of their spin motion with that of proton and molecular spins. Second, we analyze the role of other effects, such as the Nyquist-noise triggering and Dicke correlation, studying their influence on the spin dynamics of nanoclusters. We show that these effects are negligible as compared to the Purcell effect. And, third, we compare different approaches to describing spin dynamics, demonstrating the advantage of using a microscopic approach based on quantum equations of motion.

2. Phenomenological classical equations

Dynamics of the magnetic moment $\mathbf{M}$ of a magnetic particle is usually described by the classical equation

$$\frac{d\mathbf{M}}{dt} = -|\gamma_S|\mathbf{M} \times \mathbf{H}_{eff} + \mathbf{R} \,, \tag{1}$$

in which γ_S is the giromagnetic ratio of the particle with spin S and $\mathbf{R}$ is a relaxation term. The effective magnetic field is given by the variational derivative $\mathbf{H}_{eff} = -\,\delta E/\delta\mathbf{M}$ of the particle energy E. The length of the magnetic moment is conserved, when the right-hand side of the equation $d\mathbf{M}^2/dt = 2\mathbf{M}\cdot\mathbf{R}$ is zero.

Choosing the relaxation term in the form

$$\mathbf{R} = -\,\frac{\alpha|\gamma_S|}{M}\,\mathbf{M} \times (\mathbf{M} \times \mathbf{H}_{eff}) \,, \tag{2}$$

one gets the Landau-Lifshitz equation, where α is a dissipation parameter and $M \equiv |\mathbf{M}|$. Under form (2), $|\mathbf{M}|$ is conserved. The equation was initially derived [20] for describing energy dissipation in the process of magnetic domain wall motion inside bulk ferromagnetic matter. Though it is often applied for treating the dynamics of ferromagnetic particles [21].

Taking the relaxation term as

$$\mathbf{R} = \frac{\alpha}{M}\,\mathbf{M} \times \frac{d\mathbf{M}}{dt} \,, \tag{3}$$

one comes to the Gilbert equation [22]. This equation, up to a renotation of parameters, is equivalent to the Landau-Lifshitz equation. Hence, it has the same region of applicability, though it is also used for describing the magnetization rotation of magnetic particles [10].

Another form of the relaxation term has been advanced by Bloch [23] as

$$\mathbf{R} = -\,\frac{M_x - M_x^*}{T_2}\,\mathbf{e}_x - \frac{M_y - M_y^*}{T_2}\,\mathbf{e}_y - \frac{M_z - M_z^*}{T_1}\,\mathbf{e}_z \,, \tag{4}$$

where $\mathbf{M}^*$ is an equilibrium magnetization, $\mathbf{e}_\alpha$ are unit coordinate vectors, and the relaxation parameters are characterized by the longitudinal relaxation time T_1 and transverse relaxation time T_2. The latter is also called the dephasing time. For an ensemble of N magnetic particles with a large average spin polarization

$$s \equiv \frac{1}{SN}\sum_{j=1}^{N}\langle S_j^z\rangle \,, \tag{5}$$

the transverse term has to be renormalized [16,24] as $1/\widetilde{T}_2 = (1 - s^2)/T_2$.

The Landau-Lifshits equation has a single dissipation parameter α and preserves spherical symmetry, thus, describing isotropic magnetization rotation. Because of these properties, it is appropriate for bulk macroscopic ferromagnetic matter with spherical magnetic symmetry. It may also be used for magnetic clusters, possessing this symmetry, which, however, is a rather rare case.

The Bloch equation has two relaxation parameters, T_1 and T_2. Therefore it can describe more general situation of anisotropic relaxation, which is more realistic for treating nanoclusters in a medium or below the blocking temperature, when $T_2 \ll T_1$. The Bloch equations have been employed for considering the electron and nuclear spin motion in a strongly coherent regime [25-29] and for spin-polarized ^{129}Xe gas [30]. But these equations cannot describe the whole process of spin relaxation starting from an incoherent quantum stage, for which a microscopic approach is necessary [29,31,32]. The initial stage of spin relaxation is triggered by quantum spin fluctuations that can be identified with nonequilibrium spin waves [4,16,31-33].

3. Microscopic quantum approach

In a self-consistent quantum approach, we start with a microscopic spin Hamiltonian $\hat{H}$ that is a functional of spin operators $\mathbf{S}$. The evolution equations are given by the Heisenberg equations of motion

$$i\hbar \, \frac{d\mathbf{S}}{dt} = [\mathbf{S}, \, \hat{H}] \, . \tag{6}$$

The advantage of using the quantum approach is in the following. First, it takes into account quantum effects that can be important for small clusters. Hence, it is more general. Second, at the initial stage of free spin relaxation, quantum spin fluctuations are of principal importance, being the triggering mechanism for starting the spin motion. Third, being based on an explicit spin Hamiltonian makes it possible to control the used approximations and to have well defined system parameters.

We assume that a magnetic cluster is inserted into an magnetic coil, of n turns and length l, of a resonant electric circuit characterized by resistance R, inductance L, and capacity C. The coil axis is taken along the axis x. Moving magnetic moments induce in the coil the electric current j described by the Kirchhoff equation

$$L \, \frac{dj}{dt} + Rj + \frac{1}{C} \int_0^t j \, dt = - \, \frac{d\Phi}{dt} + E_f \, , \tag{7}$$

in which the magnetic flux $\Phi = 4\pi n M_x / cl$ is formed by the mean transverse magnetization $M_x = \mu_0 \sum_{j=1}^N \langle S_j^x \rangle$, where $\mu_0 \equiv \hbar\gamma_S$. Here E_f is an additional electromotive force, if any. The resonator natural frequency and circuit damping, respectively, are

$$\omega = \frac{1}{\sqrt{LC}} \, , \qquad \gamma = \frac{R}{2L} \, . \tag{8}$$

The coil current creates the magnetic field

$$H = \frac{4\pi n}{cl} \, j \tag{9}$$

that is the solution to the equation

$$\frac{dH}{dt} + 2\gamma H + \omega^2 \int_0^t H(t') \, dt' = -4\pi\eta \, \frac{dm_x}{dt} \, , \tag{10}$$

where $\eta \equiv V/V_{coil}$ is the filling factor and

$$m_x \equiv \frac{M_x}{V} = \frac{\mu_0}{V} \sum_{j=1}^{N} \langle S_j^x \rangle$$

is the transverse magnetization density. The external electromotive force is omitted. The field H is the feedback field, created by moving spins and acting back on them.

4. Dynamics of a single nanocluster

The typical Hamiltonian of a nanocluster is

$$\hat{H} = -\mu_0 \mathbf{B} \cdot \mathbf{S} - D(S^z)^2 + D_2(S^x)^2 + D_4 \left[(S^x)^2(S^y)^2 + (S^y)^2(S^z)^2 + (S^z)^2(S^x)^2 \right] , \qquad (11)$$

where the total magnetic field

$$\mathbf{B} = B_0 \mathbf{e}_z + B_1 \mathbf{e}_x + H \mathbf{e}_x \qquad (12)$$

consists of an external constant field B_0, weak transverse anisotropy field B_1, and the feedback resonator field H. The anisotropy parameters D, D_2, D_4 are defined by the particular type of considered nanoclusters.

The main attention will be payed to the investigation of spin dynamics starting from a strongly nonequilibrium initial state, where the magnetization is directed opposite to the constant external magnetic field B_0.

First, we study the influence of the thermal Nyquist noise of the coil in order to understand whether it can trigger the spin motion in a nanocluster. For the thermal-noise relaxation time, we find

$$t_T = \frac{4\gamma V_{coil}}{\hbar \gamma_S^2 \omega} \tanh\left(\frac{\omega}{2\omega_T} \right) , \qquad (13)$$

where $\omega_T \equiv k_B T/\hbar$ is the thermal frequency defined by temperature T. At low temperatures, below the blocking temperature, say at $T = 1$ K, we have $\omega_T \sim 10^{12}$ s^{-1}. Then the thermal-noise relaxation time is

$$t_T \simeq \frac{2\gamma V_{coil}}{\hbar \gamma_S^2 \omega_T} \qquad \left(\frac{\omega}{\omega_T} \ll 1 \right) . \qquad (14)$$

On the other side, for the reversal time, caused by the resonator feedback field, we have

$$t_{rev} \simeq \frac{V_{coil}}{\pi \hbar \gamma_S^2 S} . \qquad (15)$$

The ratio of the latter to the thermal time (13) is $t_{rev}/t_T \sim \omega_T/2\pi\gamma S$. For the typical values $T = 1$ K, $\gamma \sim 10^{10}$ s^{-1}, and $S \sim 10^3$, this ratio is small: $t_{rev}/t_T \sim 10^{-2}$. Therefore the thermal Nyquist noise does not play any role in the spin dynamics of a nanocluster.

We have accomplished numerical solution of the evolution equations for nanocluster parameters typical of Fe, Ni, and Co nanoclusters. The Zeeman frequency is taken as $\omega_0 \equiv 2\mu_B B_0/\hbar \sim 10^{11}$ s^{-1}. For the feedback rate, we have $\gamma_0 \equiv \pi\eta\hbar\gamma_S^2 S/V_{coil} \sim 10^{10}$ s^{-1}. The typical anisotropy parameters satisfy the relations $D/(\hbar\gamma_0) \sim 10^{-3}, D_2/(\hbar\gamma_0) \sim 10^{-3}, D_4/(\hbar\gamma_0) \sim 10^{-10}$. At the initial time, the spin is assumed to be directed along the axis z. The resonator natural frequency is taken to be in resonance with the Zeeman frequency defined by the field B_0. The behavior of the spin polarization (5) is shown in Fig. 1, where we compare the spin motion in the presence of the resonator ($h \neq 0$) and in the absence of the latter ($h = 0$). Clearly, without the resonator feedback field, the spin is blocked, while in the presence of the resonator, it reverses in short time $t_{rev} \sim 10^{-10}$ s.

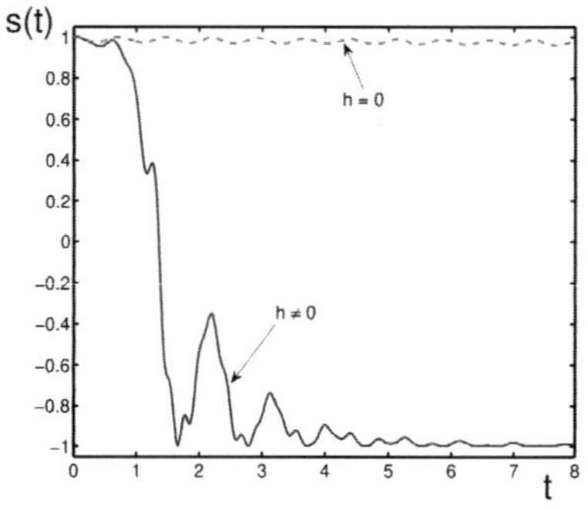

Figure 1. Spin reversal of a single nanocluster, with parameters typical of nanoclusters made of Fe, Ni, and Co.

5. Dynamics of nanocluster assemblies

The ensemble of nanoclusters is described by the Hamiltonian

$$\hat{H} = \sum_i \hat{H}_i + \frac{1}{2} \sum_{i \neq j} \hat{H}_{ij} \; , \tag{16}$$

where the indices $i, j = 1, 2, \ldots, N$ enumerate nanoclusters. The single nanocluster Hamiltonians are

$$\hat{H}_i = -\mu_0 \mathbf{B} \cdot \mathbf{S}_i - D(S_i^z)^2 + D_2(S_i^x)^2 + D_4 \left[(S_i^x)^2(S_i^y)^2 + (S_i^y)^2(S_i^z)^2 + (S_i^z)^2(S_i^x)^2 \right] \; , \tag{17}$$

with the total magnetic field

$$\mathbf{B} = B_0 \mathbf{e}_z + H \mathbf{e}_x \; . \tag{18}$$

The interaction term takes into account the dipolar spin interactions

$$\hat{H}_{ij} = \sum_{\alpha\beta} D_{ij}^{\alpha\beta} S_i^\alpha S_j^\beta \; , \tag{19}$$

through the dipolar tensor $D_{ij}^{\alpha\beta} = \mu_0^2 \left(\delta_{\alpha\beta} - 3 n_{ij}^\alpha n_{ij}^\beta \right) / r_{ij}^3$, in which $r_{ij} \equiv |\mathbf{r}_{ij}|$, $\mathbf{n}_{ij} \equiv \mathbf{r}_{ij}/r_{ij}$, and $\mathbf{r}_{ij} \equiv \mathbf{r}_i - \mathbf{r}_j$.

One sometimes says that spin systems are similar to atomic systems, where transition dipoles are correlated by means of the photon exchange through the common radiation field. This correlation leads to coherent atomic radiation called the Dicke superradiance [34]. One says that moving spins also radiate electromagnetic field that could yield the correlated spin motion, in the same way as in the Dicke effect. To check whether this is so, we need to compare the time t_{rad}, required for inducing spin correlations through the common radiation field with the spin dephasing time T_2. As the radiation time [35,36] for nanoclusters, we have

$$t_{rad} = \frac{3c^3}{2\hbar \gamma_S^2 \omega^3 S} \; , \tag{20}$$

while the spin dephasing time is

$$T_2 = \frac{1}{\hbar \rho \gamma_S^2 S} \; . \tag{21}$$

For the typical nanocluster density $\rho \sim 10^{20}$ cm^{-3} and $S \sim 10^3$, the spin dephasing time is $T_2 \sim 10^{-10}$ s. While for the radiation time (20), with $\omega \sim 10^{11}$ s^{-1}, we have $t_{rad} \sim 10^8$ s $= 10$ years. The ratio of times (20) and (21) is extremely large: $t_{rad}/T_2 = 3c^3\rho/(2\omega^3) \sim 10^{18}$. This tells us that the spin motion in no way can be correlated through electromagnetic radiation.

That is, the Dicke effect has no relation to the coherent spin motion. But spins can be correlated only through the Purcell effect requiring the presence of a feedback field caused by a resonator.

The feedback rate due to the resonator is

$$\gamma_0 = \pi\eta\rho\hbar\gamma_S^2 S \; . \tag{22}$$

The reversal time for N correlated nanoclusters becomes

$$t_{rev} = \frac{1}{\gamma_0} = \frac{V_{coil}}{\pi\hbar\gamma_S^2 SN} = \frac{t_{rev}^1}{N} \; , \tag{23}$$

where t_{rev}^1 is the relaxation time (15) for a single nanocluster inside the same coil.

We solved the evolution equations for the nanocluster assemblies involving the scale separation approach [4,14] that is a generalization of the Krylov-Bogolubov [37] averaging method. Four classes of spin objects have been investigated.

(i) Polarized nuclear materials, such as propanediol $C_3H_8O_2$, butanol C_4H_9OH, and ammonia NH_3, with the parameters: $S = 1/2, \rho = 10^{22}\text{cm}^{-3}, T = 0.1\text{K}, B_0 \sim 10^4\text{G}, \omega_0 \sim 10^8\text{s}^{-1}, \lambda \sim 10^2\text{cm}, T_1 \sim 10^5\text{s}, T_2 \sim 10^{-5}\text{s}, \tau \equiv 1/\gamma \sim 10^{-6}\text{s}$. Recall that in these nanomolecules the magnetization is due to polarized proton spins.

The following characteristic times are found: thermal-noise time $t_T \sim 10^{16}\text{s} \sim 10^9$ years, radiation time $t_{rad} \sim 10^{15}\text{s} \sim 10^8$ years, and reversal time $t_{rev} \sim 10^{-6}\text{s}$.

Therefore, neither the Nyquist thermal noise nor the photon exchange through the radiated field play any role in the relaxation process. Spin dynamics, resulting in the magnetization reversal, is completely due to the action of the resonator feedback field. As is explained above, the same concerns nanomolecules and nanoclusters

(ii) Nuclear polarized ferromagnets, where proton spins are polarized and interact through hyperfine forces with electrons participating in forming ferromagnetic order. In such materials, the electron subsystem plays the role of an additional resonator enhancing effective nuclear correlations. Being interested in the motion of nuclear spins, under a fixed mean electron magnetization, we find the reversal time $t_{rev} \sim 10^{-9}$ s.

(iii) Molecular magnets, such as Mn_{12} and Fe_8, with the typical parameters: $S = 10, \rho = 10^{20} - 10^{21}\text{cm}^{-3}, T_B = 1\text{K}, B_0 \sim 10^5\text{G}, \omega_0 \sim 10^{13}\text{s}^{-1}, \lambda \sim 10^{-2}\text{cm}, \omega_A \equiv E_A/\hbar \sim 10^{10} - 10^{12}\text{s}^{-1}, T_1 \sim 10^5 - 10^7\text{s}, T_2 \sim 10^{-10}\text{s}$. The reversal time is $t_{rev} \sim 10^{-11}$ s.

(iv) Magnetic nanoclusters composed of Fe, Ni, and Co, at $T = 1$ K, with the typical parameters: $S = 10^3, \rho = 10^{20}\text{cm}^{-3}, T_B = 10 - 40\text{K}, B_0 \sim 10^4\text{G} = 1\text{T}, \omega_0 \sim 10^{11}\text{s}^{-1}, \lambda \sim 1\text{cm}, T_1 \sim 10^{34}\text{s} \sim 10^{27}\text{years}, T_2 \sim 10^{-10}\text{s}$. The reversal time can be very small reaching the value $t_{rev} \sim 10^{-12}$ s.

In the case of magnetic molecules and, especially, nanoclusters, because of their high spins, the system of many clusters can produce quite strong coherent radiation of the maximal intensity

$$I_{max} \sim \frac{2\mu_0^2}{3c^3} S^2\omega^4 N_{coh}^2 \; , \tag{24}$$

where $N_{coh} \sim \rho\lambda^3$ is the number of clusters in a coherent packet. The intensity of radiation of magnetic molecules, with $N_{coh} \sim 10^{14}$, is of order $I_{max} \sim 10^5 W$. And for magnetic nanoclusters, with $N_{coh} \sim 10^{20}$, the radiation intensity can reach $I_{max} \sim 10^{12} W$.

There can happen several regimes of spin dynamics depending on the initial spin polarization, the strength of a triggering pulse, and the effective coupling parameter

$$g \equiv \frac{\gamma\gamma_0\omega_0}{\gamma_2(\gamma_2^2 + \Delta^2)} \; , \tag{25}$$

where $\gamma_2 \equiv 1/T_2$ and $\Delta \equiv \omega - \omega_0$ is the detuning from resonance. These regimes for nanoclusters can be classified analogously to those occurring for nuclear magnets [4,16]: incoherent free relaxation, weakly coherent free induction, weakly coherent superradiance, strongly coherent pure superradiance, strongly coherent triggered superradiance, pulsing superradiance, and punctuated superradiance [38].

It is important to stress that the existence of magnetic anisotropy in magnetic nanoclusters does not preclude the realization of fast spin reversal, provided the external magnetic field is sufficiently strong. The influence of the anisotropy energy E_A on the spin reversal of a nanocluster system is shown in Fig. 2, where $A \equiv E_A/\omega_0$. This regime corresponds to pure spin superradiance.

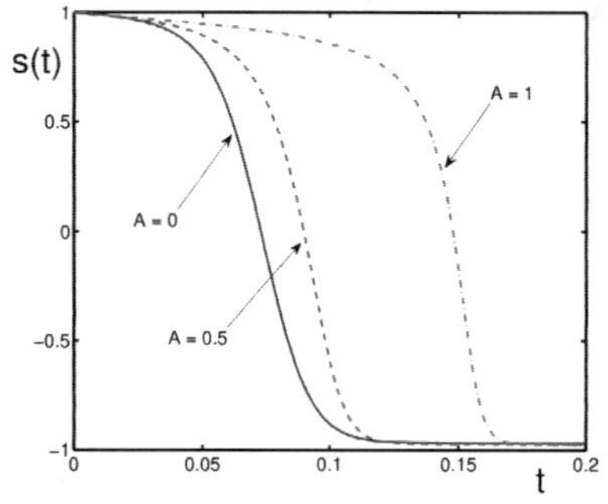

Figure 2. Influence of magnetic anisotropy on the spin reversal of an ensemble of many nanoclusters, with the parameters typical of Fe, Ni, and Co.

Coherent dynamics in the spin assemblies, formed by magnetic nanomolecules, have an important difference from the spin dynamics in an ensemble of magnetic nanoclusters. Magnetic molecules are identical and form the systems with well organized crystalline lattices. While magnetic nanoclusters vary in their shapes, sizes, and total spins, which results in an essential nonuniform broadening. Computer simulations, accomplished together with V.K. Henner and P.V. Kharebov, demonstrate that this nonuniformity does not destroy coherent spin motion. A detailed analysis of the computer simulations, with nonuniform nanocluster distributions, will be presented in a separate publication.

In conclusion, we have considered spin dynamics in magnetic nanomolecules and nanoclusters, starting from a strongly nonequilibrium state, with the magnetization directed opposite to the applied external magnetic field. We have compared several methods of describing the spin dynamics, showing that a microscopic approach, based on the quantum equations of motion, is the most accurate. We also have analyzed the influence of different effects on spin dynamics. The effects of the Nyquist-noise triggering and of Dicke correlations are found to be negligible for spin systems. This principally distinguishes spin systems from atomic systems or quantum dot systems [39], where correlations, leading to coherent radiance, are caused by the Dicke effect of interactions through the common radiation field.

The feedback field, developing in the resonator, reaches rather high values, of the order of the applied constant magnetic field. Such a strong feedback field suppresses the influence of mutual cluster interactions. Generally, in an ensemble of nanoclusters of sufficiently high density, in addition to dipole interactions, there can appear exchange interactions [40] that can influence equilibrium properties of nanoclusters. But in the considered case of strongly nonequilibrium spin dynamics, the exchange interactions are also suppressed by the self-organized resonator feedback field.

This important conclusion can be formulated as follows: *Coherent spin dynamics are completely governed by the Purcell effect that is caused by the action of the resonator feedback field.*

[1] Barbara B, Thomas L, Lionti F, Chiorescu I, and Sulpice A 1999 *J. Magn. Magn. Mater.* **200** 167
[2] Wernsdorfer W 2001 *Adv. Chem. Phys.* **118** 99
[3] Ferre J 2002 *Topics Appl. Phys.* **83** 127
[4] Yukalov V I and Yukalova E P 2004 *Phys. Part. Nucl.* **35** 348
[5] Wang J and Zeng X C 2009 in: *Nanoscale Magnetic Materials and Applications* (Springer, Berlin), p. 35
[6] Bedanta S and Kleemann W 2009 *J. Phys. D* **42** 013001
[7] Berry C C 2009 *J. Phys. D* **42** 224003
[8] Beveridge J S, Stephens J R, and Williams M E 2011 *Annu. Rev. Anal. Chem.* **4** 251
[9] Hoang V V and Ganguli D 2012 *Phys. Rep.* **518** 81
[10] Bauer M, Fassbender J, Hillebrands B, and Stamp R L 2000 *Phys. Rev. B* **61** 3410
[11] Yukalov V I and Yukalova E P 2012 *J. Appl. Phys.* **111** 023911
[12] Purcell E M 1946 *Phys. Rev.* **69** 681
[13] Bloembergen N and Pound R V 1954 *Phys. Rev.* **95** 8
[14] Yukalov V I and Yukalova E P 2000 *Phys. Part. Nucl.* **31** 561
[15] Yukalov V I 2002 *Laser Phys.* **12** 1089
[16] Yukalov V I 2005 *Phys. Rev. B* **71** 184432
[17] Yukalov V I, Henner V K, Kharebov P V and Yukalova E P 2008 *Laser Phys. Lett.* **5** 887
[18] Yukalov V I, Henner V K and Kharebov P V 2008 *Phys. Rev. B* **77** 134427
[19] Yukalov V I and Yukalova E P 2011 *Laser Phys. Lett.* **8** 804
[20] Landau L D and Lifshitz E M 1935 *Phys. Zeits. Sow.* **8** 153
[21] Henner V, Raikher Y and Kharebov P 2011 *Phys. Rev. B* **84** 144412
[22] Gilbert T L 1956 PhD Thesis (Illinois Institute of Technology, Chicago)
[23] Bloch F 1946 *Phys Rev.* **70** 460
[24] Abragam A and Goldman M 1982 *Nuclear Magnetism: Order and Disorder* (Clarendon, Oxford)
[25] Yukalov V I 1992 *Laser Phys.* **2** 559
[26] Fokina N P, Khutsishvili K O and Chkhaidze S G 1992 *J. Exp. Theor. Phys.* **102** 1013
[27] Khutsishvili K O and Chkhaidze S G 1992 *Physica B* **176** 54
[28] Fokina N P, Khutsishvili K O and Chkhaidze S G 1992 *Physica B* **179** 171
[29] Yukalov V I 1996 *Nucl. Instrum. Methods Phys. Res. A* **37** 345
[30] Yoshimi A, Asahi K, Sakai K, Tsuda M, Yogo K, Ogawa H, Suzuki T and Nakagura M 2002 *Phys. Lett. A* **304** 13
[31] Yukalov V I 1995 *Phys. Rev. Lett.* **75** 3000
[32] Yukalov V I 1996 *Phys. Rev. B* **53** 9232
[33] Rueckriegel A, Kreisel A and Kopietz P 2012 *Phys. Rev. B* **85** 054422
[34] Dicke R H 1954 *Phys. Rev.* **93** 99
[35] Yukalov V I and Yukalova E P 2005 *Laser Phys. Lett.* **2** 302
[36] Yukalov V I 2005 *Laser Phys. Lett.* **2** 356
[37] Bogolubov N N and Mitropolsky Y A 1961 *Asymptotic Methods in the Theory of Nonlinear Oscillations* (Gordon and Breach, New York)
[38] Yukalov V I and Yukalova E P 2002 *Phys. Rev. Lett.* **88** 257601
[39] Yukalov V I and Yukalova E P 2010 *Phys. Rev. B* **81** 075308
[40] Kondratyev V N and Lutz H O 1998 *Phys. Rev. Lett.* **81** 4508

Response of magnetic nanoparticle assemblies

V N Kondratyev[1, 2], V V Krylov[1], O A Bezshyyko[1], L O Golinka-Bezshyyko[1] and V A Osipov[2]
[1] Physics Department, Taras Shevchenko National University, Kiev, UA-03022 Ukraine
[2] Bogolubov Laboratory of Theoretical Physics, JINR, 141980, Dubna, Russia

E-mail: vkondra@univ.kiev.ua

Abstract. Properties of magnetic nanoparticle assemblies are analysed by employing the randomly jumping interacting moments model including quantum fluctuations, ferromagnetic inter-particle coupling and disorder. Using spatially uniform external field we find the respective magnetic state equation and phase diagram indicating an existence of spinodal regions and critical points. Possibilities to use such systems as magnetoresistive sensors are considered.

1. Introduction
Supermagnetism (SM) was invoked to specify the properties of a system containing quantum confined objects, e.g., quantum dots, atomic clusters and/or magnetic nanoparticles, see [1-5] and refs. therein. In particular, an assembly of magnetic nanoparticles (MNPs) displays ferromagnetic (FM) long range order at sufficiently large density. Such a property requires to consider the inter-dot magnetic interaction beyond dipolar components [1-5], like quadrupole, superexchange, strain fields etc. Especially, the superferromagnetic (SFM) structures with certain electronic coherent states [1] provide a system exhibiting giant magnetoresistance (GMR) [6,7] and can be employed as magnetoresistive (MR) sensors. The SFM response in external magnetic fields represents an important relevant issue.
In this contribution we use the recently developed randomly jumping interacting moments (RJIM) model, see [2,3] and refs. therein, to analyse SFM magnetic reactivity, on examples of (i) spatially uniform field changing slowly in time and (ii) spatially local perturbation, i.e., a sensor mode for detection of, e.g., a single magnetic particle. In Sect.2 we briefly recall the RJIM model and the Monte Carlo simulation technique, cf., [2,3], used to describe the system dynamics. Optimized serial algorithms and parallel CUDA programming on graphic processing units (GPU) were developed to provide acceptable level of calculation rates. Arrays of nanoparticles with multiple magnetic response anomalies are demonstrated to display strong dependence on system disorder. Possible implications of peculiarities in magnetic planar array response as sensors are discussed in Sect 2. Conclusions are given in Sect. 3.

2. SFM structure of magnetic nanoparticle assemblies with disorder
Overall magnetisation of a system composed from Π elements with moments m_i can be represented as $P=\sum_i m_i/V=\langle m\rangle/V_D$ with total volume V and an area occupied by ith nanoparticle

$V_D = V / \Pi$. For SFM with ferromagnetic coupling of a strength *J* between the nearest neighbor (nn) nanoparticles the respective Ising term $-J\sum_{ij} m_i m_j$ contributes to the Hamiltonian. Here the sum runs over the nn elements. Besides, inhomogeneity and disorder in the form of defects, grain boundaries, impurities lead to random crystalline anisotropy and varying interaction strengths in the super-crystalline heterostructure. Such effects can be accounted for by the random fields h_i. We point out also dynamical components of h_i due to inexactness of the model description with nn interaction. The central limit theorem suggests, thereby, that the random fields obey the Gaussian distribution $W(h) = \exp\{-h^2 / R^2\} / R\sqrt{\pi}$ of a width *R*, which we call the disorder. The total Hamiltonian H of an array in a field *H* can be expressed through an interaction of the nanoparticle magnetic moment m_i with local fields

$$b_i = H(t) + JV_D \sum_{j \subset nn} P_j + h_i \tag{1}$$

as $\mathrm{H} = -\sum_i m_i b_i$. We refer for this model as randomly jumping interacting moments (RJIM) model [2,3]. Hereafter, we consider an array of nanoparticles with single discontinuity in magnetic moment response, $m_i = m \cdot \mathrm{sign}(b_i)$. Due to the ferromagnetic interaction a jumping moment can cause some of the nearest neighbours to jump, which may in turn trigger some of their neighbours, and so on, generating, thereby, (re)magnetizing avalanche. As a consequence some sharp stepwise discontinuity arises on magnetization curves. Such sharp changes of magnetic induction result in a release of magnetic energy with noise signals proportional to avalanche size, similarly to the well-known Barkhausen effect.

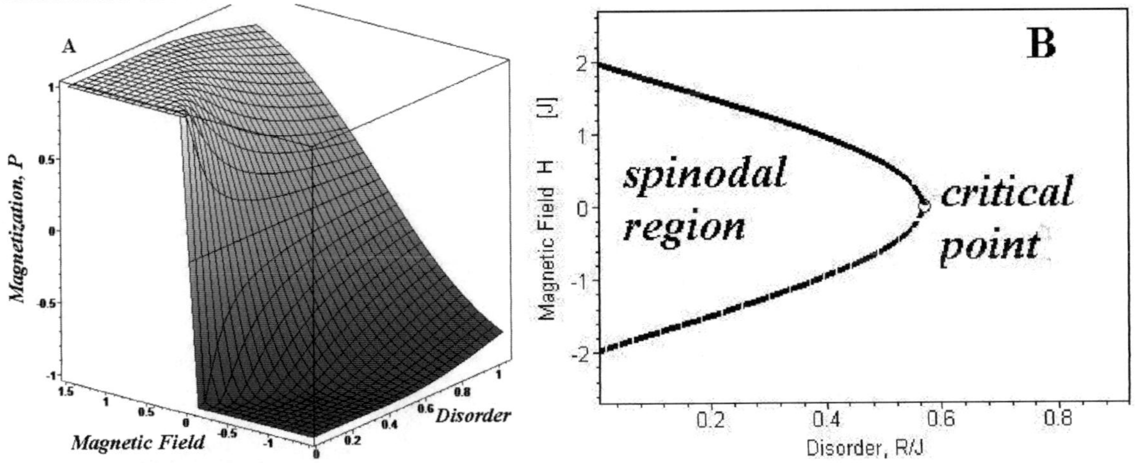

Figure 1. Magnetic state equation (Panel A) and phase diagram (Panel B) for SFM composed from MNPs with single jump at *b*=0 and disorder *R*. The values *H*, *b* and *R* are measured in units of *J*.

2.1. Magnetic state equation and phase diagram of SFM

The basic features of the nonequilibrium system corresponding to the Hamiltonian Eq. (1) can be analysed by employing the mean-field approach, in which one assumes an equal interaction strength between MNPs with the coupling constant $J_{ij} = J / \Pi$. The local field in Eq. (1) can be then simplified to the form $b_i^{mf} = H(t) + JP + h_i$ with an averaged over a sample magnetization *P*, see above. We see, therefore, that random fields can be viewed as mean-field fluctuations (cf, e.g., [2,3,8]).

In the thermodynamic limit $\Pi \to \infty$ for spatially uniform field *H* we calculate the magnetic state equation (MSE) $P = \int dhW(h)m(b)$ and find for the magnetic susceptibility

$$\chi = -dP/dH = [\chi_{\mathrm{NI}}^{-1} - J]^{-1} \tag{2}$$

with $\chi_{NI}=W(b)$ representing the susceptibility of an array without inter-particle interaction (i.e. J=0). The case of negatively defined susceptibility, see Eq. (2), defines spinodal region for an array. Figure 1 shows MSE and phase diagram. For coupled MNPs the value $J\chi_{NI}$ measures an average number of induced jumps per single jumping moment. When the number of induced moment jumps exceeds 1 the system favours to evolve in an avalanche spanning almost entire sample with a macro-(de)magnetization discontinuity. Evidently, the relation $\chi_{NI} \geq J^{-1}$ corresponds to the instability condition. Such spinodal regions are located on $\{H,R\}$-plane between the low and up critical fields as indicated in Fig. 1B. The field-lines meet at the critical point $\{ H_c = b_0 =0,\ R_c = J/\sqrt{\pi} \}$. The difference between the number of induced jumps and 1, $d = J\chi_{NI} - 1$, provides, therefore, a measure for a vicinity of self-organized criticality.

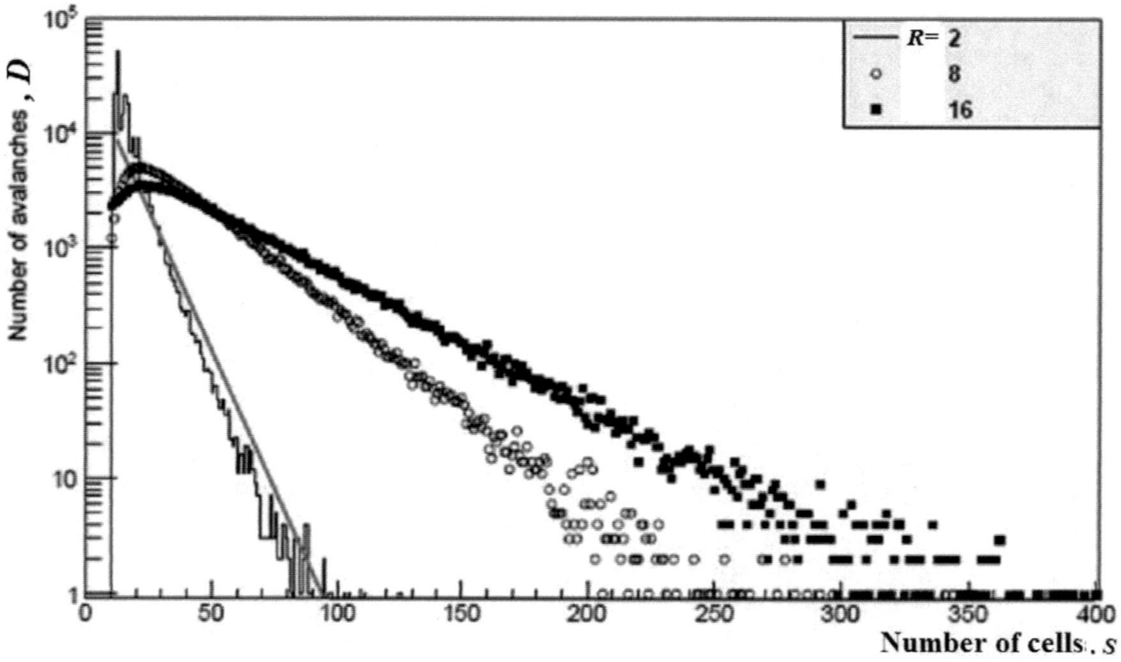

Figure 2. Size distribution of avalanches induced by local fields for 2×10^5 randomly distributed positions (see text) at different disorders R= 2 (line), 8(open circles), 16(solid circles). Solid line shows the fit according to Eq. (3).

2.2. *Sensing magnetic nanoparticles by SFM structures*

To analyse possibilities for using the realistic SFM systems as MR sensors for detecting, e.g., a single magnetic particle we consider spatially local field inducing jumps of several assembly elements. Such an event creates an avalanche of moment jumps due to ferromagnetic coupling and the GMR signal proportional to a number S of particles in an avalanche. We perform the simulations for SFM of simple quadric planar lattice of a size $(1000)^2$ with random distribution of such avalanche *grains*. As seen in Fig. 2 the avalanche size distribution $D(S)$ shows nearly exponential behaviour with increasing width at growing disorder. To understand such a trend we employ the mean-field arguments. Since within the mean-field treatment an average number of the moments induced to jump by a single jumping *seed* is site independent for the avalanche size distribution at $1 \ll S \ll \Pi$ one obtains [3]

$$D_{mf}(S) \sim S^{-3/2} \exp\{-Sd^2/2\} \tag{3}$$

For growing disorder in the considered range the parameter d decreases. Therefore, the distribution *D(S)* becomes wider according to Eq. (3). As a consequence, such conditions correspond to stronger MR signal of a sensor.

3. Conclusions

As seen above the magnetic nanoparticle assemblies display the SFM structure effects at sufficiently dense packing. Magnetic state equation and phase diagram of SFM are demonstrated to exhibit spinodal regions on {*disorder, magnetic field*}-plane and the critical points. We argue that such SFM features are well suited for magnetoresistive sensors. The disorder in magnetic nanoparticle assemblies improves the detection ability at considered parameters. We note that quantum fluctuations due to the nanoparticle discrete levels can bring additional anomalies of magnetic response [9] and new phases [3]. Such properties could provide further possibilities for sensors.

Acknowledgments

One of the authors (V.N.K.) thanks JINR (Dubna, Russia) for the warm hospitality and the financial support.

References

[1] Kondratyev V N and Lutz H O 1998 *Phys. Rev. Lett.* **81** 4508; 1999 *Eur. Phys. J.* D **9** 483
[2] Kondratyev V N 2002 *Phys. Rev. Lett.* **88** 221101; 2002 *JAERI-Research* 2001–057
[3] Kondratyev V N 2006 *Phys. Lett.* A **354** 217; 2010 *J. Phys.CS* **248** 012027
[4] Hutten A et al. 2004 *J. Biotechnology* **112** 47
[5] Bedanta S and Kleemann W 2009 *J. Phys. D: Appl. Phys.* **42** 013001
[6] Gruenberg P et al 1986 *Phys. Rev. Lett.* **57** 2442
[7] Babich M N et al 1988 *Phys. Rev. Lett.* **61** 2472
[8] Kondratyev V N 1993 *Phys. Lett.* A **179** 209; 1994 *ibid.* **190** 465; 1996 *Z. Phys.* B **99** 473
[9] Kondratyev V N 2002 *J. Nucl. Sci. Technol.* **1** Suppl. 2. 550; 2002 *J. Nucl. Radiochem. Sci.* **3** 205

About applicability of thermodynamic parameters to small drops and clusters

G V Kharlamov[1], A A Onischuk[2], S V Vosel[2], P A Purtov[2]

[1]Novosibirsk State Technical University, 20 K. Marx Prospekt, 630092, Novosibirsk, Russia,
[2]Institute of Chemical Kinetics and Combustion of RAS, 3 Institutskaya str., 630090, Novosibirsk, Russia

E-mail: g_kharlamov@ngs.ru

Abstract. In our paper the results of molecular dynamics calculations of nanodrops are presented. The density profiles, the Irving-Kirkwood pressure tensors, the chemical potentials of the systems, the equimolar radii of the drops and the radii of tension, the mechanical and thermodynamic surface tensions have been calculated.
It is shown that both the mechanical and thermodynamic surface tensions decrease with the decrease of the equimolar radius of the drop and reach zero at the same R_0 depending on temperature. With the further equimolar radius decrease the surface tension becomes negative. It means that such droplet is metastable and that the notion of the macroscopic surface tension cannot be applied to such small drops. The dependence of the ratio of the drop surface tension to the surface tension of the flat liquid-vapor interface on the ratio of the drop equimolar radius to R_0 is a universal function.

1. Introduction

Thermodynamic description of nanodrops is a very important physical problem. There is a problem of applicability of thermodynamic parameters to small drops and clusters as the parameters are macroscopic thermodynamic notions, but nanodrops are microscopic systems. One of such parameters is the surface tension. J. Gibbs pointed out that the surface tension of a droplet (σ) must decrease with the decrease of the droplet tension surface radius (R_s) so that at $R_s = 0$ the surface tension must be also equal to zero ($\sigma = 0$) [1]. At the same time the equimolar radius of the droplet (R_e) is not equal to zero. Consequently, the notion of the surface tension cannot be applied to such droplets.

2. Thermodynamics of drops

The surface tension concept in thermodynamics is propounded by J. Gibbs and is conventional [1]. In accordance with J. Gibbs the additional term σdA (σ is the surface tension, dA is a change of a liquid surface area) is introduced in fundamental thermodynamic equations. For example, such equation for the internal energy of the system consisting of a liquid drop and vapor has the usual form

$$dU = T\,dS - P_l\,dV_l - P_v\,dV_v + \sigma\,dA + \mu\,dm. \tag{1}$$

Here dU, dS, and dm are changes of the internal energy, entropy, and mass of the whole system, dV_l and dV_v are changes of the volumes of the liquid droplet and the vapor, dA is a change of the drop surface area, P_l, P_v are the pressures in the drop and in the vapor, T is the system temperature, μ is the chemical potential of the system.

Thus, such surface tension σ determines additional to the thermodynamic potentials energy related to a liquid surface area change and is called the thermodynamic surface tension. In addition to surface tension, J. Gibbs introduced a dividing surface which is termed as a tension surface. It is a mathematical surface, to which the forces of the surface tension are applied. At the thermodynamic description of the systems with flat interfaces of liquid – vapor, the position of the tension surface has no importance, but there is a problem of choice of the tension surface for curved interfaces.

For the system, consisting of a liquid drop surrounded by natural vapor, the value of the surface tension depends on the choice of the tension surface position. J. Gibbs suggested choosing the tension surface so that the surface tension had the minimum value. The Gibbs model assumes that the droplet is a sphere with a radius of the tension surface. Inside the sphere there is a liquid medium with the properties of a bulk liquid. Sometimes this medium is called the comparison phase. The chemical potential of this phase coincides with the chemical potential of the vapor surrounding the sphere. Thus, the pressure, density, and temperature in the droplet are equal to the pressure, density, and temperature in the bulk liquid at the same chemical potential. The tension surface of the drop generally does not coincide with the equimolar surface; therefore the part of the system mass does not belong to any of two phases and is the part of the substance adsorbed on the interface. The difference between the tension surface radius and the equimolar surface radius is determined by the parameter $\delta = R_e - R_s$ which is called the Tolman length [2]. The difference of the pressures between the comparison phase and the surrounding vapor obeys the Laplace formula

$$P_l - P_g = 2\sigma/R_s \,, \tag{2}$$

where P_l and P_g are the pressures of the liquid in the comparison phase and the vapor respectively. The choice of the pressure and density of the liquid in the comparison phase is arbitrary. The values of σ, R_s and also the equimolar radius R_e will be changed according to this choice.

On the other hand, there is a mechanical definition of the surface tension [3, 4] connected with the substitution of the continuous droplet pressure profile by the step one. The pressure tensor of a drop has the spherical symmetry

$$P(r) = \mathrm{e}_r\mathrm{e}_r P_N(r) + (\mathrm{e}_\theta\mathrm{e}_\theta + \mathrm{e}_\phi\mathrm{e}_\phi) P_T(r) \,, \tag{3}$$

where $P_N(r)$ and $P_T(r)$ are the normal and tangential pressures.

The mechanical surface tension is obtained from the pressure tensor by [3, 4]

$$\sigma = \int_0^\infty (r/R_s)^2 (P_N(r) - P_T(r))\, dr \,. \tag{4}$$

Since R_s is unknown beforehand in this formula, it is necessary to have one more equation determining R_s. In papers [4, 5] it is suggested to use the equation

$$\sigma = \int_0^\infty (R_s/r)(P_N(r) - P_T(r))\, dr \tag{5}$$

which is derived from the Buff's equations based on the mechanical equilibrium conditions of the system. Using the equations (4) and (5), it is possible to calculate R_s and σ if we know the profiles of the pressure tensor components $P_N(r)$ and $P_T(r)$.

It is not obvious in advance that the surface tension calculated by formulae (4) and (5) will coincide with the thermodynamic surface tension used in (1). Here there is a problem of relation of the surface tension determined by formulae (4) and (5) to the Gibbs model of the surface tension.

3. Simulation and results

In this work, the thermodynamic properties of droplets were calculated by the molecular dynamics method. The calculations were made for the system containing 60 – 4500 molecules in a cubic cell with periodic boundary conditions. Interaction between the molecules was specified by the Lennard-Jones potential. A special procedure was used to obtain an equilibrium system composed of a liquid drop in the center of the cell and vapor occupying the remaining space. The size of the drop depended on the number of particles in the cell and the mean density of the system. The Irving-Kirkwood pressure tensor was calculated by the formulae of statistical mechanics [4]. The density profiles, chemical potential of the system, equimolar and surface tension radii of drops, mechanical and thermodynamic surface tensions, and Tolman length, δ, were calculated.

Reduced variables were used: distance $r = r^*/r_0$, temperature $T = kT^*/\varepsilon$, energy $U = U^*/\varepsilon$, density $\rho = \rho^* r_0^3$, time $t = t^*/r_0(\varepsilon/m_0)^{1/2}$, pressure $p = p^* r_0^3/\varepsilon$, surface tension $\sigma = \sigma^* r_0^2/\varepsilon$, and chemical potential $\mu = \mu^* m_0/\varepsilon$. Here ε and r_0 are the Lennard-Jones potential parameters, m_0 is the molecule mass. The variables labeled by an asterisk are dimensional. The calculations methods of the mechanical and thermodynamic surface tensions were discussed in our papers [6 – 8].

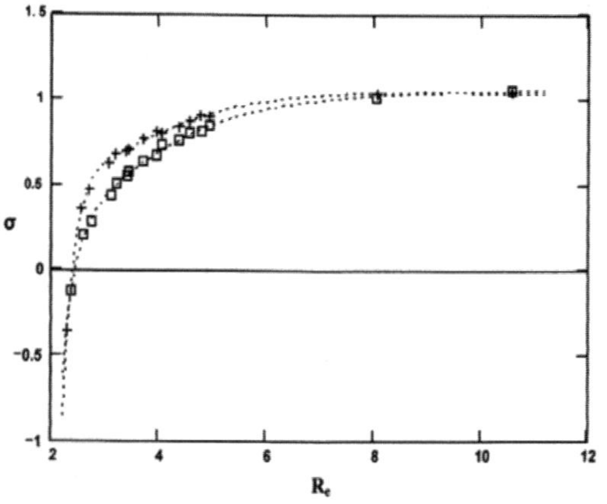

Figure 1. The dependence of the mechanical (□) and thermodynamic (+) surface tensions on the equimolar radius R_e at the temperature $T = 0.65$

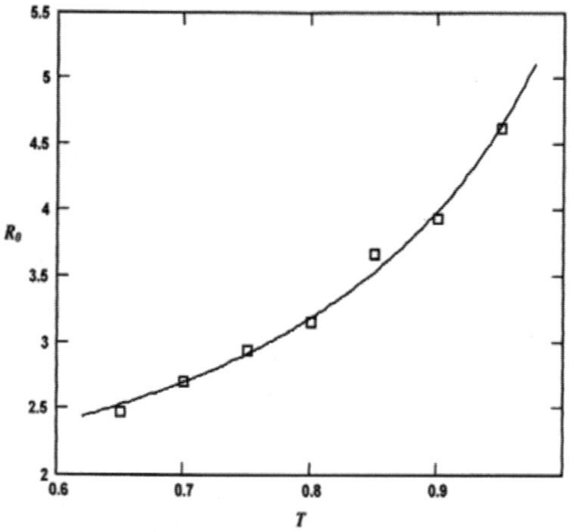

Figure 2. The dependence of R_0 on the temperature. The solid line is the dependence (7)

Figure 1 presents the mechanical and thermodynamic surface tension dependences on the drop equimolar radius. Significant decrease of both mechanical and thermodynamic surface tension can be observed with R_e decrease. Both surface tensions equal zero at a certain equimolar radius R_0. This radius corresponds to the drop containing about 50 molecules. The radius of the tension surface R_s also equals zero. The surface tension becomes negative at the further equimolar radius decrease though the drop holds stability in the conditions of the numerical experiment. Figure 1 shows that the thermodynamic surface tension coincides with the mechanical one for relatively big drops containing

2000 and more molecules. For the drops containing smaller quantity of molecules, the thermodynamic surface tension exceeds the mechanical one up to their disappearance at $R_e = R_0$.

It has been established in our papers [6 - 8] that the dependence of the nanodrops mechanical surface tension on the equimolar radius is a universal function relative to the temperature, at which the numerical experiment is conducted. It has been found that approximation of this dependence by the polynomial of the ratio R_0/R_e can be expressed as follows

$$\sigma(R_e) = \sigma_\infty\left[1 - 0.519 \cdot R_0/R_e + 0.426 \cdot (R_0/R_e)^2 - 0.907 \cdot (R_0/R_e)^3\right]. \quad (6)$$

In addition, the drop equimolar radius R_0, at which $\sigma = 0$, greatly depends on the temperature and becomes infinite in the point equal 1.199 (see figure 2). The approximation of the dependence of R_0 on the temperature can be expressed by the fit

$$R_0 = 0.768 + \frac{0.963}{1.199 - T}. \quad (7)$$

The surface tension of the flat dividing surface of liquid – vapor σ_∞ also depends on the temperature and becomes zero in the same point.

The thermodynamic surface tension dependence on the equimolar radius R_e is similar in its quality to that of the mechanical surface tension though it differs a little in its quantity. For relatively large drops, the surface tension approaches the value of the surface tension of the flat dividing surface. For quite small drops, the surface tension becomes zero or even negative. But the drop does not evaporate at these conditions. If the number of molecules in the simulation cell is increased at the same pressure, the drop will evaporate after all. It leads to the increase of fluctuations of the force acting on the nanodrop from the vapor. That's why the state of a nanodrop with a negative surface tension can be considered as a metastable one in respect of small fluctuations.

On the other hand, the negative surface tension and negative radius of tension surface R_s have no physical meaning and become formal mathematical parameters of the model. Apparently, it testifies that the notion of surface tension, which is used in macroscopic theory, cannot be applied to such small drops. The limit of applicability of the surface tension concept to the droplets changes from about 50 molecules at $T = 0.65$ to about 300 molecules at $T = 0.95$.

The authors are grateful to the Russian Foundation for Basic Research (RFBR) (project № 11-08-01204a) for financial support.

References

[1] Gibbs J W 1928 *Thermodynamics. The Collected Works Vol. I.* (Longmans, Green. New York)
[2] Tolman R C 1949 *J. Chem. Phys.* **17** 333.
[3] Ono S, Kondo S 1960 *Molecular Theory of Surface Tension in Liquids* (Springer – Verlag, Berlin – Gottingen – Heidelberg)
[4] Rowlinson J S, Widom B 1982 *Molecular theory of capillarity* (Clarendon Press, Oxford)
[5] Thompson S M, Gubbins K E, Walton J P R B, Chantry R A R, Rowlinson J S 1984 *J. Chem. Phys.* **81** 530.
[6] Kharlamov G V, Onischuk A A, Purtov P A, Vosel S V, Bolesta A V 2008 *Atmospheric and Oceanic Optics,* **21** 679 (in Russian).
[7] Kharlamov G V, Onischuk A A, Purtov P A, Vosel S V, Bolesta A V 2010 *e-Journal of Surface Science and Nanotechnology,* **8** 197.
[8] Kharlamov G V, Onischuk A A, Vosel S V, Purtov P A 2011 *Colloids and Surfaces* **A379** 10.

Journal of Physics: Conference Series **393** (2012) 012007 doi:10.1088/1742-6596/393/1/012007

Narrow optical band-pass filters and nanoplasmonics

Y.B. Martynov[1], R.G. Nazmitdinov [2,3], P.P. Gladyshev[1,4], and I.A. Tanachev[1,4]

[1] Institute of Applied Acoustics, 141980 Dubna, Russia
[2] Departament de Física, Universitat de les Illes Balears, 07122 Palma de Mallorca, Spain
[3] BLTP, Joint Institute for Nuclear Research, 141980 Dubna, Russia
[4] Dubna University, 141980 Dubna, Russia

E-mail: `rashid@theor.jinr.ru`

Abstract. We show that nanoparticles consisting of a dielectric core and metallic shell layer can be designed to produce a narrow optical window by employing a Fano resonance phenomenon. A mechanical model provides a remarkable description of the plasmon interactions in the dipole approximation of the Mie theory for optical response of this nanoparticle.

1. Introduction

Narrow optical band-pass filters are widely used in the systems with optical processing of information, color displays development and optical computers. Existing optical band-pass filters are based on the interference of optical modes from multiple thin dielectric layers which are manufactured by different types of expensive layer deposition techniques. The cost of such optical filters can be greatly reduced if one can apply simple non-vacuum technology for their manufacturing. The advent of nanotechnology gives impetus to the field of plasmonics which enables one to operate with light at the nanoscale, well below the scale accessible for the classical techniques. Indeed, nowadays there is an enormous experimental and theoretical activity focussed on the study of properties localized surface plasmon resonances (SRR) in metal nanoparticles. This interest is due to enhanced light absorption, spectral tuning of the SPR by controlling the size, shape, composition of metal nanoparticles and surrounding medium [1, 2, 3]. Moreover, these studies are motivated by various applications of metal nanoparticles in biosensing technology [4], for nanoscale switches [5], imaging below the diffraction limit [6], materials with negative refractive index [7]. Nanoplasmonics offers to employ various remarkable optical properties of nanoparticles for design of novel narrow band pass filters.

Recently the structural tunability of plasmon resonances in metal-based nanoshells was demonstrated [8]. Nanoshell is a nanoparticle consisting of a dielectric core which is covered by a thin metallic shell. It can be shown that such particles may have a few plasmonic resonances, depending on the number of layers in the complex nanonostructure. In particular, the interaction of the plasmons on the inner and outer surfaces of the shell gives rise to two hybridized plasmon modes which energies can be varied widely with the variation of the inner to outer shell radius ratio. The original plasmon modes could be associated with the sphere and cavity plasmon modes. To enhance the transparency in the optical band-pass we suggested to choose nanoshell material permittivity in such a way that a nanoshell dipole polarizability equals to zero in the center of the optical band-pass [9]. In this contribution we will demonstrate that a narrow optical window with the enhanced transparency can be explained by Fano resonance in strongly coupled systems, which gives rise to *a plasmon-induced transparency.*

2. The Model

Let us consider a spherical scatterer of radius a_2, composed of a homogeneous material with permittivity ε_2 and permeability μ_2, surrounded by a metallic shell with a radius a_1 with constitutive parameters ε_1 and μ_1. We assume that a nanoparticle is embedded in a typical optical plastics with the permittivity ε_m. We do not discuss a particular device but rather focus on the filtering effect produced by small nanoparticle with metallic nanoshell.

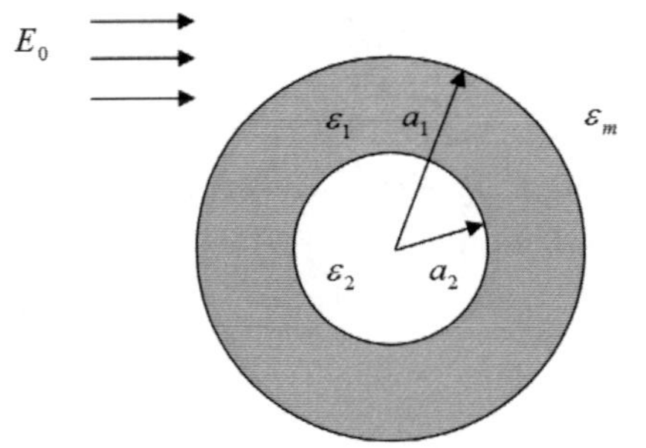

Figure 1. Cross section of a spherical scatterer composed of two concentric layers of different isotropic materials in a suitable spherical reference system.

The geometry of the problem is depicted in Fig. 1. Let the illuminating wave be represented by an $E_0e^{-i\omega t}$-monochromatic plane wave, and consider a given object centered without loss of generality at the origin of a spherical coordinate system. Adopting a standard Mie expansion, one can write the scattered field from a plane wave as the well-known discrete sum of spherical harmonics with complex amplitudes c_n^{TE} and c_n^{TM} ($n \geq 1$ being integer), respectively for TE and TM spherical waves (see, e.g. [10]). It was shown that the dipolar term ($n = 1$) is dominant in the Mie expansion for light scattering from a spherical small plasmonic particle with a radius $a \leq \lambda/10$, where λ is a wavelength [11]. Below we employ this fact to elucidate the effect of nanoparticle composed of a dielectric spherical core and a homogenous metallic shell on filtering phenomena at nanoscale.

In the dipole approximation the scattering cross section is given by

$$\sigma_{sc} = \frac{8\pi}{3}k^4|\alpha|^2 \,. \tag{1}$$

Here, $k = 2\pi\sqrt{\varepsilon_m}f/c$ is a wave number and f is a frequency of incident photon, and a dipole polarizability reads as (see, for example, [12])

$$\alpha = a_1^3\frac{(\varepsilon_1 - 1)(\varepsilon_2 + 2\varepsilon_1) + \beta(1 + 2\varepsilon_1)(\varepsilon_2 - \varepsilon_1)}{(\varepsilon_1 + 2)(\varepsilon_2 + 2\varepsilon_1) + 2\beta(\varepsilon_1 - 1)(\varepsilon_2 - \varepsilon_1)} \,. \tag{2}$$

We introduced the following notations: $\varepsilon_{1,2} \to \varepsilon_{1,2}/\varepsilon_m$, $\beta = (a_1/a_2)^3$. Hereafter, for the sake of discussion we present results for the cross sections in conventional units (c.u.).

To illustrate the validity of the dipole approximation we present results for the composite nanoparticle which consists of a dielectric core CdS with a silver shell. For the small composite nanoparticle there is a remarkable agreement between results obtained in the dipole approximation (defined by Eqs.(1), (2)), and a total cross section which contains all multipoles in the Mie expansion [10] up to $l = 10$ (see Fig.2). The filtering effect is observed between two resonance peaks. It will preserve when one considers the extinction cross section $\sigma_{ext} = \sigma_{sc} + \sigma_{ab}$ [9].

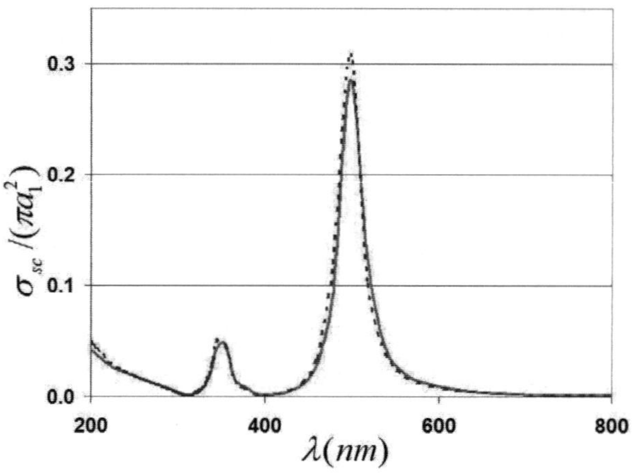

Figure 2. The comparison of the total scattering cross section calculated in the standard Mie approach [10] (dashed line) with the one calculated in the dipole approximation (solid line). The parameters for the composite nanoparticle are following: $a_1 = 10$ nm for the silver shell, the ratio $\beta = 0.12$, the dielectric core CdS with the permittivity $\varepsilon_2 = 6.5$. The permittivity of the surrounding was set to be $\varepsilon_m = 2.25$. The real and imaginary contributions to the silver permittivity ε_1 as a function of the wavelength are taken from [13].

These results can be understood from the following arguments. Evidently, if the polarization is zero $\alpha = 0$, the scattering cross section of nanoparticle becomes zero as well. It results in the drastic increase of the transparency of a medium composed of such nanoparticles. The requirement $\alpha = 0$ (which is fulfilled as soon as the numerator is zero) determines an analytical dependence of the shell properties on the size and dielectric characteristics of the nanoparticle core [9]. One obtains two solutions which we denote as $\varepsilon^t_{1,2}$. Once the denominator becomes equal to zero, the conditions are realized for the two surface plasmons (produced by the inner and outer surfaces of the shell) to be excited by incident light (an electromagnetic wave with a wave frequency $\omega = 2\pi f$). From this condition one obtains for the shell permittivity two solutions which we denote as $\varepsilon^p_{1,2}$. Taking into account these results, with the aid of the Drude model for the dielectric permittivity of the metal

$$\varepsilon_1 = \varepsilon_\infty - \frac{\omega_p^2}{\omega^2 + i\gamma\omega}, \tag{3}$$

where $\varepsilon_\infty \to \varepsilon_\infty/\varepsilon_m$ is a metal dielectric constant at high frequency, $\omega_p \to \omega_p/\varepsilon_m$ is a plasmon frequency, one obtains for the polarizability

$$\alpha = k_0 a_1^3 \frac{(x_1+i)(x_2+i)}{(y_1+i)(y_2+i)}, \quad k_0 = \frac{(\varepsilon_\infty - \varepsilon_1^t)(\varepsilon_\infty - \varepsilon_2^t)}{(\varepsilon_\infty - \varepsilon_1^p)(\varepsilon_\infty - \varepsilon_2^p)} \tag{4}$$

Here, we introduced the following notations: $x_{1,2} = (\omega^2 - {E^{(t)}_{1,2}}^2)/\gamma\omega$, $y_{1,2} = (\omega^2 - {E^{(r)}_{1,2}}^2)/\gamma\omega$, $E^{(t)}_{1,2} = \omega_p^2/(\varepsilon_\infty - \varepsilon^t_{1,2})$, $E^{(r)}_{1,2} = \omega_p^2/(\varepsilon_\infty - \varepsilon^p_{1,2})$. With the aid of the above equations, we transform the scattering cross section (1) to the form

$$\sigma_{sc} = \frac{8\pi}{3} k^4 k_0^2 a_1^6 \frac{(x_1^2+1)(x_2^2+1)}{(y_1^2+1)(y_2^2+1)} \tag{5}$$

In order to make an optical window to be narrow let us assume that the conditions for the maximum of transparency $x_1 = 0$ and for the plasmon resonance $y_1 = 0$ should be close to each other. This assumption leads to the following relation

$$x_1 = y_1 + q\,, \quad q = \frac{E_1^{(r)^2} - E_1^{(t)^2}}{\gamma\omega} \tag{6}$$

Introducing the following notations:

$$\Gamma = \frac{\gamma\omega}{\omega + \sqrt{E_1^{(r)}}}\,, \quad \epsilon = \frac{(\omega - \sqrt{E_1^{(r)}})}{\Gamma}\,, \tag{7}$$

by means of Eq.(6) we transform Eq.(5) to the form

$$\sigma_{sc} = \frac{8\pi}{3} k^4 k_0^2 a_1^6 \frac{(x_2^2+1)}{(y_2^2+1)} \frac{(\epsilon+q)^2+1}{\epsilon^2+1}. \tag{8}$$

This form corresponds to a general formula for the asymmetric resonance (Fano resonance) in a nonconservative system [14]. Thus, the basic mechanism, responsible for the appearance of the transparency windows, is due to Fano-type resonance in a complex nanoparticle.

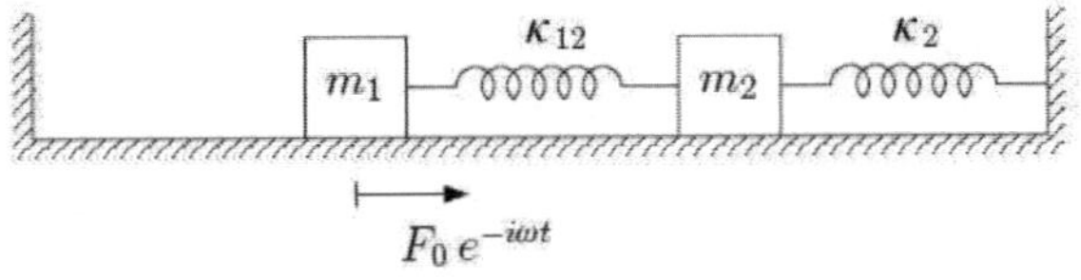

Figure 3. Two-body classical oscillator model with masses m_1, m_2 coupled with springs. Oscillator (1) is driven by the periodic harmonic force $F_0 \exp(-i\omega t)$.

Indeed, the major feature of the Fano resonance is the interference of two resonances (see for review [15]). To elucidate the connection of Eq.(2) to this mechanism let us consider a simple case $\varepsilon_2 = \varepsilon_\infty = 1$. In the Drude approximation, with the aid of Eq.(3), we obtain for the dipole polarizability

$$\alpha = \frac{1}{3}\omega_p^2(1-\beta)a_1^3 \frac{(\omega^2 - \frac{2}{3}\omega_p^2 + i\gamma\omega)}{(\frac{1}{3}\omega_p^2 - \omega^2 - i\gamma\omega)(\omega^2 - \frac{2}{3}\omega_p^2 + i\gamma\omega) + \frac{2}{9}\beta\omega_p^4}. \tag{9}$$

To shed light on the optical response of the nanoshell defined by the dipole polarizability (9) let us employ a model of two coupled oscillators [15]. Consider a mechanical system consisting of two masses m_1 and m_2, which slide over a horizontal surface with friction parameters γ_1 and γ_2, respectively. Suppose that the masses are attached to one another, and to one immovable wall, by means of two light horizontal springs of spring constant k_2 and k_{12}, as shown in Fig.3. Left mass m_1 is driven by a periodic harmonic force $F(t) = F_0 e^{-i\omega t}$. The instantaneous state of the system is conveniently specified by the displacements of the left and right masses, $q_1(t)$ and $q_2(t)$, respectively. The extensions of the middle and right springs are thus $q_1 - q_2$ and q_2, respectively, assuming that $q_1 = q_2 = 0$ corresponds to the equilibrium configuration in which the springs are all unextended. The equations of motion of the first and the second oscillators are thus

$$\begin{cases} \ddot{q}_1 + \gamma\dot{q}_1 + \omega_1^2 q_1 - \Omega_1^2 q_2 = (F_0/m_1)e^{-i\omega t} \\ \ddot{q}_2 + \gamma\dot{q}_2 + \omega_2^2 q_2 - \Omega_2^2 q_1 = 0 \end{cases} \tag{10}$$

Here, $\omega_1^2 = \Omega_1^2 = k_{12}/m_1$, $\omega_2^2 = (k_2 + k_{12})/m_2$, $\Omega_2^2 = k_{12}/m_2$. To make a connection to Eq.(9), without losing generality, we have introduced $\gamma = \gamma_1/m_1 = \gamma_2/m_2$. Assuming the solutions in the form $q_{1,2} = c_{1,2}e^{-i\omega t}$, one finds

$$c_1 = -\frac{F_0}{m_1} \frac{\omega^2 - \omega_2^2 + i\gamma\omega}{(\omega_1^2 - \omega^2 - i\gamma\omega)(\omega^2 - \omega_2^2 + i\gamma\omega) + \Omega_1^2\Omega_2^2}, \quad c_2 = -c_1 \frac{\Omega_2^2}{\omega^2 - \omega_2^2 + i\gamma\omega} \tag{11}$$

Thus, in the Drude approximation the dipole polazability Eq.(9) coincides exactly with the definition of the coefficient c_1, Eq.(11), if we assume that $(F_0/m_1) = -\omega_p^2/3(1-\beta)a_1^3$, $\omega_1 = \omega_p/\sqrt{3}$, $\omega_2 = \omega_p \times \sqrt{2/3}$, $\Omega_2^2 = \beta\omega_2^2$. In this case ω_1 and ω_2 correspond to the sphere and cavity plasmon modes, respectively.

There is a total suppression of the amplitude of the system at the eigenfrequency ω_2 wihout the dissipation ($\gamma = 0$). The amplitude of the first oscillator becomes zero as a result of destructive interference of oscillations from the external force and the second oscillator. This model demonstrates one of the basic properties of the Fano resonance, namely, resonant destructive interference at a particular optical frequency.

3. Summary

In conclusion, we have shown that nanoshell can be design with a highly tunable Fano resonance. Varying the size, the ratio $\beta = (a_2/a_1)^3$, the permitttivities of the core and the shell it is possible to reach a narrow optical window which can be used to create effective narrow band pass filters.

Acknowledgments

This work is partly supported by Russian Federal Program Grant No 2011-1.3-513-054-006, RFBR Grant No.11-02-00086 (Russia), and by Grant No. FIS2008-00781/FIS (Spain).

References

[1] Schuller J A, *et al.*, 2010 *Nat. Mat.* **9**, 193
[2] Luk'yanchuk B, *et al.*, 2010 *Nat. Mat.* **9**, 707
[3] Klimov V, 2011 *Nanoplasmonics* (Singapore: Pan Stanford Publishing)
[4] Kabashin A V, *et al.*, 2009 *Nat. Mat.* **8**, 867
[5] Krasavin A V and Zheludev N I, 2004 *Appl. Phys. Lett.* **84**, 1416
[6] Pendry J B, 2000 *Phys. Rev. Lett.* **85**, 3966
[7] Fang N, Lee H, Sun C, and Zhang X, 2005 *Science* **308**, 534
[8] Prodan E, Radloff C, Halas H J, and Nordlander P, 2003 *Science* **302**, 419
[9] Martynov Y B, Nazmitdinov R G, Tanachev I A, and Gladyshev P P, 2012 *JETP Lett.* **95**, 122
[10] Jackson J D, 1999 *Classical Electrodynamics* (NewYork: Wiley) 3rd ed.
[11] Alú A and Engheta N, 2005 *Phys. Rev. E* **72**, 016623
[12] Kerker M, 1975 *J. Opt. Soc. Am.* **65**, 376; Chew H and Kerker M, 1976 *ibid* **66**, 445
[13] Palik E D, 1998 *Handbook of Optical Constants of Solids* vol.1
[14] Gallinet B and Martin O J F, 2011 *Phys. Rev. B* **83**, 235427
[15] Miroshnichenko A E, Flach S, Kivshar Yu S, 2010 *Rev. Mod. Phys.* **82**, 2257

Magnetoresistance of two-dimensional electrons with spin-orbit coupling disorder

E Ya Sherman[1,2], V K Dugaev[3,4,5], M M Glazov [6], J Barnaś[7,8]

[1] Department of Physical Chemistry, Universidad del País Vasco UPV-EHU, 48080, Bilbao, Spain
[2] IKERBASQUE Basque Foundation for Science, Alameda Urquijo 36-5, 48011 Bilbao, Spain
[3] Department of Physics, Rzeszów University of Technology, al. Powstańców Warszawy 6, 35-959 Rzeszów, Poland
[4] Department of Physics and CFIF, Instituto Superior Técnico, TU Lisbon, Av. Rovisco Pais, 1049-001 Lisbon, Portugal
[5] Institut für Physik, Martin-Luther-Universität Halle-Wittenberg, Heinrich-Damerow-Str. 4, 06120 Halle (Saale), Germany
[6] Ioffe Physical-Technical Institute RAS, 194021 St.-Petersburg, Russia
[7] Department of Physics, Adam Mickiewicz University, Umultowska 85, 61-614 Poznań, Poland
[8] Institute of Molecular Physics, Polish Academy of Sciences, Smoluchowskiego 17, 60-179 Poznań, Poland

E-mail: evgeny_sherman@ehu.es

Abstract. We present analysis of random spin-orbit coupling in two-dimensional electron systems. This randomness leads to several interesting order-from-disorder phenomena, where this random interaction having nanoscale spatial size produces or modifies macroscopic spin and charge dynamics. Anisotropic negative magnetoresistance related solely to the spin dynamics of electrons is one of the examples of these effects and will be considered here in detail.

1. Introduction

Spin-dependent transport phenomena in solids are in the basis of spintronics, a rapidly developing field of fundamental and applied physics. One of the spintronics keywords is spin-orbit coupling, usually considered as the coupling between electron spin and momentum. The spin-orbit coupling leads to at least two classes of phenomena [1, 2, 3] such as (1) spin relaxation and (2) ability of spin manipulation by external electric field. In addition, spin-orbit coupling determines two novel transport effects: the spin-Hall- [4, 5] and the anomalous Hall effect [6, 7]. In the former the electric field applied to a sample causes spin current while in the latter it leads to a transversal current in a system with finite magnetization even if the external magnetic field is zero.

On the other hand, an interesting aspect of the influence of magnetic field on electric current is the magnetoresistance, that is the dependence of the system conductivity on the applied magnetic field. This effect is usually attributed to the change in the electron motion in external magnetic field. However, in the systems with spin-orbit coupling one expects that the magnetization of the system influencing electron spin will also somehow influence the orbital motion and, as a result, lead to a nonzero magnetoresistance.

This paper focuses on two major points. First, for completeness, we discuss spin-orbit coupling as a random function of the electron position. Second, we demonstrate that this randomness leads to a very unusual negative magnetoresistance, where the conductivity increases when the magnetic field is applied. We estimate the magnitude of this effect and show that it can be observable in narrow-gap semiconductor structures such as InSb-based quantum wells.

2. Physics and Hamiltonian of random spin-orbit coupling

We begin with the conventional Rashba Hamiltonian of spin-orbit coupling in the form [8]:

$$H_R = \alpha \left(\sigma_x k_y - \sigma_y k_x \right), \tag{1}$$

where σ_i and k_j are the Cartesian components of spin and momentum (we use $\hbar \equiv 1$), respectively. The coupling constant α is the material and structure-dependent parameter, which appears due to the structure inversion asymmetry with respect to the $z \to -z$ operation, where z is perpendicular to the structure, and vanishes if the structure is fully symmetric. If the asymmetry with respect to the reflection in the structure plane is caused by the electric field applied across the quantum well, the coupling can be expressed in terms of the field as [9]:

$$\alpha = \xi e \mathcal{E}_z, \tag{2}$$

where ξ is a constant, e is the fundamental charge, and $\mathcal{E}_z$ is the field component. The value of ξ rapidly increases with the atomic mass of the compound. The numerical examples are: $\xi = 0.05$ nm^2 in GaAs, $\xi \sim 1$ nm^2 in typical InGaAs alloys, and $\xi = 5$ nm^2 for InSb systems. Random spatial variations in the electric field cause random spatial variations in the spin-orbit coupling. The Rashba Hamiltonian requires modification to restore the Hermitian form and becomes [10]:

$$H_R = \frac{1}{2} \left(\sigma_x \left\{ k_y, \alpha \right\} - \sigma_y \left\{ k_x, \alpha \right\} \right), \tag{3}$$

where $\{...\}$ stands for anticommutator.

We consider an example of random spin-orbit coupling in semiconductor structure, where the details can be understood in terms of a simple model. The structure consists of a two-dimensional conducting channel and two dopant layers positioned at $z = \pm R_d$ with the two-dimensional concentration of dopants per layer n_d (Fig.1). The average electric field in the well is, therefore, zero. Figure 1 shows, however that inevitable fluctuations in the dopant density break the symmetry locally. As a result, random electric field appears in the system and leads to the random Rashba field. The z-component of the Coulomb field of the dopant ions is given by the sum over the dopant layers

$$\mathcal{E}_z(\boldsymbol{\rho}) = -\frac{e}{\varkappa} \sum_j f(\boldsymbol{\rho}, \mathbf{r}_j), \tag{4}$$

where the dopant j position $\mathbf{r}_j = (\mathbf{r}_{\parallel}^j, z)$ and $\varkappa$ is the dielectric constant. Function $f(\boldsymbol{\rho}, \mathbf{r}_j)$ is the Coulomb field in the form:

$$f(\boldsymbol{\rho}, \mathbf{r}_j) = \frac{z}{\left[\left(\boldsymbol{\rho} - \mathbf{r}_{\parallel}^j \right)^2 + R_d^2 \right]^{3/2}}. \tag{5}$$

We assume that the correlation function of the dopant concentration is the "white noise"

$$\langle (n(\mathbf{r}_1) - n_d)(n(\mathbf{r}_2) - n_d) \rangle = n_d \delta(\mathbf{r}_1 - \mathbf{r}_2), \tag{6}$$

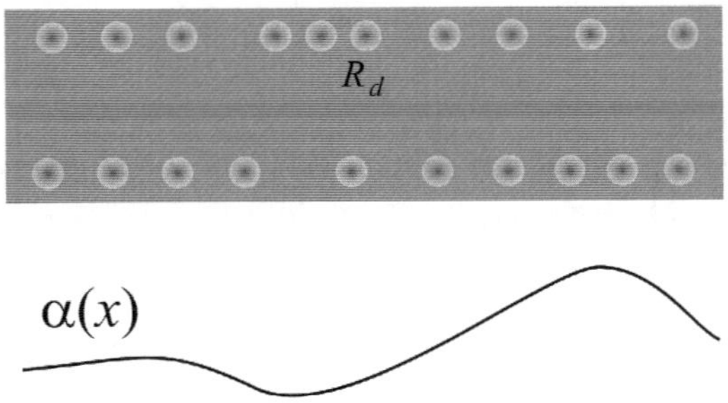

Figure 1. (Upper panel) Single quantum well surrounded by two dopant layers. Filled circles schematically show positions of dopant ions. (Lower panel) A typical profile of a random spin-orbit coupling produced by given distribution of dopants.

and the fluctuations obey the Gaussian statistics, where the higher-order correlatives can be reduced to the lower-order ones. Under these assumptions the correlation function of random electric field z-components, $\langle \mathcal{E}_{z,\mathrm{r}}(0)\mathcal{E}_{z,\mathrm{r}}(\boldsymbol{\rho})\rangle$, where subscript r corresponds to the random contribution, depends on the distance between the points only. By using Eqs.(4)-(6) we obtain the correlation function of the random electric field:

$$\langle \mathcal{E}_{z,\mathrm{r}}(0)\mathcal{E}_{z,\mathrm{r}}(\boldsymbol{\rho})\rangle = \langle \mathcal{E}^2_{z,\mathrm{r}}\rangle F_{\mathrm{corr}}(\rho), \tag{7}$$

with

$$\langle \mathcal{E}^2_{z,\mathrm{r}}\rangle = 2\pi \left(\frac{e}{\varkappa}\right)^2 \frac{n_d}{R_d^2}. \tag{8}$$

The resulting fluctuation in the Rashba coupling is of the order of

$$\alpha_{\mathrm{r}} \equiv e\xi\sqrt{\langle \mathcal{E}^2_{z,\mathrm{r}}\rangle} = \sqrt{2\pi}\xi\frac{e^2}{\varkappa}\frac{\sqrt{n_d}}{R_d}. \tag{9}$$

To obtain a numerical estimate for a narrow-gap semiconductor such as InSb, we assume typical parameters $n_d = 5\times 10^{11}$ cm^{-2}, $R_d = 20$ nm and obtain α_{r} of the order of $0.5 - 1\times 10^{-6}$ meVcm.

With the increase in the distance to the dopant layers, the random variations in the electric field $\mathcal{E}_{z,\mathrm{r}}(\rho)$ become weaker and smoother since the electric field becomes only weakly sensitive to the details of the charge distribution. It is interesting to mention that correlation length of the random electric field is solely determined by the distance to the dopant layer rather than by the distance between the dopant ions since the white-noise distribution does not have a characteristic spatial scale and the correlations pick-up the scale of the distance to the dopant layer.

This randomness can be crucial for systems where spin dynamics is prohibited otherwise. An important example is given by symmetric GaAs quantum wells grown along the [110]-direction, where the symmetry of the Dresselhaus coupling prohibits evolution of the spin component along the growth direction. On the other hand, the random Rashba coupling in these systems leads to a variety of interesting phenomena, including spin relaxation [11, 12] and nonlinear spin-Hall effect at finite frequency of the electric field [13], which can be seen as the "order-from-disorder" effect, where the macroscopic current is solely due to randomness in the driving spin-orbit field. Another kind of the "order-from-disorder" is the conventional spin-Hall effect, which, in contrast to common wisdom, does not vanish in the presence of disorder [14].

To make the picture of random spin-orbit coupling in low-dimensional systems more complete, we mention several other examples where the randomness of spin-orbit coupling is important.

First example is given by Si/Ge interfaces in the Si_xGe_{1-x}/Si quantum wells, where random alignment of the bonds leads to the random spin-orbit coupling of the Rashba or Dresselhaus type [15]. Second example is given by graphene, where adatoms and long-range ripples with randomly coordinate-dependent curvature form a short-range and long-range random spin-orbit fields, respectively [16, 17]. Third example is topological insulators, including HgTe quantum wells, where random spin-orbit coupling combined with electron-electron interaction leads to a variety of nontrivial transport phenomena [18]. Fourth, recent developments on the tunneling spectroscopy of the depletion layers in Cs-covered InSb systems clearly demonstrate the random coordinate dependence of the Rashba coupling [19] seen as coordinate-dependent spin splitting of the Landau levels. Finally, we mention recent developments on the physics of thin Bi films showing the random distribution of the spin-orbit coupling there [20].

In the momentum representation Eq. (3) can be rewritten as

$$H^{(so)}_{\mathbf{kk}'} = \frac{\alpha_{\mathbf{kk}'}}{2}\left[\sigma_x(k_y + k'_y) - \sigma_y(k_x + k'_x)\right]. \tag{10}$$

The Fourier transform of the correlation function of fields in Eq. 10 is given by:

$$\langle\alpha^2\rangle_q = 2\pi\langle\alpha^2\rangle\mathcal{F}(q) \tag{11}$$

where $\mathcal{F}(q)$ is the system-dependent range function. A simple generic approximation for this function can be given by $\mathcal{F}(q) = R^2 e^{-qR}$; the result becomes exact in the case of random Rashba field produced by the Coulomb field charged dopants shown in Fig.1 with $R = 2R_d$. This equation demonstrates two most important features of the random spin-orbit coupling: variation $\langle\alpha^2\rangle$ and correlation length R. The knowledge of these two parameters allows one in general to study the related to the random spin-orbit coupling effects.

3. Magnetoresistance due to random spin-orbit coupling

3.1. Hamiltonian for kinetic phenomena in magnetic field.

We describe two-dimensional electron gas of interest by the Hamiltonian $\hat{H} = \hat{H}_0 + \hat{H}_R$, where (we use the units with $\hbar \equiv 1$)

$$\hat{H}_0 = -\frac{\nabla^2}{2m} + U_{\text{rnd}}(\mathbf{r}) + \sigma_x\beta B, \tag{12}$$

corresponding to the magnetic field along the x-axis in the quantum well plane with m being the electron effective mass. The term $U_{\text{rnd}}(\mathbf{r})$ describes the spin-diagonal disorder assumed to be of white-noise type, and $\beta = g\mu_B/2$, where g is the electron Landé factor. Assuming in-plane magnetic field we avoid the diamagnetic orbital effects, and therefore take into account only the Zeeman term. The magnetization leads to splitting of the Fermi surface, and in the limit of small magnetization $\Delta K = 2|\beta|Bm/k_F$, as shown in Fig. 2, where k_F is the Fermi momentum at $B = 0$. Here the weak magnetization condition implies $\Delta K \ll k_F$.

The random spin-orbit coupling leads to spin-flip scattering (Fig.2(a)) between states with opposite spin orientations with the minimum momentum transfer equal to $|\Delta K|$ as well as to spin-conserving scattering (Fig.2(b)) with the zero minimum momentum transfer. Probabilities of both these scatterings depend on the direction of the electron momentum. We will see in the next subsection that these two features of the random spin-orbit coupling lead to the negative anisotropic magnetoresistance.

3.2. Negative anisotropic magnetoresistance.

Conventional dependence of the conductivity of two-dimensional electron gas on magnetic field perpendicular to the structure is due to the orbital motion and can be expressed as:

$$\sigma(B) = \frac{\sigma(0)}{1 + \omega_c^2\tau^2}, \tag{13}$$

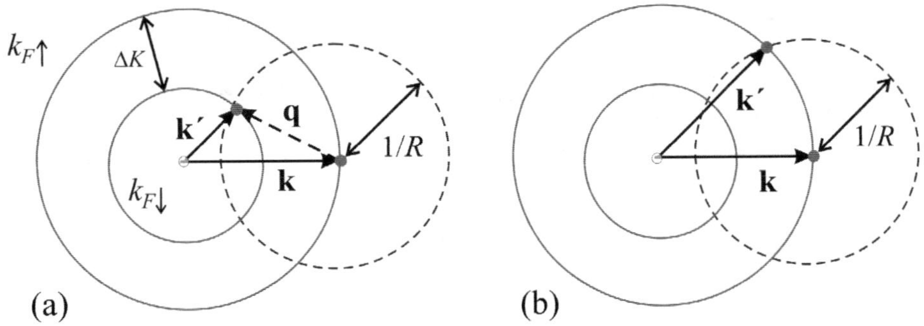

Figure 2. Spin-split Fermi surface. $1/R$ is the possible range of momentum change, $q = |\mathbf{k}' - \mathbf{k}|$, in the spin-flip scattering process between two Fermi surfaces. In the limit of small Fermi surface splitting, $\Delta K = 2|\beta|Bm/k_F$ (spin orientation is taken with respect to the x -axis). This geometry with $k_{F\uparrow} > k_{F\downarrow}$ corresponds to a negative g-factor of electron. Panel (a) shows spin-flip transitions while panel (b) corresponds to the spin-conserving ones.

where $\omega_c = eB/mc$ is the cyclotron frequency, and $\sigma(0)$ is the zero-field conductivity. As can be seen in Eq.13, usually, the conductivity decreases in a magnetic field the Lorenz force prevents straightforward motion on electrons, and, as result, works against the electron propagation necessary to carry electric current. However, the orbital motion in some special kinds of disorder can be more complicated and the magnetic–field dependence of the conductivity can correspond to the increase rather than to the decrease with the magnetic field [21]. Also, quantum effects such as weak antilocalization can lead to negative magnetoresistance in a certain interval of magnetic fields [22, 23]. However, in the presence of spin-orbit coupling, the orbital motion, and, as a result, the conductivity, becomes spin-dependent. Here we explore this possibility for the random spin-orbit coupling and show its unusual consequences.

In magnetic field the conductivity can be expressed as the sum of spin-dependent terms:

$$\sigma(B) = \frac{e^2}{m}\left(n_\uparrow\tau_\uparrow + n_\downarrow\tau_\downarrow\right), \tag{14}$$

where $n_\uparrow$ and $n_\downarrow$ are corresponding concentrations, and $\tau_\uparrow$ and $\tau_\downarrow$ are the momentum relaxation times, dependent on magnetic field. These times can demonstrate anisotropy, which will be of our interest below. To present the analysis in a simple and general form, we use the conventional Boltzmann kinetic equation approach for the momentum- and spin-dependent distribution functions $f_{\mathbf{k}\downarrow}$ and $f_{\mathbf{k}\uparrow}$:

$$\frac{e}{m}(\mathbf{E}\cdot\mathbf{k})\frac{\partial f_{\mathbf{k}\uparrow}}{\partial\varepsilon_k} = \int\frac{d^2k'}{(2\pi)^2}\left[W_{\mathbf{kk}'}\left(f_{\mathbf{k}\uparrow} - f_{\mathbf{k}'\uparrow}\right) + W^f_{\mathbf{kk}'}\left(f_{\mathbf{k}\uparrow} - f_{\mathbf{k}'\downarrow}\right)\right] \tag{15}$$

$$\frac{e}{m}(\mathbf{E}\cdot\mathbf{k})\frac{\partial f_{\mathbf{k}\downarrow}}{\partial\varepsilon_k} = \int\frac{d^2k'}{(2\pi)^2}\left[W_{\mathbf{kk}'}\left(f_{\mathbf{k}\downarrow} - f_{\mathbf{k}'\downarrow}\right) + W^f_{\mathbf{kk}'}\left(f_{\mathbf{k}\downarrow} - f_{\mathbf{k}'\uparrow}\right)\right], \tag{16}$$

where $\varepsilon_k = k^2/2m$. The scattering probabilities are

$$W_{\mathbf{kk}'} = 2\pi\left[w_{\mathbf{kk}'} + \frac{1}{4}|\alpha_{\mathbf{kk}'}|^2(k_y + k'_y)^2\right]\delta(\varepsilon_k - \varepsilon_{k'}), \tag{17}$$

$$W^f_{\mathbf{kk}'} = 2\pi\frac{|\alpha_{\mathbf{kk}'}|^2}{4}(k_x + k'_x)^2\,\delta(\varepsilon_k - \varepsilon_{k'} \pm 2\beta B), \tag{18}$$

for spin-conserving ($W_{\mathbf{kk}'}$) and spin-flip $\left(W^f_{\mathbf{kk}'}\right)$ processes, respectively. The $w_{\mathbf{kk}'}$- term is due to spin-diagonal disorder described by the $U_{\rm rnd}(\mathbf{r})$ term in Eq.10 and is assumed to be a constant w_0 below. Change in the electron momentum: $q = |\mathbf{k}-\mathbf{k}'| = \sqrt{(\Delta K)^2 + k_F^2\gamma^2}$ in the spin-flip process, and $q = k_F|\gamma|$ in the spin-conserving process, where γ is the scattering angle.

Equations 17 and 18 show two features of the spin-dependent scattering: anisotropy and strong dependence on the magnetization. We concentrate on the semiclassical regime $k_F R \gg 1$, realizable in experimentally available quantum wells. To get an idea of time scales involved in the spin-dependent scattering, we introduce the spin relaxation time τ_s due to the spin-orbit coupling [10]:

$$\frac{1}{\tau_s} = 4m\langle\alpha^2\rangle k_F R, \tag{19}$$

which will be used below for comparison. This expression corresponds to the generalized Dyakonov-Perel' spin relaxation mechanism due to the random spin precession in the random spin-orbit field.

We begin with the zero magnetic field limit, where the splitting of the Fermi surface is very small, that is $R\Delta K \ll 1$. For $k_F R \gg 1$, the spin-dependent scattering occurs only at small angles $\leq 1/k_F R$ and effectively becomes independent of the direction of the electron momentum since the sum $W_{\mathbf{kk}'} + W^f_{\mathbf{kk}'}$ does not depend on the direction of $\mathbf{k}$ here. In this approximation we can write

$$\frac{eE\cos\varphi}{m}\frac{\partial f_k^0}{\partial\varepsilon_k} = 2\pi\nu\int\frac{d\varphi'}{2\pi}\left(w_0 + 2\pi k^2R^2\left\langle\alpha^2\right\rangle e^{-kR|\varphi'-\varphi|}\right)(f_{\mathbf{k}} - f_{\mathbf{k}'}) \tag{20}$$

where φ is the angle between $\mathbf{E}$ and $\mathbf{k}$ (Fig.3) whereas φ' is the angle between $\mathbf{E}$ and $\mathbf{k}'$, $\nu = m/2\pi$ in the density of states per single spin component, and f_k^0 is the equilibrium distribution function.

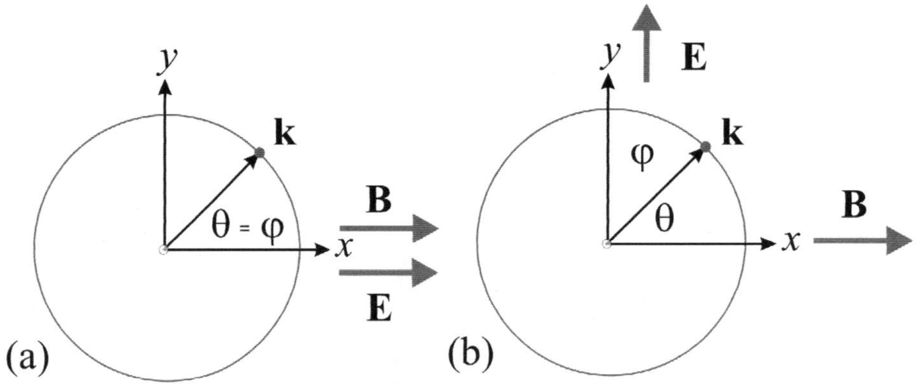

Figure 3. Geometry of magnetoresistance. (a) longitudinal magnetoresistance, (b) transversal magnetoresistance.

For $B = 0$, where the scattering anisotropy is unimportant, we obtain a conventional Drude conductivity

$$\sigma(0) = \frac{ne^2\tau}{m}, \tag{21}$$

with the momentum relaxation time modified by the spin-orbit coupling disorder as:

$$\frac{1}{\tau} = \frac{1}{\tau_0} + \frac{1}{\tau_1},\quad \frac{1}{\tau_1} = \frac{2m\left\langle\alpha^2\right\rangle}{k_F R} = \frac{1}{\tau_s}\times\frac{1}{2\left(k_F R\right)^2}. \tag{22}$$

Here τ_0 is the relaxation time due to scattering from short-range spin-diagonal disorder. The correction to the conductivity can be expressed in terms of the spin relaxation time and the $k_F R$ parameter, which suppresses the correction to the conductivity since the spin-flip scattering occurs at small angles only.

Now we can study the effect of strong magnetic field, where spin-flip scattering is almost suppressed, and only spin-conserving terms (Fig.2(b)) remain. Since one of contributions (spin-flip one) in the electron scattering is completely suppressed, and the remaining spin-conserving one is anisotropic, one can expect that the conductivity of the electron gas will increase and become, in turn, anisotropic. In these strong fields $|\beta| BmR/k_F \gg 1$, the kinetic equation reads

$$\frac{eE\cos\varphi}{m}\frac{\partial f_k^0}{\partial \varepsilon_k} = 2\pi\nu \int \frac{d\theta'}{2\pi}\left(w_0 + 2\pi k_F^2 R^2 \left\langle \alpha^2 \right\rangle e^{-kR|\theta-\theta'|}\sin^2\theta\right)(f_{\mathbf{k}} - f_{\mathbf{k}'}), \tag{23}$$

where $\mathbf{k} = k(\cos\theta, \sin\theta)$ and $\mathbf{k}' = k'(\cos\theta', \sin\theta')$. We begin with the longitudinal magnetoresistance for $\mathbf{E} \parallel \mathbf{B}$. In this case the angle φ between $\mathbf{E}$ and $\mathbf{k}$ is equal to θ, as shown in Fig.3(a). To account for the scattering anisotropy, we seek for solution of the Boltzmann equation in the form:

$$f_{\mathbf{k}} = f_k^0 + kE(g_1\cos\varphi + g_2\cos^3\varphi) \tag{24}$$

with g_1, g_2 being scalar functions of $\mathbf{k}$ and obtain

$$g_1 = \frac{e\tau}{m}\frac{\partial f_k^0}{\partial \varepsilon_k}, \qquad g_2 = \frac{\tau_0}{\tau_1} g_1. \tag{25}$$

The term $g_2\cos^3\varphi$ in the distribution function leads to the increase and anisotropy in the conductivity due to suppressed spin-flip scattering. The resulting conductivity is

$$\sigma_l(B) = \sigma(0)\left(1 + \frac{3\tau_0}{4\tau_1}\right), \tag{26}$$

and the longitudinal magnetoresistance is negative.

Now we calculate the transversal magnetoresistance taking into account that $\theta + \varphi = \pi/2$ for $\mathbf{B} \perp \mathbf{E}$ (Fig.3(b)). Using the solution similar to Eq.24, we obtain:

$$\sigma_t(B) = \sigma(0)\left(1 + \frac{\tau_0}{4\tau_1}\right) < \sigma_l(B). \tag{27}$$

The transversal conductivity increases, however, less than the longitudinal one, leading to anisotropic magnetoresistance. This anisotropy is due to the properties of the electron scattering in the random Rashba field, namely due to the $\sin^2\theta$ in the scattering probability in Eq.23. This factor vanishes for electron momentum along the x-axis and has maximum for $\mathbf{k}$ parallel to the y-axis. To understand the scale of the effect in InSb we estimate τ_1 for $k_F R = 5$ as 5×10^{-11} s. Taking into account that for the mobility $\mu = 10^5$ cm^2/Vs, the relaxation time τ_0 is of the order of 10^{-12} s, the decrease in the resistivity and the corresponding anisotropy can of the order of few percent, sufficient for the experimental observation. The required magnetic field to split the Fermi surface and observe this effect is of the order of one Tesla, dependent on the system parameters.

Here we mention that this effect is not related to the negative magnetoresistance in the weak antilocalization regime [22, 23] and can be related to the spin-dependent phenomena in the negative magnetoresistance recently observed experimentally [24].

4. Conclusions.

We considered two-dimensional electron gas with a spatially random spin-orbit coupling. This randomness having the correlation length of the order of 10...100 nm arises due to a conventional disorder always present in these systems. The scattering of electrons by the randomly fluctuating spin-orbit coupling is strongly sensitive to the external magnetic field and leads to new interesting "order-from-disorder" kinetic phenomena in the charge and spin channels. As an example, we studied the negative anisotropic magnetoresistance solely related to this random spin-orbit coupling. These effects can be observed experimentally and used in spintronics applications.

Acknowledgments

This work is partly supported by the DFG in Germany and by National Science Center in Poland as a research project in years 2011 – 2014. The work of EYS was supported by the MCINN of Spain grant FIS 2009-12773-C02-01,d "Grupos Consolidados UPV/EHU del Gobierno Vasco" grant IT-472-10, and by the UPV/EHU under program UFI 11/55. We are grateful to M. Morgenstern, M. Oestreich, R. Winkler, and M.-W. Wu for valuable discussions.

References

[1] Zutić I, Fabian J and Das Sarma S 2004 Rev. Mod. Phys. **76** 323
[2] Fabian J, Matos-Abiague A, Ertler C, Stano P and Zutić I 2007 Acta Phys. Slovaca **57** 565
[3] Wu M W, Jiang J H and Weng M Q 2010 Phys. Rep. **493** 61
[4] Engel H A, Rashba E I and Halperin B I 2007 in *Handbook of Magnetism and Advanced Magnetic Materials, v 5: Spintronics and Magnetoelectronics* (eds. Kronmüller H and Parkin S, John Wiley & Sons, New York) p. 2858
[5] *Spin Physics in Semiconductors* 2008 (ed. Dyakonov M I, Springer, Berlin)
[6] Nagaosa N, Sinova J, Onoda S, MacDonald A H and Ong N P 2010 Rev. Mod. Phys. **82** 1539
[7] Nunner T S *et al.* 2007 *Phys. Rev.* B **76** 235312
[8] Rashba E I 1964 *Sov. Phys. Solid State* **2** 1874; Bychkov Yu A and Rashba E I 198) *JETP Lett.* **39** 79
[9] Winkler R 2003 *Spin-Orbit Coupling Effects in Two-Dimensional Electron and Hole Systems* (Springer, Berlin)
[10] Glazov M M, Sherman E Ya and Dugaev V K 2010 Physica E **42** 2157
[11] Müller G M, Römer M, Schuh D, Wegscheider W, Hübner J and Oestreich M 2008 *Phys. Rev. Lett.* **101** 206601
[12] Griesbeck M, Glazov M M, Sherman E Ya, Schuh D, Wegscheider W, Schüller C and Korn T 2012 *Phys. Rev.* B **85** 085313
[13] Ivanov V I, Dugaev V K, Sherman E Ya and Barnaś J 2011 *Phys. Rev.* B **84** 085326
[14] Dugaev V K, Inglot M, Sherman E Ya and Barnaś J 2010 *Phys. Rev.* B **82** 121310
[15] Golub L E and Ivchenko E L 2004 *Phys. Rev.* B **69** 115333
[16] Ertler C, Konschuh S, Gmitra M and Fabian J 2009 *Phys. Rev.* B **80** 041405
[17] Zhang P and Wu M W 2012 *New J. Phys.* **14** 033015
[18] Ström A, Johannesson H and Japaridze G I 2010 *Phys. Rev. Lett.* **104** 256804
[19] Morgenstern M, Georgi A, Straßer C, Ast C R, Becker S and Liebmann M 2012 Physica E **44** 1795
[20] Rudolph M and Heremans J J 2012 Appl. Phys. Lett. **100** 241601
[21] Mirlin A D, Polyakov D G, Evers F and Wölfle P 2001 *Phys. Rev. Lett.* **87** 126805; Mirlin A D, Wilke J, Evers F, Polyakov D G and Wölfle P 1999 *Phys. Rev. Lett.* **83** 2801
[22] Faniel S, Matsuura T, Mineshige S, Sekine Y and Koga T 2011 *Phys. Rev.* B **83** 115309 and references therein.
[23] M. Kohda *et al.* 2012 arXiv:1205.5630.
[24] Hatke A T, Zudov M A, Reno J L, Pfeiffer L N and West K W 2012 *Phys. Rev.* B **85** 081304

Dubna-Nano 2012 IOP Publishing
Journal of Physics: Conference Series **393** (2012) 012009 doi:10.1088/1742-6596/393/1/012009

A geometrical crossover in excited states of two-electron quantum dots in a magnetic field

R.G. Nazmitdinov [1,2], N. S. Simonović [3], A. R. Plastino [4,5], and A. V. Chizhov [2]

[1] Departament de Física, Universitat de les Illes Balears, 07122 Palma de Mallorca, Spain
[2] BLTP, Joint Institute for Nuclear Research, 141980 Dubna, Russia
[3] Institute of Physics, University of Belgrade, 11001 Belgrade, Serbia
[4] Instituto Carlos I de Fisica Teorica y Computacional, Universidad de Granada, 18071 Granada, Spain
[5] National University La Plata, UNLP-CREG-CONICET, C.C. 727, La Plata 1900, Argentina

E-mail: rashid@theor.jinr.ru

Abstract. We use the entanglement measure to study the evolution of quantum correlations in two-electron axially-symmetric parabolic quantum dots under a perpendicular magnetic field. We found that the entanglement indicates on the shape transition in the density distribution of two electrons in the lowest state with zero angular momentum projection at the specific value of the applied magnetic field.

1. Introduction

Nowadays there is a growing interest in using quantum entanglement measures for study of quantum correlations in topologically ordered systems [1]. The analysis of these systems is a highly non-trivial task due to the absence of the order parameter. Particular examples are the integer and fractional quantum Hall liquids which cannot be understood in terms of the traditional description of phases based on symmetry breaking and local order parameters. The main stream of such an analysis is focused on many-particle one-dimensional systems (cf [2]). Recently, topological phases were studied in finite systems such as two-dimensional electrons at very high magnetic field (Laughlin states) [3] and weakly interacting two-dimensional rotating Bose-Einstein condensate [4]. There are attempts to find a relation between topological orders of different topological phases and quantum phase transitions, driven by quantum fluctuations at zero temperature in many-body systems (see a textbook [5]).

It is believed that various quantum phases could exist in quantum dots (QDs) at different strengths of the applied perpendicular magnetic field [6]. At small magnetic field strengths one observes the orbital momentum and spin oscillations of the ground state of a QD by increasing the field strength. At certain field range the oscillations disappear and electrons form a fully polarized state called the maximum density droplet. It is expected that a further increase of the magnetic field should lead to the formation of the Wigner molecule, a finite-size analogue of the Wigner crystallization of the homogeneous electron gas. A natural question arises: if QDs can be considered as a finite-size analogy of conventional condensed matter systems what are signatures of quantum phase transitions in QDs ?

To shed light on this question we will employ the entanglement and compare its evolution with the evolution of quantum spectra of QDs as a function of the magnetic field. Evidently, finite systems can only show precursors of the QPT behaviour. However, they are also important for the development of the concept. Two-electron QDs being realictic tractable nontrivial systems are, in particular, attractive because their eigenstates can be obtained very accurately, or in some cases, exactly (cf [7, 8]). Moreover, it was found that at certain values of the magnetic field quantum spectra of two-electron QDs become degenerate due to onset of the spherical symmetry [9, 10]. The goal objective of the present paper is to demonstrate that a quantum entanglement can be used to indicate this transition in a three-dimensional (3D) quantum dot under a magnetic field.

2. Basics

Our analysis is carried out by means of the numerical diagonalization of the Hamiltonian

$$H = \sum_{j=1}^{2} \left[\frac{1}{2m^*} \left(\mathbf{p}_j - \frac{e}{c}\mathbf{A}_j \right)^2 + U(\mathbf{r}_j) \right] + \frac{k}{|\mathbf{r}_1 - \mathbf{r}_2|} + H_{spin}. \tag{1}$$

Here $k = e^2/4\pi\varepsilon_0\varepsilon_r$ and $H_{spin} = g^*\mu_B(\mathbf{s}_1 + \mathbf{s}_2)\cdot\mathbf{B}$ describes the Zeeman term, where $\mu_B = e\hbar/2m_e c$ is the Bohr magneton. As an example, we will use the effective mass $m^* = 0.067m_e$, the relative dielectric constant $\varepsilon_r = 12$ and the effective Landé factor $g^* = -0.44$ (bulk GaAs values). For the perpendicular magnetic field we choose the vector potential with gauge $\mathbf{A} = \frac{1}{2}\mathbf{B}\times\mathbf{r} = \frac{1}{2}B(-y, x, 0)$. The confining potential is approximated by a 3D axially-symmetric harmonic oscillator $U(\mathbf{r}) = m^*[\omega_0^2(x^2+y^2) + \omega_z^2 z^2]/2$, where $\hbar\omega_z$ and $\hbar\omega_0$ are the energy scales of confinement in the z-direction and in the xy-plane, respectively.

By introducing the center of mass (CM) and relative coordinates: $\mathbf{R} = \frac{1}{2}(\mathbf{r}_1 + \mathbf{r}_2)$ and $\mathbf{r}_{12} = \mathbf{r}_1 - \mathbf{r}_2$, – the Hamiltonian (1), in agreement with the Kohn theorem [11], separates into the CM and relative-motion terms $H = H_{\rm CM} + H_{\rm rel}$ (see details in [8]). The CM term is described by the oscillator Hamiltonian with the mass $\mathcal{M} = 2m^*$ and frequencies of the one-particle confining potential U. The Hamiltonian for relative motion in cylindrical coordinates takes the form

$$H_{\rm rel} = \frac{1}{2\mu}\left(p_{\rho_{12}}^2 + \frac{l_z^2}{\rho_{12}^2} + p_{z_{12}}^2\right) + \frac{\mu}{2}(\Omega^2\rho_{12}^2 + \omega_z^2 z_{12}^2) + \frac{k}{r_{12}} - \omega_L l_z, \tag{2}$$

where $\mu = m^*/2$ is the reduced mass, l_z ($\to -i\hbar\partial/\partial\varphi_{12}$) is the projection of angular momentum for relative motion and $\rho_{12} = (x_{12}^2 + y_{12}^2)^{1/2}$, $\varphi_{12} = \arctan(y_{12}/x_{12})$, $r_{12} = (\rho_{12}^2 + z_{12}^2)^{1/2}$. Here, $\omega_L = eB/2m^*c$ is the Larmor frequency, and the effective lateral confinement frequency $\Omega = (\omega_L^2 + \omega_0^2)^{1/2}$ depends through ω_L on the magnetic field.

The total two-electron wave function $\Psi(\mathbf{r}_1, \mathbf{r}_2) = \psi(\mathbf{r}_1, \mathbf{r}_2)\chi(\sigma_1, \sigma_2)$ is a product of the orbital $\psi(\mathbf{r}_1, \mathbf{r}_2)$ and spin $\chi(\sigma_1, \sigma_2)$ wave functions. Due to the Kohn theorem, the orbital wave function is factorized as a product of the CM and the relative motion wave functions

$$\psi(\mathbf{r}_1, \mathbf{r}_2) = \psi_{\rm CM}(\mathbf{R})\,\psi_{\rm rel}(\mathbf{r}_{12}). \tag{3}$$

According to the Pauli principle, the orbital wave function must be symmetric (or, equivalently, $\psi_{\rm rel}(\mathbf{r}_{12})$ must be even) for the antisymmetric (singlet: $S = M_S = 0$) spin state, and it must be antisymmetric ($\psi_{\rm rel}(\mathbf{r}_{12})$ must be odd) for the symmetric (triplet: $S = 1$, $M_S = 0, \pm 1$) spin states. Thus, for the relative motion the parity of $\psi_{\rm rel}(\mathbf{r}_{12})$ is a good quantum number as well as the magnetic quantum number m, since l_z is the integral of motion.

The CM eigenfunction is a product of the Fock-Darwin state (the eigenstate of electron in an isotropic 2D harmonic oscillator potential under a perpendicular magnetic field) [12] in the (X, Y)-plane and the oscillator function in the Z-direction (both sets for a particle of mass $\mathcal{M}$). In this paper we consider the lowest CM eigenstate which has the form $\psi_{\rm CM}(\mathbf{R}) = \psi_{\rm CM}^{(xy)}(X,Y)\,\psi_{\rm CM}^{(z)}(Z)$, where $\psi_{\rm CM}^{(xy)}(X,Y) = \sqrt{2\bar{\Omega}/\pi}\,e^{-\bar{\Omega}(X^2+Y^2)}$ and $\psi_{\rm CM}^{(z)}(Z) = (2\bar{\omega}_z/\pi)^{1/4}e^{-\bar{\omega}_z Z^2}$ (i.e. zero principle quantum numbers), with $\bar{\Omega} = m^*\Omega/\hbar$ and $\bar{\omega}_z = m^*\omega_z/\hbar$.

Since the Coulomb interaction mixes the eigenstates of non-interacting electrons, the eigenfunctions of the Hamiltonian for relative motion (2) are expanded in the basis of the Fock-Darwin states $\Phi_{n,m}(\rho_{12}, \varphi_{12})$ and oscillator functions in the z_{12}-direction $\phi_{n_z}(z_{12})$ (for a particle of mass μ), i.e.

$$\psi_{\rm rel}(\mathbf{r}_{12}) = \sum_{n,n_z} c_{n,n_z}^{(m)}\,\Phi_{n,m}(\rho_{12},\varphi_{12})\,\phi_{n_z}(z_{12}). \tag{4}$$

The coefficients $c_{n,n_z}^{(m)}$ can be determined by diagonalizing the Hamiltonian (2) in the same basis. Evidently, in numerical analysis the basis is restricted to a finite set $\{\Phi_{n,m}\,\phi_{n_z} | n = 0,\ldots,n_{\rm max};\ n_z = 0,\ldots,n_z^{\rm max}\}$. It must be, however, large enough to provide a good convergence for the numerical results. Since the function $\psi_{\rm rel}(\mathbf{r}_{12})$ has a definite parity and the parity of the functions $\Phi_{n,m}\,\phi_{n_z}$ is $(-1)^{m+n_z}$, the index n_z in the expansion (4) takes either even or odd values.

For non-interacting electrons ($k = 0$) the eigenfunctions $\psi_{\rm rel}$ are simply the basis functions $\Phi_{n,m}\,\phi_{n_z}$, and, therefore, the ground state is described by the wave function $\psi_{\rm rel} = \Phi_{0,0}\,\phi_0$. When two interacting electron move in the external field created by the confining potential and the applied, varying steadily, magnetic field, the quantum number m of the ground state (in the form (4)) evolves from zero to higher values as the magnetic field strength increases. It results in the well known singlet-triplet transitions [13]. Namely, for a given m the dominant term in the expansion (4) will be $\Phi_{0,m}\,\phi_0$ ($\Rightarrow$ all n_z are even) and the parity of the ground state is $(-1)^m$, which determines the total spin to be $S = \frac{1}{2}[1-(-1)^m]$. Note that the quantum number M_S associated with the spin wave function evolves as following: for even m the total spin $S = 0$ and, thus, $M_S = 0$; for odd m the total spin $S = 1$ and M_S can be -1, 0 or 1. The Zeeman splitting (with $g^* < 0$) will lower the energy of the $M_S = 1$ component of the triplet states, while leaving the singlet states unchanged. As a consequence, the ground state will be characterized by $M_S = S$. With the increasing magnetic field the intervals of the triplet states will increase at the cost of the singlet ones, and eventually, the singlet ground states will be totally suppressed. The increase of the magnetic field leads to the formation of a ring and a torus of maximal density in 2D- and 3D-densities, respectively (see Fig. 4 in Ref.[14]).

At the value $\omega_L^{\rm sph} = (\omega_z^2 - \omega_0^2)^{1/2}$ the magnetic field gives rise to the *spherical symmetry* ($\omega_z/\Omega = 1$) (with $\omega_z > \omega_0$) in the *axially-symmetric* two-electron QD [9, 10]. This phenomenon was also recognized in the results for many interacting electrons in self-assembled QDs [15]. In the later case it was interpreted as an approximate symmetry that had survived from the non-interacting case due to the dominance of the confinement energy over a relatively small Coulomb interaction energy. However, the symmetry is not approximate but *exact* even for strongly interacting electrons, because the radial electron-electron repulsion does not break the rotational symmetry. A natural question arises how to detect such a transition looking on the density distribution only. The related question is, if such a transition occurs, what are the concomitant structural changes?

To this end we employ the entanglement measure based on the linear entropy of reduced density matrices (cf [16])

$$\mathcal{E} = 1 - 2\,{\rm Tr}[\rho_r^{(orb)^2}]\,{\rm Tr}[\rho_r^{(spin)^2}], \tag{5}$$

where $\rho_r^{(orb)}$ and $\rho_r^{(spin)}$ are the single-particle reduced density matrices in the orbital and spin spaces, respectively. This measure is quite popular for the analysis of the entanglement of two-fermion systems, in particular, two electrons confined in the parabolic potential in the absence of the magnetic field [17].

The trace $\mathrm{Tr}[\rho_r^{(spin)^2}]$ of the two-electron spin states with a definite symmetry χ_{S,M_S} has two values: (i) 1/2 if $M_S = 0$ (anti-parallel spins of two electrons); (ii) 1 if $M_S = \pm 1$ (parallel spins). The condition $M_S = S = \frac{1}{2}[1-(-1)^m]$ yields

$$\mathrm{Tr}[\rho_r^{(spin)^2}] = \tfrac{1}{2}(1+|M_S|) = \frac{3-(-1)^m}{4}. \tag{6}$$

The trace of the orbital part $\mathrm{Tr}[\rho_r^{(orb)^2}]$

$$\begin{aligned}\mathrm{Tr}[\rho_r^{(orb)^2}] &= \int d\mathbf{r}_1\, d\mathbf{r}_1'\, d\mathbf{r}_2\, d\mathbf{r}_2'\, \psi(\mathbf{r}_1,\mathbf{r}_2)\, \psi^*(\mathbf{r}_1',\mathbf{r}_2) \\ &\qquad \psi^*(\mathbf{r}_1,\mathbf{r}_2')\, \psi(\mathbf{r}_1',\mathbf{r}_2').\end{aligned} \tag{7}$$

is more involved. Indeed, in virtue of Eqs.(3), (4), one obtains

$$\begin{aligned}\mathrm{Tr}[\rho_r^{(orb)^2}] = &\sum_{n_1=0}^{n_{\max}}\sum_{n_2=0}^{n_{\max}}\sum_{n_3=0}^{n_{\max}}\sum_{n_4=0}^{n_{\max}}\sum_{n_{z_1}=0}^{n_z^{\max}}\sum_{n_{z_2}=0}^{n_z^{\max}}\sum_{n_{z_3}=0}^{n_z^{\max}}\sum_{n_{z_4}=0}^{n_z^{\max}} \\ &c^{(m)}_{n_1,n_{z_1}}\, c^{(m)}_{n_2,n_{z_2}}\, c^{(m)}_{n_3,n_{z_3}}\, c^{(m)}_{n_4,n_{z_4}} \\ &I(n_1,n_2,n_3,n_4;m)\, J(n_{z_1},n_{z_2},n_{z_3},n_{z_4}),\end{aligned} \tag{8}$$

where

$$\begin{aligned}I(n_1,n_2,n_3,n_4;m) = &\int d\mathbf{r}_1\, d\mathbf{r}_1'\, d\mathbf{r}_2\, d\mathbf{r}_2' \\ &\psi^{(xy)}_{\mathrm{CM}}(\tfrac{\mathbf{r}_1+\mathbf{r}_2}{2})\, \psi^{(xy)^*}_{\mathrm{CM}}(\tfrac{\mathbf{r}_1'+\mathbf{r}_2}{2})\, \psi^{(xy)^*}_{\mathrm{CM}}(\tfrac{\mathbf{r}_1+\mathbf{r}_2'}{2})\, \psi_{\mathrm{CM}}(\tfrac{\mathbf{r}_1'+\mathbf{r}_2'}{2}) \\ &\Phi_{n_1,m}(\mathbf{r}_1-\mathbf{r}_2)\, \Phi^*_{n_2,m}(\mathbf{r}_1'-\mathbf{r}_2)\, \Phi^*_{n_3,m}(\mathbf{r}_1-\mathbf{r}_2')\, \Phi_{n_4,m}(\mathbf{r}_1'-\mathbf{r}_2')\end{aligned} \tag{9}$$

(here $\mathbf{r}_i$ are vectors in the xy-plane) and

$$\begin{aligned}J(n_{z_1},n_{z_2},n_{z_3},n_{z_4}) = &\int dz_1\, dz_1'\, dz_2\, dz_2' \\ &\psi^{(z)}_{\mathrm{CM}}(\tfrac{z_1+z_2}{2})\, \psi^{(z)^*}_{\mathrm{CM}}(\tfrac{z_1'+z_2}{2})\, \psi^{(z)^*}_{\mathrm{CM}}(\tfrac{z_1+z_2'}{2})\, \psi^{(z)}_{\mathrm{CM}}(\tfrac{z_1'+z_2'}{2}) \\ &\phi_{n_{z_1}}(z_1-z_2)\, \phi^*_{n_{z_2}}(z_1'-z_2)\, \phi^*_{n_{z_3}}(z_1-z_2')\, \phi_{n_{z_4}}(z_1'-z_2').\end{aligned} \tag{10}$$

The magnetic field dependence of the entanglement $\mathcal{E}$ naturally occurs via inherent variability of the expansion coefficients. The values of the I and J integrals for any choice of indices can be determined analytically, which simplifies the numerical calculations.

For our analysis it is convenient to use the so-called Wigner parameter $R_W = (k/l_0)/\hbar\omega_0 = l_0/a^*$, a measure of the Coulomb interaction strength relative to the confinement strength (cf [8]). Here, $l_0 = \sqrt{\hbar/m^*\omega_0}$ is the oscillator length and $a^* = \hbar^2/km^*$ is the effective Bohr radius. For our choice of the parameters (GaAs) and for the confinement frequency $\hbar\omega_0 \approx 2.8$ meV we have $R_W \approx 2$. The numerical analysis demonstrates a good convergency for the basis with $n_{max} = n_z^{max} = 4$.

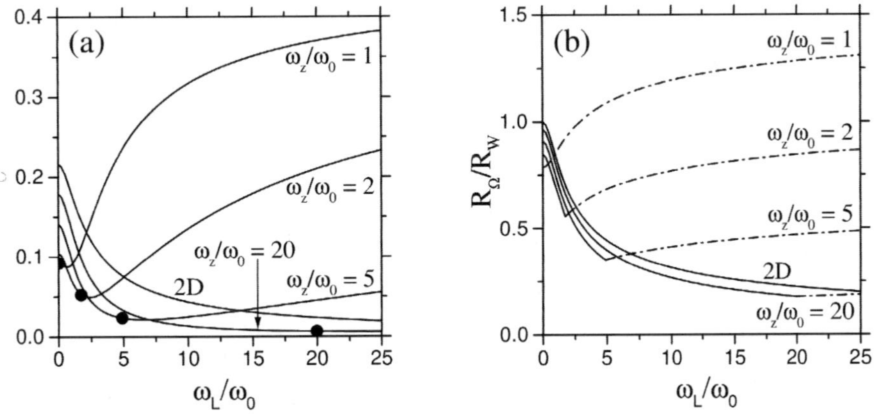

Figure 1. (a) Entanglement of the lowest state with $m = 0$ at $R_W = 2$ and various ratios ω_z/ω_0 as functions of the parameter ω_L/ω_0. The circles denote the values of ω_L/ω_0 when QDs with given ratios ω_z/ω_0 become spherically symmetric. (b) The relative strength of the Coulomb interaction $R_\Omega^{(2D)}/R_W$ (solid line) and $R_\Omega^{(1D)}/R_W$ (dash-dotted line) for the lowest state with $m = 0$ at various ratios ω_z/ω_0 as functions of the parameter ω_L/ω_0.

3. Analysis of results

The strongest correlations are expected for the states with $m = 0$, since they are not affected by the orbital electron motion which weakens the Coulomb interaction. In the absence of the magnetic field ($B = 0$) the entanglement decreases if the ratio ω_z/ω_0 decreases from ∞ (2D model) to 1 (spherically symmetric 3D model); see Fig. 1(a) at $\omega_L/\omega_0 = 0$. This effect could be explained by introducing the effective charge k_{eff} [9, 18] which determines the effective electron-electron interaction $V_C = k_{\text{eff}}/\rho_{12}$ in the QD. In the 3D dot the electrons can avoid each other more effectively than in the 2D one. Therefore, the Coulomb interaction has a smaller effect on the 3D spectrum (the ratio $k_{\text{eff}}/k \sim 0.5$) in contrast to the 2D case when $k_{\text{eff}}/k = 1$. Thus, a decreasing of the ratio ω_z/ω_0 yields an analogous effect as the reduction of the electron-electron interaction.

Fig. 1(a) shows the entanglement measure $\mathcal{E}$ of the lowest angular momentum state $m = 0$ as a function of the magnetic field (the parameter ω_L/ω_0) at a fixed value of R_W and for different ratios ω_z/ω_0. In the 2D case the entanglement decreases monotonically with the increase of the magnetic field. The constant electron-electron interaction becomes relatively weaker, since the effective lateral confinement ($\hbar\Omega$) increases with the magnetic field. If we introduce the characteristic length of the effective confinement $l_\Omega = \sqrt{\hbar/m^*\Omega}$, the parameter $R_\Omega = l_\Omega/a^*$ (which is equal R_W at $B = 0$) determines the relative strength of Coulomb interaction at a given effective confinement. Evidently, R_Ω decreases with the increase of the magnetic field B (see Fig. 1(b), the line labelled by '2D'). In the 3D case, however, the entanglement decreases until $\omega_L = \omega_L^{\text{sph}}$, when the spherical symmetry occurs. After this point the entanglement starts to increase (see Fig. 1(a)).

This behaviour can be explained by the influence of magnetic field on the effective strength R_Ω, which is twofold here. Indeed, in the 3D case the magnetic field affects the effective charge as well as the effective confinement. For the quasi-2D system of electrons ($\Omega \ll \omega_z$) the effective charge is $k_{\text{eff}}^{(2D)} = \langle \rho_{12} V_C \rangle$ (see Eq. (18) in Ref. [18]), where $V_C = k/\sqrt{\rho_{12}^2 + z_{12}^2}$ is the full 3D Coulomb interaction. The mean value $\langle \rho_{12} V_C \rangle$ is calculated by means of the eigenstates of H_{rel} in the approximation of non-interacting electrons. Here, the eigenstate is $\Phi_{0,m}\,\phi_0$ (for

Journal of Physics: Conference Series **393** (2012) 012009 doi:10.1088/1742-6596/393/1/012009

explicit expressions see Eqs. (19),(20) in Ref. [18]). Thus, for the quasi-2D case the parameter $R_\Omega^{(2D)} = (m^*/\hbar^3\Omega)^{1/2}\, k_{\rm eff}^{(2D)}$ can be used as a measure for the relative strength of the Coulomb interaction.

For $\Omega \gg \omega_z$ (very strong magnetic field) the electrons are pushed laterally towards the dot's center. The magnetic field, however, does not affect the vertical confinement. As a consequence the electrons practically can move only in the z-direction and the QD becomes a quasi-1D system. In this case a measure for the relative strength of Coulomb interaction can be defined as $R_\Omega^{(1D)} = (m^*/\hbar^3\omega_z)^{1/2}\, k_{\rm eff}^{(1D)}$, where the effective charge for a quasi-1D system is $k_{\rm eff}^{(1D)} = \langle |z_{12}| V_C\rangle$. It can be shown that for the lowest state with $m = 0$ one obtains $k_{\rm eff}^{(1D)}/k = (1+\sqrt{\omega_z/\Omega})^{-1}$.

The quantities $R_\Omega^{(2D)}$ and $R_\Omega^{(1D)}$ for the lowest state with $m = 0$, as functions of the parameter ω_L/ω_0 (in the domains $0 < \omega_L < \omega_L^{\rm sph}$ and $\omega_L > \omega_L^{\rm sph}$, respectively), are shown in Fig. 1(b) for different ratios ω_z/ω_0. One observes that the effective strength $R_\Omega^{(2D)}$ decreases with the increase of the magnetic field for different ratios ω_z/ω_0, similar to the 2D case. The oppositely ordered confinement $\Omega^{(1D)}$ (which is not defined for the 2D case) increases with ω_L and, therefore, the effective strength $R_\Omega^{(1D)}$ increases as well. In order to match $R_\Omega^{(1D)} = R_\Omega^{(2D)}$ at $\omega_L = \omega_L^{\rm sph}$ (i.e. when $\Omega = \omega_z$) the strength $R_\Omega^{(1D)}$ is scaled by the factor $\pi/2$. Although at this point the 3D system is far from the 2D model and from the 1D model and, as a consequence, $R_\Omega^{(2D)}$ and $R_\Omega^{(1D)}$ do not match smoothly, these two functions taken together give a qualitative picture how the effective electron-electron interaction in a 3D QD changes with the magnetic field.

The minimum of entanglement for the lowest state with $m = 0$ at $\Omega = \omega_z$ can be associated with the condition of equivalence of the oscillator scales in the lateral and vertical confinements: $l_\Omega \equiv l_z = \sqrt{\hbar/m^*\omega_z}$. These quantities can be understood as the amplitudes of electron oscillations in the ρ and z-directions, respectively, for single-particle states with $n_\rho = 0$ and $n_z = 0$ [18]. At this point the effective Coulomb interaction becomes isotropic, which results in the small mixing of single-particle states, in contrast to the 2D and 1D cases. This result can be readily extended for the infinite square well potential in the z-direction. In this case $l_z \approx d/2$; d is the thickness of semiconductor layer where the dot is created (see Fig.1 in Ref. [18])). Therefore, the minimum of entanglement is expected at $l_\Omega = d/2$ which defines the value of the magnetic field by means of the Larmor frequency $\omega_L = \sqrt{(4\hbar/m^*d^2)^2 - \omega_0^2}$.

To get deep insight into this transition we calculate the probability density $|\psi(\mathbf{r}_{12})|^2$ and potential surfaces for various values of the magnetic field (see Fig. 2). Since the symmetry is exact for any strength of the electron-electron interaction at the transition point, in order to illuminate the effect, we use $R_W = 10$. For the magnetic field $\omega_L < \omega_L^{\rm sph}$ the density maximum is located in the (x_{12}, y_{12})-plane ($z_{12} = 0$, see Fig. 2(a)). For $\omega_L > \omega_L^{\rm sph}$, however, there are two maxima located symmetrically along the z_{12}-axis ($\rho_{12} = 0$, see Fig. 2(c)). The analysis of the behaviour of the stationary point of the potential $V = \frac{1}{2}\mu(\Omega^2\rho_{12}^2 + \omega_z^2 z_{12}^2) + k/r_{12}$ as a function of the magnetic field provides the explanation. For $\omega_L < \omega_L^{\rm sph}$ ($\Omega < \omega_z$) the stationary point $\rho_{12} = \rho_0$, $z_{12} = 0$ is the minimum of the potential surface (see Fig. 2(d)). Here $\rho_0 = (k/\mu\Omega^2)^{1/3}$ [19]. By increasing the magnetic field over the value $B_{\rm sph}$ ($\Omega > \omega_z$) the stationary point transforms to the saddle point and two new minima appear, divided by a potential barrier (see Fig. 2(f)). In other words, for $m = 0$ a bifurcation of the stationary point located at $(\rho_0, 0)$ occurs at the value of magnetic field when $\omega_L = \omega_L^{\rm sph}$ (see Fig. 2(b,e)). In the domain $\Omega > \omega_z$, for $m = 0$, the minima are located at $z_{12} = \pm z_0$ in the z_{12}-axis ($\rho_{12} = 0$), where $z_0 = (k/\mu\omega_z^2)^{1/3}$. Similar behavior is observed for the states with $m \neq 0$. However, the effect is less prominent due to weakening of correlations by the orbital electron motion.

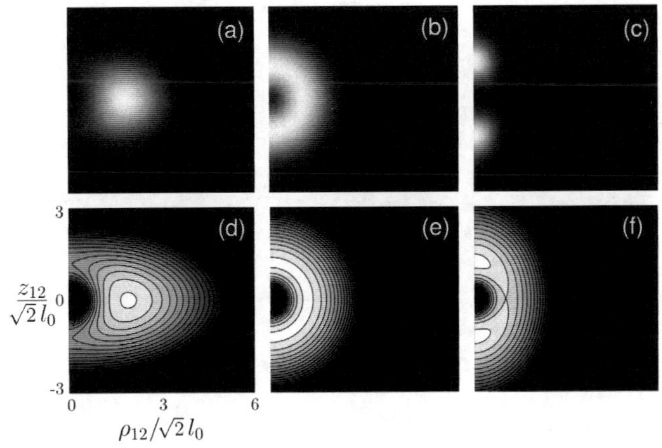

Figure 2. The probability density $|\psi(\mathbf{r}_{12})|^2$ of the lowest $m = 0$ state (top) and the contour plots of the potential surface (bottom) for the QD with $\omega_z/\omega_0 = 2$ and $R_W = 10$ shown in (ρ_{12}, z_{12})-plane for the cases: (a,d) $\omega_L/\omega_0 = 0$ $(\omega_z/\Omega = 2)$, (b,e) $\omega_L/\omega_0 = \omega_L^{\rm sph}/\omega_0 = 1.73205$ $(\omega_z/\Omega = 1)$ and (c,f) $\omega_L/\omega_0 = 2.29129$ $(\omega_z/\Omega = 0.8)$.

4. Conclusions

Summarizing, we have shown that the 3D approach provides a consistent description of the shape transition in the excited states of two-electron QDs under the magnetic field. The entanglement of the lowest state with $m = 0$, being first a decreasing function of the magnetic field, starts to increase after the transition point with the increase of the magnetic field. This behaviour is understood as the transition from the lateral to the vertical localization of the two-electron probability density for this state, which becomes prominent at strong Coulomb interaction $(R_W \gg 2)$.

Acknowledgments

This work is partly supported by RFBR Grant No.11-02-00086 (Russia), Project 171020 of Ministry of Education and Science of Serbia, Spanish MICINN Grant No. FIS2008-00781, and Project FQM-2445 of the Junta de Andalucia (Spain).

References

[1] Amico L, Fazio R, Osterloh A and Vedral V 2008 *Rev. Mod. Phys.* **80** 517
[2] Turner A M, Pollmann F and Berg E 2011 *Phys. Rev. B* **83** 075102; Fidkowski L and Kitaev A 2011 *Phys. Rev. B* **83** 075103
[3] Haque M, Zozulya O and Schoutens K 2007 *Phys. Rev. Lett.* **98** 060401
[4] Liu Z, Guo H-L, Vedral V and Fan H 2011 *Phys. Rev. A* **83** 013620
[5] Sachdev S *Quantum Phase Transitions* 2011 (Cambridge: Cambridge University Press) 2nd Edition
[6] Reimann S M and Manninen M 2002 *Rev. Mod. Phys.* **74** 1283
[7] Kais S, Herschbach D R and Levine R D 1989 *J. Chem. Phys.* **91** 7791; Taut M 1994 *J. Phys. A* **27** 1045
[8] Nazmitdinov R G 2009 *Physics of Particles and Nuclei* **40** 71
[9] Nazmitdinov R G, Simonović N S and Rost J M 2002 *Phys. Rev. B* **65** 155307
[10] Simonović N S and Nazmitdinov R G 2003 *Phys. Rev. B* **67** 041305(R)
[11] Kohn W 1961 *Phys. Rev.* **123** 1242
[12] Fock V 1928 *Z. Phys.* **47** 446; Darwin C G 1930 *Proc. Cambridge Philos. Soc.* **27** 86
[13] Wagner M, Merkt U and Chaplik A V 1992 *Phys. Rev. B* **45** 1951
[14] Nazmitdinov R G and Simonović N S 2007 *Phys. Rev. B* **76** 193306
[15] Wojs A, Hawrylak P, Fafard S and Jacak L 1996 *Phys. Rev. B* **54** 5604
[16] Coleman A and Yukalov V 2000 *Reduced Density Matrices* (Berlin: Springer-Verlag)

[17] Buscemi F, Bordone P and Bertoni A 2007 *Phys. Rev. A* **75** 032301; Naudts J and Verhulst T 2007 *Phys. Rev. A* **75** 062104; Coe J P, Sudbery A and Amico I D' 2008 *Phys. Rev. B* **77** 205122; Pipek J and Nagy I 2009 *Phys. Rev. A* **79** 052501; Yañez R J, Plastino A R and Dehesa J S 2010 *Eur. Phys. J. D* **56** 141; Kościk P and Okopińska A 2010 *Phys. Lett. A* **374** 3841
[18] Simonović N S and Nazmitdinov R G 2008 *Phys. Rev. A* **78** 032115
[19] Puente A, Serra L and Nazmitdinov R G 2004 *Phys. Rev. B* **69** 125315

Journal of Physics: Conference Series **393** (2012) 012010 doi:10.1088/1742-6596/393/1/012010

Entanglement of electron orbital motion in open two-electron quantum dots

A. V. Chizhov

Bogoliubov Laboratory of Theoretical Physics, Joint Institute for Nuclear Research, 141980 Dubna, Russia

E-mail: chizhov@theor.jinr.ru

Abstract. In the present report, the effects of decoherence in a two-electron quantum dot coupled with its environment are investigated. Environmental effects on the quantum dynamics of electrons are modeled by coupling the dot with a thermo-bath at fixed temperatures. In such an approach, the special emphasis is placed on the analysis of the entanglement degradation of electron orbital motion depending on the quantum dot parameters (the strength of electron interaction and quantum dot shapes) and on the temperature of environment.

1. Introduction

Quantum dots (QDs) represent mesoscopic systems of artificially trapped electrons in a localized space of about a few hundreds Angstroms of a linear size. All characteristics of QDs (size, number of electrons, dielectric environment) as well as their external parameters (electric and magnetic fields, temperature) can be kept under control that allows one to use QDs as handy laboratories to study quantum effects at low-dimensional scale. In particular, they are expected to give a natural realization of a quantum bit or "qubit" and can be used in quantum computing and information storage. In systems of electrons the realization of qubits is usually associated with spins of electrons [1, 2], since they have just two projections onto any axe. At the same time, a quantized character of QD energy spectra allows one to represent qubits in terms of pseudospin based on the orbital motion of electrons [3, 4, 5]. Therefore, the study of quantum correlations of the spatial function in the dynamics of electrons, i.e., orbital entanglement of electrons, may be of great importance for the realization of multi-qubit logic gates [6]. Creation of such logic gates is needed for the accomplishment of general algorithms in the quantum information theory [7]. Thus, the question how to efficiently produce and control the orbital entanglement is one of fundamental as well as technological problems.

Recently, an exactly solvable model of an isolated two-electron QD was treated for the study of quantum correlation properties of electrons [8]. A two-electron QD is the simplest pattern with the essential features of more complex systems. In this model, electrons were assumed to move in a two-dimensional parabolic confining potential, whereas the interaction between electrons was approximated by the effective harmonic potential of the Johnson-Payne type. Within the bounds of such a model it was possible to get an analytical description of the electron orbital motion and the electron energy spectrum in accordance with the shell structure. In the presence of a perpendicular magnetic field a detailed study of the degree of the orbital entanglement of the QD ground state and its first excited states was performed for various values of a QD shape, the strength of the interaction and the magnetic field [9].

Entanglement is known to be very sensitive to the environmental effects which lead to the phenomenon of decoherence. Time of decoherence is one of the crucial problems appeared in quantum systems. The aim of this communication is to evaluate the entanglement degradation of electron orbital states in a two-electron QD coupled to its environment. The analysis is given with a special emphasis to the effect of finite temperatures of the environment.

2. The Model

Decoherence effects in QDs arise from the interaction of electrons with phonon excitations in the QD structural materials [10, 11, 12]. In theoretical treatment, an open QD can be modeled by way of coupling the dot with a boson reservoir at fixed temperatures. Consider the motion of electrons being restricted to the xy plane by a confining potential and identifying the real coupling constants as $\kappa_{x,y\,j}$, the Hamiltonian of the whole system can thus be written as

$$\hat{\mathcal{H}} = \hat{H}_{2e} + \sum_j \Bigg[\left(\hat{P}_{xj}^2 + \hat{P}_{yj}^2\right)/2M_j + M_j\left(\Xi_{xj}^2\hat{Q}_{xj}^2 + \Xi_{yj}^2\hat{Q}_{yj}^2\right)/2$$
$$+\kappa_{xj}(\hat{x}_1+\hat{x}_2)\hat{Q}_{xj} + \kappa_{yj}(\hat{y}_1+\hat{y}_2)\hat{Q}_{yj} + \frac{\kappa_{xj}(\hat{p}_{x1}+\hat{p}_{x2})\hat{P}_{xj}}{m^* M_j\omega_x\Xi_{xj}} + \frac{\kappa_{yj}(\hat{p}_{y1}+\hat{p}_{y2})\hat{P}_{yj}}{m^* M_j\omega_y\Xi_{yj}}\Bigg]. \quad (1)$$

In Eq. (1) $\hat{H}_{2e}$ denotes the Hamiltonian of a system of two electrons. $\hat{Q}_{x,y\,j}$ and $\hat{P}_{x,y\,j}$ stand for position and momentum operators of the reservoir oscillators with the mode frequencies $\Xi_{x,y\,j}$, whereas $\hat{x}_{1,2}$ ($\hat{y}_{1,2}$) and $\hat{p}_{x\,1,2}$ ($\hat{p}_{y\,1,2}$) describe projections of position and momentum operators of QD electrons, respectively. Position and momentum operators of the two electrons enter into the interaction terms in the forms of sums in order to comply with the symmetry demands of the full Hamiltonian with respect to the mutual permutation of electrons. The confining potential in few-electron QDs is usually assumed, to a good approximation, as parabolic [13] with two confining frequencies ω_x and ω_y, which are not equal in a general case, so that

$$\hat{H}_{2e} = \sum_{i=1}^{2}\left[\frac{\hat{p}_{xi}^2+\hat{p}_{yi}^2}{2m^*} + \frac{m^*}{2}\left(\omega_x^2x_i^2+\omega_y^2y_i^2\right)\right] + \hat{V}(\mathbf{r}_1,\mathbf{r}_2). \quad (2)$$

Here m^* is an effective mass for the conduction electrons, and operator $\hat{V}(\mathbf{r}_1,\mathbf{r}_2)$ describes the electron-electron interaction which is taken in the approximation of the Johnson-Payne type [14]

$$\hat{V}(\mathbf{r}_1-\mathbf{r}_2)=V_0-\lambda^2\frac{m^*}{2}(\mathbf{r}_1-\mathbf{r}_2)^2, \quad (3)$$

where V_0 and λ are positive parameters within the model.

Introducing new variables for QD electrons

$$\mathbf{R}=\frac{\mathbf{r}_1+\mathbf{r}_2}{2}, \qquad \mathbf{r}=\mathbf{r}_1-\mathbf{r}_2, \quad (4)$$

Hamiltonian $\hat{H}_{2e}$ is divided into the the center-of-mass and relative motion terms

$$\hat{H}_{2e}=\hat{H}_{\rm CM}+\hat{H}_{\rm rel}+V_0, \quad (5)$$

$$\hat{H}_{\rm CM} = \frac{\hat{\mathbf{P}}^2}{2M}+\frac{M}{2}[\omega_x^2\hat{X}^2+\omega_y^2\hat{Y}^2],$$
$$\hat{H}_{\rm rel} = \frac{\hat{\mathbf{p}}^2}{2\mu}+\frac{\mu}{2}[\omega_1^2\hat{x}^2+\omega_2^2\hat{y}^2], \quad (6)$$

where

$$\omega_{1,2}^2 = \omega_{x,y}^2 - 2\lambda^2, \quad M = 2m^*, \quad \mu = \frac{m^*}{2}. \tag{7}$$

In these notations the full Hamiltonian (1) takes the form

$$\hat{\mathcal{H}} = \hat{H}_{\rm CM}^{dis} + \hat{H}_{\rm rel} + V_0 \tag{8}$$

with

$$\begin{aligned} \hat{H}_{\rm CM}^{dis} &= \hat{H}_{\rm CM} + \sum_j \Big[\left(\hat{P}_{xj}^2 + \hat{P}_{yj}^2\right)/2M_j + M_j\left(\Xi_{xj}^2\hat{Q}_{xj}^2 + \Xi_{yj}^2\hat{Q}_{yj}^2\right)/2 \\ &+2\kappa_{xj}\hat{X}\hat{Q}_{xj} + 2\kappa_{yj}\hat{Y}\hat{Q}_{yj} + 2\frac{\kappa_{xj}\hat{P}_x\hat{P}_{xj}}{MM_j\omega_x\Xi_{xj}} + 2\frac{\kappa_{yj}\hat{P}_y\hat{P}_{yj}}{MM_j\omega_y\Xi_{yj}}\Big]. \end{aligned} \tag{9}$$

Hamiltonians $H_{\rm rel}$ and $\hat{H}_{\rm CM}^{dis}$ can be written in the secondary quantized representation

$$\hat{H}_{\rm rel} = \hbar\omega_1(\hat{c}_1^\dagger\hat{c}_1 + 1/2) + \hbar\omega_2(\hat{c}_2^\dagger\hat{c}_2 + 1/2), \tag{10}$$

$$\begin{aligned} \hat{H}_{\rm CM}^{dis} &= \hbar\omega_x(\hat{C}_1^\dagger\hat{C}_1 + 1/2) + \hbar\omega_y(\hat{C}_2^\dagger\hat{C}_2 + 1/2) + \hbar\sum_j \Big[\Xi_{xj}\left(\hat{f}_{xj}^\dagger\hat{f}_{xj} + 1/2\right) + \Xi_{yj}\left(\hat{f}_{yj}^\dagger\hat{f}_{yj} + 1/2\right) \\ &+\varsigma_{xj}\left(\hat{C}_1^\dagger\hat{f}_{xj} + \hat{C}_1\hat{f}_{xj}^\dagger\right) + \varsigma_{yj}\left(\hat{C}_2^\dagger\hat{f}_{yj} + \hat{C}_2\hat{f}_{yj}^\dagger\right)\Big] \end{aligned} \tag{11}$$

with annihilation operators $\hat{c}_{1,2}$, $\hat{C}_{1,2}$, and $\hat{f}_{x,y\,j}$ for relative and center-of-mass motion of electrons, and reservoir bosons, respectively, and the transformed coupling constants $\varsigma_{x,y\,j} = 2\kappa_{x,y\,j}/sqrtMM_j\omega_{x,y}\Xi_{x,y\,j}$.

3. Quantum dynamics of electron subsystem

The operator dynamics of electrons can be found in the Heisenberg-Langevin approximation [15]. In view of the fact that the relative-motion operators are not coupled to the reservoir (see Eq. (10)), they follow the free evolution in time

$$\hat{c}_1(t) = \hat{c}_1(0)\exp(-i\omega_1 t), \quad c_2(t) = \hat{c}_2(0)\exp(-i\omega_2 t), \tag{12}$$

whereas the center-of-mass motion operators are subject to damping

$$\hat{C}_1(t) = u_1(t)\hat{C}_1(0) + \sum_j w_{xj}(t)\hat{f}_{xj}(0), \quad \hat{C}_2(t) = u_2(t)\hat{C}_2(0) + \sum_j w_{yj}(t)\hat{f}_{yj}(0), \tag{13}$$

with

$$u_1(t) = \exp(-i\omega_x t - \gamma_1 t/2), \quad u_2(t) = \exp(-i\omega_y t - \gamma_2 t/2), \tag{14}$$

$\gamma_{1,2} = 2\pi|\varsigma(\omega_{x,y})|^2\varrho(\omega_{x,y})$ are the damping constants ($\varrho(\omega)$ being the density distribution of the reservoir oscillators), and functions $w_{x,y\,j}(t)$ obey the equalities

$$\sum_j |w_{xj}(t)|^2 = 1 - \exp(-\gamma_1 t), \quad \sum_j |w_{yj}(t)|^2 = 1 - \exp(-\gamma_2 t) \tag{15}$$

in order to fulfill the commutation relations for operators $\hat{C}_{1,2}$. Then the time evolution of two-mode annihilation operators of the two QD electrons confined in the xy plane

$$\hat{d}_{x\,1,2} = \frac{m^*\omega_x\hat{x}_{1,2} + i\hat{p}_{x1,2}}{\sqrt{2\hbar m^*\omega_x}}, \quad \hat{d}_{y\,1,2} = \frac{m^*\omega_y\hat{y}_{1,2} + i\hat{p}_{y1,2}}{\sqrt{2\hbar m^*\omega_y}}, \tag{16}$$

can be obtained in accordance with Eqs. (4)

$$\hat{d}_{x\,1,2}(t) = \frac{1}{2\sqrt{2}}\left[2\hat{C}_1(t) \pm \left(\sqrt{\frac{\omega_1}{\omega_x}} + \sqrt{\frac{\omega_x}{\omega_1}}\right)\hat{c}_1(t) \mp \left(\sqrt{\frac{\omega_1}{\omega_x}} - \sqrt{\frac{\omega_x}{\omega_1}}\right)\hat{c}_1^\dagger(t)\right], \quad (17)$$

$$\hat{d}_{y\,1,2}(t) = \frac{1}{2\sqrt{2}}\left[2\hat{C}_2(t) \pm \left(\sqrt{\frac{\omega_2}{\omega_y}} + \sqrt{\frac{\omega_y}{\omega_2}}\right)\hat{c}_2(t) \mp \left(\sqrt{\frac{\omega_2}{\omega_y}} - \sqrt{\frac{\omega_y}{\omega_2}}\right)\hat{c}_2^\dagger(t)\right] \quad (18)$$

by using Eqs. (12) and (13).

Eqs. (17) and (18) allow one to study the state evolution of orbital motion of QD electrons. In case the whole system of electrons and reservoir bosons at the initial moment is described by the total density matrix $\hat{\varrho}(0)$, a four-mode quantum state of the electron subsystem in time can be represented in the phase space by the Wigner function

$$W(\alpha_{x1}, \alpha_{y1}, \alpha_{x2}, \alpha_{y2}|t) = \frac{1}{\pi^8}\int \chi(\beta_{x1}, \beta_{y1}, \beta_{x2}, \beta_{y2}|t)\exp\{\sum_{k=1}^{2}[\alpha_{xk}\beta_{xk}^* + \alpha_{yk}\beta_{yk}^* - \text{c.c.}]\}\text{d}^8\beta, \quad (19)$$

based on a symmetric characteristic function

$$\chi(\beta_{x1}, \beta_{y1}, \beta_{x2}, \beta_{y2}|t) = \text{Tr}\{\hat{\varrho}(0)\hat{D}(\beta_{x1}, \beta_{y1}, \beta_{x2}, \beta_{y2}|t)\} \quad (20)$$

with the displacement operator

$$\hat{D}(\beta_{x1}, \beta_{y1}, \beta_{x2}, \beta_{y2}|t) = \exp\{\sum_{k=1}^{2}[\beta_{xk}\hat{d}_{xk}^\dagger(t) + \beta_{yk}\hat{d}_{yk}^\dagger(t) - \text{h.c.}]\}.$$

4. Degradation of QD charge entanglement

In order to take account of temperature effects of the QD environment on the orbital entanglement of electrons (i.e., the charge entanglement), the state of reservoir bosons might be chosen as chaotic, so that their mean boson numbers

$$\langle n_1\rangle = \langle \hat{f}_{xj}^\dagger(0)\hat{f}_{xj}(0)\rangle = \left\{\text{e}^{\hbar\omega_x/k_BT} - 1\right\}^{-1}, \quad \langle n_2\rangle = \langle \hat{f}_{yj}^\dagger(0)\hat{f}_{yj}(0)\rangle = \left\{\text{e}^{\hbar\omega_y/k_BT} - 1\right\}^{-1} \quad (21)$$

are appeared to be dependent on the temperature T (k_B is the Boltzman constant).

Regarding the initial state of electrons, which is the eigenstate of the Hamiltonian $\hat{H}_{2e}$, in this case it should be one of the entangled states. In this connection, it is convenient to regard electrons at the initial moment to be in the QD vacuum state $|0\rangle_c$ (see [9]), that means $\hat{c}_{1,2}(0)|0\rangle_c = \hat{C}_{1,2}(0)|0\rangle_c = 0$.

Under these conditions, the time evolution of the state of the electron subsystem is described by the Wigner function of a Gaussian type

$$W(\alpha_{x1}, \alpha_{y1}, \alpha_{x2}, \alpha_{y2}|t) = \frac{16}{\pi^4[2\langle n_1\rangle(1-\text{e}^{-\gamma_1 t}) + 1][2\langle n_2\rangle(1-\text{e}^{-\gamma_2 t}) + 1]}$$
$$\times \exp\left\{-2\left[|\sigma_1|^2 + |\sigma_2|^2 - \frac{|\Sigma_1|^2}{2\langle n_1\rangle(1-\text{e}^{-\gamma_1 t}) + 1} - \frac{|\Sigma_2|^2}{2\langle n_2\rangle(1-\text{e}^{-\gamma_2 t}) + 1}\right]\right\}, \quad (22)$$

where $\sigma_{1,2} = \left[\sqrt{\omega_{1,2}/\omega_{x,y}}\text{Re}(\alpha_{x,y\,1} - \alpha_{x,y\,2}) + i\sqrt{\omega_{x,y}/\omega_{1,2}}\text{Im}(\alpha_{x,y\,1} - \alpha_{x,y\,2})\right]/\sqrt{2}$ and $\Sigma_{1,2} = (\alpha_{x,y\,1} + \alpha_{x,y\,2})/\sqrt{2}$.

Introducing the quadrature-component variables defined as $q_{x,y\,k} = \sqrt{2}\text{Re}\alpha_{x,y\,k}$ and $p_{x,y\,k} = \sqrt{2}\text{Im}\alpha_{x,y\,k}$, the Wigner function (22) can be represented in terms of the covariance matrix $\mathbf{V}(t)$

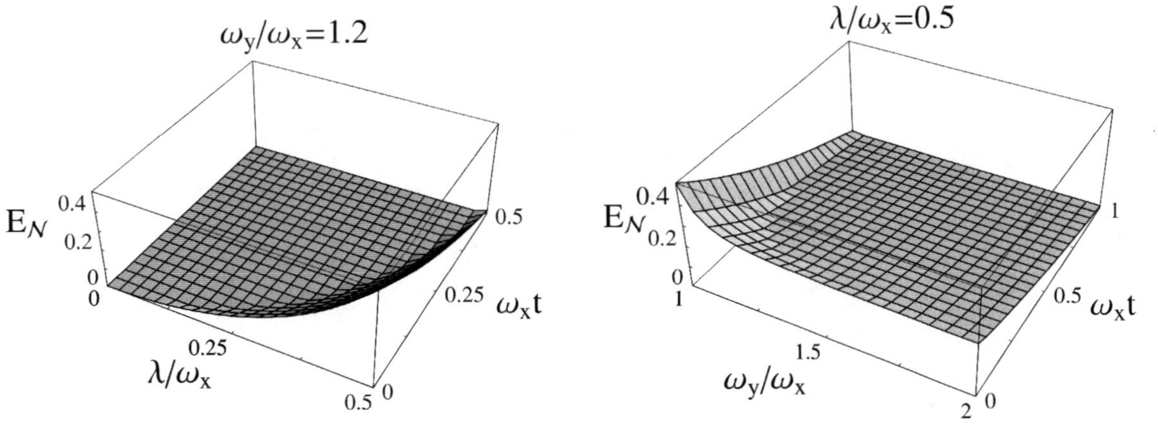

Figure 1. The logarithmic negativity $E_\mathcal{N}$ as a function of time and the strength of the electron-electron interaction (left), the QD deformation (right), when $\langle n_1 \rangle = \langle n_2 \rangle = 0.5$ and the damping constants $\gamma_1/\omega_x = \gamma_2/\omega_x = 1$.

$$W(\xi|t) = (16\pi^4\sqrt{\det V(t)})^{-1} \exp\left\{ -\frac{1}{2}\zeta^T \mathbf{V}^{-1}(t)\zeta \right\}. \tag{23}$$

Here $\zeta^T = (q_{x1}, p_{x1}, q_{y1}, p_{y1}, q_{x2}, p_{x2}, q_{y2}, p_{y2})$ is a transposed eight-vector whose elements are the quadrature-component variables.

For the entanglement measure of the two QD electrons it is appropriate to use the logarithmic negativity based on the trace norm of the partial transpose of the density matrix of a bipartite state [16]. In the case of Gaussian states, the logarithmic negativity can be calculated by the symplectic spectrum of the partial transpose of covariance matrix $\mathbf{V}^{T_1}(t)$, that is $\mathbf{V}(t)$ in which all matrix elements connecting momenta $\mathbf{p_1}$ of the first electron are multiplied by -1. The spectrum $\{\eta_k\}$ $(k = 1, \ldots, 4)$ can be found by the following expression

$$-\mathbf{J}\mathbf{V}^{T_1}(t)\mathbf{J}\mathbf{V}^{T_1}(t) = \mathrm{diag}(\eta_1, \eta_1, \eta_2, \eta_2, \eta_3, \eta_3, \eta_4, \eta_4), \tag{24}$$

where the (antisymmetric) symplectic 8×8 matrix

$$\mathbf{J} = \bigoplus_{k=1}^{4} \begin{pmatrix} 0 & 1 \\ -1 & 0 \end{pmatrix}.$$

Then the logarithmic negativity defined as

$$E_\mathcal{N} = -\sum_{j=1}^{4} \log_2(\min(1, 2\eta_j)) \tag{25}$$

determines the degree of entanglement of QD electrons for $E_\mathcal{N} > 0$.

In the report, it is given an analysis of the behavior of the logarithmic negativity for the orbital motion of QD electrons as a function of time and QD parameters. All model parameters are given with respect to the confining frequency ω_x, which serves as a dimensional parameter. In QDs the Coulomb interaction is substantially screened, so that the effective interaction (3) should be considered as a perturbation. It is why in the numerical analysis the strength of

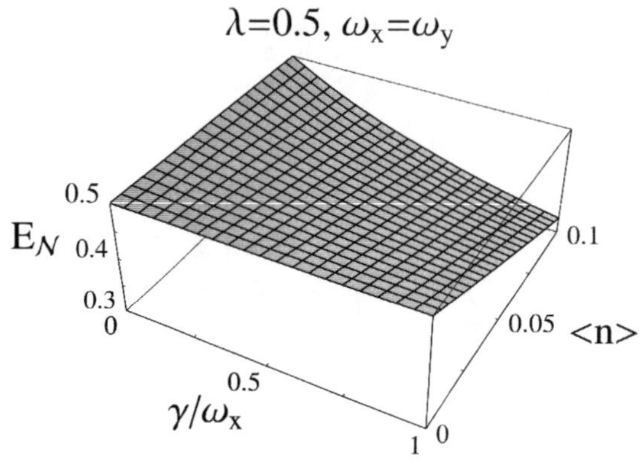

Figure 2. The logarithmic negativity $E_{\mathcal{N}}$ as a function of the damping constants $\gamma \equiv \gamma_1 = \gamma_2$ and the mean boson numbers $\langle n \rangle \equiv \langle n_1 \rangle = \langle n_2 \rangle$.

interaction is assumed to obey the condition $\lambda/\omega_x \leq 0.5$. Regarding the thermal noises that enter into the model in the form of the mean boson numbers (21), they are held as $\langle n \rangle \leq 0.1$. Taking into account the typical mean spacing of QD energy levels $\hbar\omega_x \sim 3$ meV, it corresponds to the temperatures $T \leq 10$ K according to relations (21).

It is seen in Fig. 1 that entangled states of electrons in time are noticeably subjected to the degradation caused by environmental noises. Moreover, the degradation in nearly circular QDs occurs much faster than that in strongly deformed QDs (left plot). It means that deformed QDs are more firm to temperature effects than the circular ones, although the latter can create more strong entanglement.

The dependence of the entanglement degree of QD electrons at the moment $\omega_x t = 1$ on the damping constants (QD openness) and environment thermal excitations is plotted in Fig. 2. One observes the pronounced decrease of the logarithmic negativity for $\gamma > 0$ with the increase of $\langle n \rangle$. This decrease becomes more sudden as the damping constant gains.

5. Summary

A model for a two-electron two-dimensional open QD with the anisotropic parabolic confinement is considered. The effects of the QD openness are taken into account by coupling a QD with a boson reservoir at fixed temperatures. An analysis of the dynamics of the electron orbital motion shows the appreciable degradation of the entanglement of QD electrons caused by quantum noises of reservoir excitations. The degradation is turn out to be most pronounced for circular QDs with a non-negligible degree of openness. The obtained results are of particular interest from the view point of estimation of the coherence time in quantum logic gates based on pseudospin of QD electrons.

Acknowledgments

I thank Prof. R.G. Nazmitdinov for fruitful discussions.
This work was partly supported by RFBR Grant No. 11-02-00086 (Russia).

References

[1] Loss D and DiVincenzo D P 1998 *Phys. Rev. A* **57** 120
[2] Burkard G, Loss D and DiVincenzo D P 1999 *Phys. Rev. B* **59** 2070

[3] Barenco A, Deutsch D, Ekert A and Jozsa R 1995 *Phys. Rev. Lett.* **74** 4083
[4] Tanamoto T 2000 *Phys. Rev. A* **61** 022305
[5] Hollenberg L C L, Dzurak A S, Wellard C, Hamilton A R, Reilly D J, Milburn G J and Clark R G 2004 *Phys. Rev. B* **69** 113301
[6] Koiller B, Hu X and Das Sarma S 2002 *Phys. Rev. Lett.* **88** 027903
[7] Nielsen M A and Chuang I L 2001 *Quantum Computation and Quantum Information* (Cambridge: Cambridge University Press)
[8] Nazmitdinov R G and Chizhov A V 2012 *Optics and Spectroscopy* **112** 319
[9] Chizhov A V and Nazmitdinov R G 2012 *J. Phys.: Conf. Ser.* **343** 012023
[10] Hayashi T, Fujisawa T, Cheong H D, Jeong Y H and Hirayama Y 2003 *Phys. Rev. Lett.* **91** 226804
[11] Petta J R, Johnson A C, Marcus C M, Hanson M P and Gossard A C 2004 *Phys. Rev. Lett.* **93** 186802
[12] Hu X, Koiller B and Das Sarma S 2005 *Phys. Rev. B* **71** 235332
[13] Kouwenhoven L P, Austing D G and Tarucha S 2001 *Rep. Prog. Phys.* **64** 701
[14] Johnson N F and Payne M C 1991 *Phys. Rev. Lett.* **67** 1157
[15] Louisell W H 1973 *Quantum Statistical Properties of Radiation* (New York: Wiley)
[16] Vidal G and Werner R F 2002 *Phys. Rev. A* **65** 032314

Spectral and optical characteristics of spheroidal quantum dots

A A Gusev[1], O Chuluunbaatar[1], L L Hai[1], S I Vinitsky[1], E M Kazaryan [2], H A Sarkisyan [2] and V L Derbov [3]

[1] Joint Institute for Nuclear Research, Dubna, Russia
[2] Russian-Armenian (Slavonic) University, Yerevan, Armenia
[3] Saratov State University, Saratov, Russia

E-mail: gooseff@jinr.ru

Abstract. In the framework of effective mass approximation and in strong size quantization regime the absorption coefficients for ensembles of spheroidal quantum dots (SQDs) are analyzed using the eigenvalues and eigenfunctions, calculated by means of Kantorovich and adiabatic methods. The comparison of absorption coefficients for oblate and prolate SQDs with parabolic and non-parabolic dispersion laws reveals different behavior depending on the aspect ratio (ratio of minor to major semiaxis) and the external homogeneous electric fields. The possibility of verification of the considered models is discussed.

1. Introduction

Quantum dots (QDs) are considered to be promising as the elementary basis for the new generation of semiconductor devices [1]. The unique opportunity to perform the energy level control and flexible manipulation in QDs is due to the full quantization of charge carrier energy spectra in these systems. This allows design and manufacturing of artificial structures with prescribed quantum physical characteristics [2]. That is why the scope of QDs potential applications is very wide, from heterostructure lasers to nanomedicine and nanobiology. An impressive example of such application is represented by QD lasers possessing low threshold current and high efficiency [2].

The peculiarities of physical processes in QDs are caused by both their composition and geometry. Electronic, kinetic, optical and other properties of QDs have been investigated experimentally and theoretically in many papers [3]–[15]. Particularly, the optical absorption characteristics of QDs have been shown to be strongly correlated with their geometry, on one hand, and their physical–chemical properties, on the other hand. In one of the first publications on optical transitions in QD [16] the interband absorption of light was considered in the ensemble of weakly interacting spherical QDs implanted in a dielectric matrix. The dispersion of QD sizes was characterized in the framework of Lifshitz–Slezov theory [17]. It was shown that in the absence of size dispersion, due to the full quantization of charge carriers energy spectra in QD, the absorption coefficient behaves like a delta function, and the absorption threshold frequencies depend on the peculiarities of electron and hole energy spectra. When the QD size dispersion is taken into account, the averaging procedure yields the absorption profile having finite width and height.

Recently several reports concerning the experimental implementation of narrow-band InSb QDs have appeared [18, 19], in which the dispersion law for electrons and light holes is non-parabolic and described according to the double-band mirror Kane model [20, 21]. For non-interacting band of heavy holes the dispersion law is considered as quadratic. The investigation of optical absorption peculiarities in InSb QDs is an interesting problem, taking the electron transitions from light and heavy hole bands induced by the absorbed photon energy to the conduction band into account. Interband transitions in an ensemble of cylindrical or spherical InSb QDs were considered theoretically in dipole approximation with and without magnetic field, including excitonic effects, by means of perturbation and adiabatic methods [22]–[24]. The realization of ellipsoidal QDs were done recently and experiments are intensified just nowadays. Unfortunately, the experiments of growth of ellipsoidal QDs we are aware of, are about InAsSbP compounds the zone structure of which is investigated poorly.

Quantum dots with ellipsoidal geometry are difficult to describe analytically. Therefore, one has to use numerical description of electronic and optical properties of such systems [25]–[31]. It is clear that for the first stage of investigation it is reasonable to discuss one particle states (strong size quantization regime), as far as in that case the problem becomes relatively simple. The criteria of applicability of such approximation are conditioned by the correctness of the following inequalities: $d_{max} \ll a_B^e$ and $d_{max} \ll a_B^h$, where d_{max} is the biggest linear size of localization area of electron and hole with respect to the corresponded Bohr radius,$a_B^{e,h} = (\hbar^2\kappa)/(\mu_{e,h}e^2)$, in the effective mass approximation of the $\mathbf{k}\cdot\mathbf{p}$-theory. Here, κ is the dc permittivity, $\mu_{e(h)} = \beta_{e(h)}m_0$ is the effective mass of electron or hole, m_0 is the mass of electron. As known, a comparison of theory with experiment show, if a sample thickness in a direction of size quantization contains more then ten crystal layers, i.e. start from 30-40 angstrom, then one can use effective mass approximation [1]. It should be mention, that in paper [7] in the same approximation the authors discussed the absorption of light in ellipsoidal quantum dot with asymmetric parabolic confinement potential in the presence of magnetic field. The problem of optical absorption was discussed on the base of normal mode approach [8] and exact expressions for interband transition threshold frequencies were found.

The consideration of excitonic effects is possible to do in several steps. In the case of weak size quantization regime the main energy contribution of the system is defined by the Coulomb energy of electron-hole interaction, and the presence of size quantization it is possible to define by the influence of walls on the center of mass of exciton (see, for example, [16]). Thereby, in this case the relative motion can be separated from the motion of center of mass and the shift (displacement) of excitonic line is taking place at the expense of size quantization of the motion of exciton as a whole. In the case of intermediate size quantization when for the heavy hole the size quantization is not significant, but for the electron it is strong enough, it is necessary to apply adiabatic approximation, when the motion of the hole is described in an effective field of a fast moving electron. The solution of excitonic problems of QD with ellipsoidal geometry needs separate and detailed study [9, 10]. Therefore within the frameworks of this paper we would not discuss excitonic effects, as well as electron-electron and hole-hole interactions.

Earlier we elaborated calculation schemes, symbolic-numerical algorithms (SNAs) and programs for numerical solution of boundary-value problems (BVPs) in axially-symmetric models of quantum wells, quantum wires and QDs in external fields, using the effective mass approximation [30]–[41]. The programs were based on the generalized Kantorovich method and the adiabatic method, allowed calculation of both discrete and continuous spectra and provided the prescribed accuracy.

In the present paper we apply the developed approach to the analysis of spectral and optical characteristics of oblate and prolate spheroidal QDs in strong size quantization regime with parabolic and non-parabolic dispersion laws under the influence of homogeneous electric fields (HEFs), i.e., the quantum-confined Stark effect.

2. Statement of the problem

Let us consider in the effective mass approximation of the $\mathbf{k}\cdot\mathbf{p}$-theory spheroidal QDs in strong size quantization regime with possible influence of the uniform electric field $\mathbf{F} = (0, 0, F)$. Then the Schrödinger equation for the slow-varying envelope wave function $\tilde{\Psi}^i(\tilde{\mathbf{r}}_i)$ of an electron ($i = e$) or a hole ($i = h$) reads as [10]:

$$\left\{\frac{1}{2\mu_i}\tilde{\mathbf{p}}_i^{\,2} - q_i(\mathbf{F}\cdot\tilde{\mathbf{r}}_i) + \tilde{U}_{conf}(\tilde{\mathbf{r}}_i) - \tilde{E}^i\right\}\tilde{\Psi}^i(\tilde{\mathbf{r}}_i) = 0. \qquad (1)$$

Here $\tilde{\mathbf{r}}_i$ is the radius-vector, $\tilde{\mathbf{r}}_i = (\tilde{x}_i, \tilde{y}_i, \tilde{z}_i)$, $\tilde{\mathbf{p}}_i = -\imath\hbar\nabla_{\tilde{\mathbf{r}}_i}$ is the momentum, $\tilde{E}^i$ is the energy of particles, $q_e = -e$ and $q_h = +e$ are the Coulomb charges of the electron and the hole, For the model under consideration, $\tilde{U}_{conf}(\tilde{\mathbf{r}})$ is the potential of a spherical or axially-symmetric well

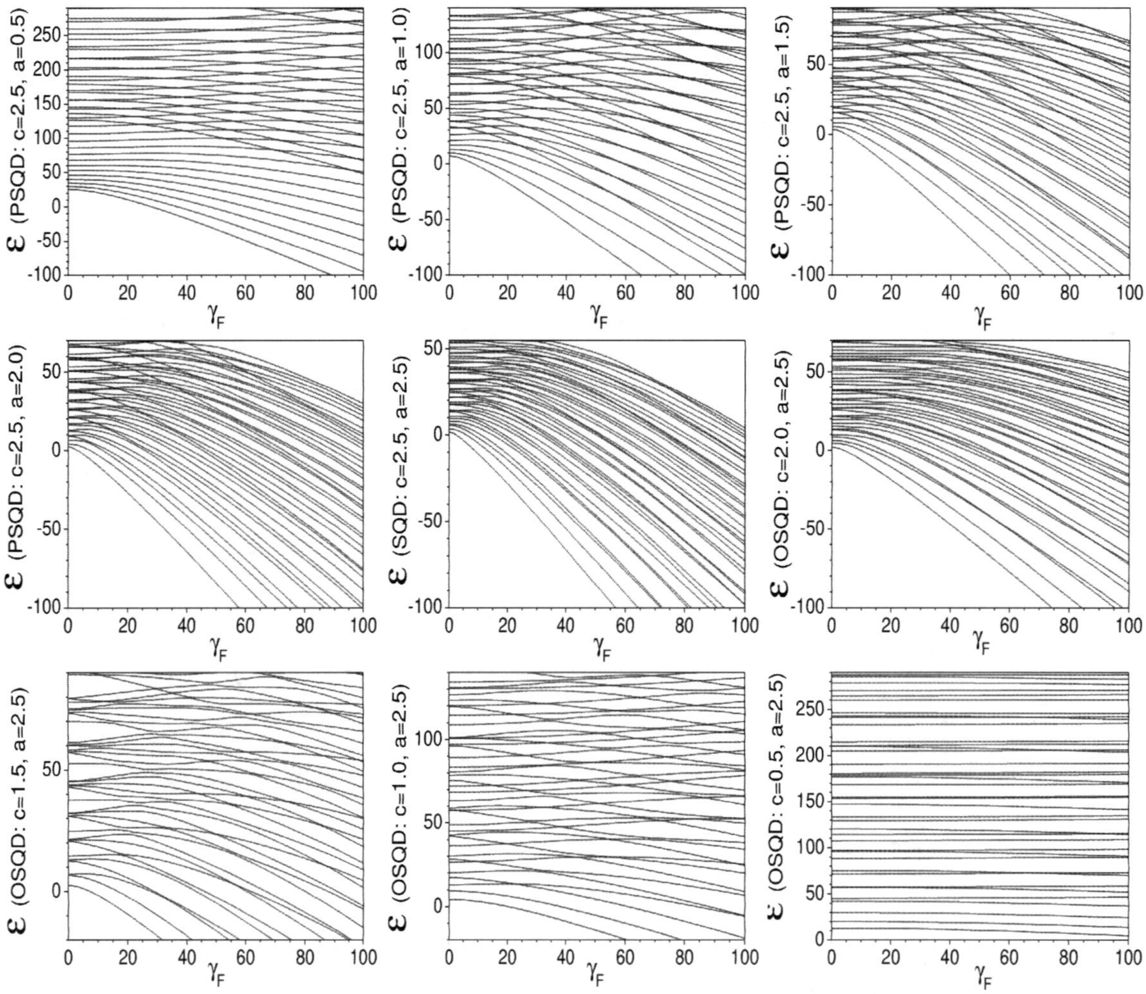

Figure 1. Dependence of energy eigenvalues $\mathcal{E}$ (in units of E_R^e) of the lower part of the spectrum of electronic states of QDs at $m = 0$ upon the electric field strength γ_F (in units of F_0^*): for spherical quantum dot (SQD) with radius $a = c = 2.5$, oblate and prolate spheroidal quantum dots (OSQD and PSQD) at different minor semiaxis (for OSQD $c = 0.5, 1, 1.5, 2$, $a = 2.5$, for PSQD $c = 2.5$, $a = 0.5, 1, 1.5, 2$).

$\tilde{U}_{conf}(\tilde{\mathbf{r}}_i) = \{0, S(\tilde{\mathbf{r}}_i) < 0; \tilde{U}_0, S(\tilde{\mathbf{r}}_i) \geq 0\}$, bounded by the surface $S(\tilde{\mathbf{r}}_i) = 0$ with walls of infinite height (infinite potential barrier model, IPBM) or finite height $1 \ll \tilde{U}_0 < \infty$ (finite potential barrier model, FPBM). Here $S(\tilde{\mathbf{r}}_i)$ depends on the parameters $\tilde{a}$, $\tilde{c}$, which are semiaxes of a spheroidal QD, $S(\tilde{\mathbf{r}}_i) \equiv (\tilde{x}_i^2 + \tilde{y}_i^2)/\tilde{a}^2 + \tilde{z}_i^2/\tilde{c}^2 - 1$. The eigenvalues and eigenfunctions needed to evaluate the absorption coefficients (ACs) were calculated with prescribed accuracy by means of program packages ODPEVP and KANTBP [34, 35, 36]. For example, the energy eigenvalues $\mathcal{E} = 2E^e = \tilde{E}^e/E_R^e$ of the lower part of the electronic spectrum of the QDs at $m = 0$ versus the electric field strength $2\gamma_F = F/F_0^*$ (in the reduced atomic units [4]) for the IPBMs of spherical, oblate, and prolate spheroidal QDs with parabolic dispersion law at different values of minor semiaxis are shown in Fig. 1. Detailed explanation of these dependencies is given in the forthcoming paper [41]. The dependence of the IPBMs and FPBMs of spheroidal and dumbbell QDs on the minor semiaxis was studied in [30, 31].

3. Absorption coefficient for an ensemble of QDs

One can use differences in the energy spectra to verify the considered models of QDs by calculating the absorption coefficient $K(\tilde{\omega}^{ph}, \tilde{a}, \tilde{c},)$ of an ensemble of identical semiconductor QDs [16]. Since we do not discuss exciton effects in the present paper, the absorption coefficient may be approximately expressed as

$$\tilde{K}(\tilde{\omega}^{ph}, \tilde{a}, \tilde{c}, u) = \sum_{\nu,\nu'} \tilde{K}_{\nu,\nu'}(\tilde{\omega}^{ph}, \tilde{a}, \tilde{c}, u) = \tilde{A} \sum_{\nu,\nu'} \tilde{I}_{\nu,\nu'}(u)\delta(\hbar\tilde{\omega}^{ph} - \tilde{W}_{\nu\nu'}), \qquad (2)$$

$$\tilde{I}_{\nu,\nu'}(u) = \left| \int \tilde{\Psi}_\nu^e(\tilde{\mathbf{r}}; \tilde{a}, \tilde{c}, F, \mu_e)\tilde{\Psi}_{\nu'}^h(\tilde{\mathbf{r}}; \tilde{a}, \tilde{c}, F, \mu_h) d^3\tilde{\mathbf{r}} \right|^2,$$

where $\tilde{A}$ is proportional to the square of the matrix element in the Bloch decomposition, $\tilde{\Psi}_\nu^e(u)$ and $\tilde{\Psi}_{\nu'}^h$ are the eigenfunctions of the electron (e) and the heavy hole (h), $\tilde{E}_\nu^e$ and $\tilde{E}_{\nu'}^h$ are the energy eigenvalues for the electron (e) and the heavy hole (h), depending on the semiaxis size $\tilde{c}, \tilde{a}$ for OSQD (or $\tilde{a}, \tilde{c}$ for PSQD) and the adiabatic set of quantum numbers $\nu = [n_{zo}, n_{po}, m]$ and $\nu' = [n'_{zo}, n_{po'}, m']$ ($\nu = [n_{pp}, n_{zp}, m]$ and $\nu' = [n'_{pp}, n'_{zp}, m']$), where $m' = -m$, $\tilde{E}_g$ is the band gap width in the bulk semiconductor, $\tilde{\omega}^{ph}$ is the incident light frequency, $\tilde{W}_{\nu\nu'} = \tilde{E}_g + \tilde{E}_\nu^e(\tilde{a}, \tilde{c}) + \tilde{E}_{\nu'}^h(\tilde{a}, \tilde{c})$ is the inter-band transition energy for which $\tilde{K}(\tilde{\omega}^{ph})$ has the maximal value. We rewrite the expression (2) in the terms of frequency shift of the incident light $\Delta\omega^{ph}/(2\pi) = (\hbar\tilde{\omega}^{ph} - \tilde{E}_g)/(2\pi\hbar)$ corresponding to the inter-band transition energy shift $\Delta\tilde{W}_{\nu\nu'} = \tilde{W}_{\nu\nu'} - \tilde{E}_g = \tilde{E}_\nu^e(\tilde{a}, \tilde{c}) + \tilde{E}_{\nu'}^h(\tilde{a}, \tilde{c})$ for which $\tilde{K}(\Delta\tilde{\omega}^{ph})$ has the maximal value, using dimensionless variables in reduced atomic units

$$\tilde{K}(\Delta\tilde{\omega}^{ph}, \tilde{a}, \tilde{c}) = \tilde{A}\tilde{E}_g^{-1} \sum_{\nu,\nu'} \tilde{I}_{\nu,\nu'}(u)\delta[f_{\nu,\nu'}(u)], \quad f_{\nu,\nu'}(u) = \lambda_1 - \frac{2E_\nu^e(a,c) + 2E_{\nu'}^h(a,c)(\mu_h/\mu_e)}{2E_g}, \qquad (3)$$

Here the parameter u will be defined below, $\lambda_1 = (\hbar\tilde{\omega}^{ph} - \tilde{E}_g)/\tilde{E}_g$ is the energy of the optical interband transitions scaled to $\tilde{E}_g$, $2E_g = \tilde{E}_g/\tilde{E}_R^e$ is the dimensionless band gap width.

For GaAs the functions $f_{\nu,\nu'}^{h\to e}(u)$ describing the ($h \to e$) interband transitions have the form

$$f_{\nu,\nu'}^{h\to e}(u) = \lambda_1 - (2E_g)^{-1}(2E_\nu^e(a,c) + 2E_{\nu'}^e(a,c)(\mu_e/\mu_h)), \qquad (4)$$

where $\mu_e = 0.067m_0$ and $\mu_h = 0.558m_0$ are the masses of electron and holes, respectively, $\tilde{E}_g =$ 1430meV is the band gap width and $\kappa = 13.18$ is the dc permittivity and $\tilde{E}_R^e = e^2/(2\kappa a_B^e) =$ 5.275 meV, $a_B^e = \hbar^2\kappa/(\mu_e e^2) = 104$Å, $E_R^h = e^2/(2\kappa a_B^h) = 49$ meV, $a_B^h = \hbar^2\kappa/(\mu_h e^2) = 15$Å and $F_0^* = E_R^e/(e a_B^e) = e/(2\kappa(a_B^e)^2) = 5.04$kV/cm.

For InSb the dispersion law for heavy holes (hh) is parabolic while for electrons (e) and light holes (lh) it is non-parabolic and may be described by the Kane model [20, 21, 24]. The energy values in our notation are:

$$2\tilde{E}_{\nu}^{hh}(InSb) = 2\tilde{E}_{\nu'}^{h}(\tilde{a},\tilde{c}), \tag{5}$$

$$2\tilde{E}_{\nu}^{e}(InSb) = 2\tilde{E}_{\nu}^{lh}(InSb) = -\tilde{E}_g/2 + \sqrt{\tilde{E}_g^2/4 + \tilde{E}_g(2\tilde{E}_{\nu}^{e}(\tilde{a},\tilde{c}))}. \tag{6}$$

As follows from Eqs. (5) and (8), to determine the energy spectrum and the wave function of the light hole and the electron one should solve the Klein-Gordon equation [42, 43], while for heavy hole the Schrödinger equation is applicable. The functions $f_{\nu,\nu'}^{hh\to e}(u)$ and $f_{\nu,\nu'}^{hh\to e}(u)$ describing the $(hh\to e)$ and the $(lh\to e)$ interband transitions have the forms

$$f_{\nu,\nu'}^{hh\to e}(u) = \lambda_1 - (1/2 + \sqrt{1/4 + (2E_{\nu}^{e}(a,c)/(2E_g))} + (2E_g)^{-1}2E_{\nu'}^{e}(a,c)(\mu_e/\mu_h)), \tag{7}$$

$$f_{\nu,\nu'}^{lh\to e}(u) = \lambda_1 - 2(\sqrt{1/4 + (2E_{\nu}^{e}(a,c)/(2E_g))}), \tag{8}$$

where $\mu_e = \mu_{lh} = 0.15m_0$ and $\mu_h \equiv \mu_{hh} = 0.5m_0$ are the masses of electron, light and heavy holes, respectively, $\tilde{E}_g = 180$ meV is the band gap width, $\kappa = 16$ is the dc permittivity, and $E_R^e = E_R^{lh} = e^2/(2\kappa a_B^e) = 7.972$ meV, $a_B^e = a_B^{lh} = \hbar^2\kappa/(\mu_e e^2) = 56.44\text{Å}$, $E_R^h = E_R^{hh} = e^2/(2\kappa a_B^{hh}) = 26.57$ meV, $a_B^h = a_B^{hh} = \hbar^2\kappa/(\mu_h e^2) = 16.93\text{Å}$.

For both electron and hole carriers the dimensionless energies $2E_{\nu}^{e} = \tilde{E}_{\nu}^{e}/\tilde{E}_R^{e}$ and $2E_{\nu}^{h}(\mu_h/\mu_e) = \tilde{E}_{\nu}^{h}/\tilde{E}_R^{e}$ are expressed in the same reduced atomic units $\tilde{E}_R^{e}$, and the overlap integral (2) between the eigenfunctions, corresponding to $E_{\nu}^{e}(\gamma_F)$ and $E_{\nu}^{h}(\gamma_F) = (\mu_e/\mu_h)E_{\nu}^{e}(-(\mu_h/\mu_e)\gamma_F)$, takes the form

$$\tilde{I}_{\nu,\nu'}(u) = \left|\int \Psi_{\nu}^{e}(\mathbf{r};a,c,\gamma_F,\mu_e)\Psi_{\nu'}^{e}(\mathbf{r};a,c,-(\mu_h/\mu_e)\gamma_F,\mu_e)d^3\mathbf{r}\right|^2. \tag{9}$$

Now consider an ensemble of OSQDs (or PSQDs), differing in the minor semiaxis values $c = u_o\bar{c}$ (or $a = u_p\bar{a}$), determined by the random parameter $u = u_o$ (or $u = u_p$). The corresponding minor semiaxis mean value is $\bar{c}$ at fixed major semiaxis a (or $\bar{a}$ at fixed major semiaxis c), and the appropriate distribution function is $P(u_o)$ (or $P(u_p)$). Commonly, in this case the normalized Lifshits-Slezov distribution function [17] is used:

$$P(u) = \{3^4eu^2\exp(-1/(1-2u/3))/2^{5/3}/(u+3)^{7/3}/(3/2-u)^{11/3}, u\in(0,3/2); 0, \text{otherwise}\}$$

having conventional properties $\int P(u)du = 1$, $\bar{u} = \int uP(u)du = 1$. The absorption coefficients $\tilde{K}^o(\tilde{\omega}^{ph},\tilde{a},\tilde{c})$ or $\tilde{K}^p(\tilde{\omega}^{ph},\tilde{a},\tilde{\bar{c}})$ of an ensemble of semiconductor OSQDs or PSQDs with different dimensions of minor semiaxes are expressed as

$$\tilde{K}^o(\tilde{\omega}^{ph},\tilde{\bar{a}},\tilde{c}) = \int \tilde{K}(\tilde{\omega}^{ph},\tilde{\bar{a}},\tilde{c},u_o)P(u_o)du_o,\quad \tilde{K}^p(\tilde{\omega}^{ph},\tilde{a},\tilde{\bar{c}}) = \int \tilde{K}(\tilde{\omega}^{ph},\tilde{a},\tilde{\bar{c}},u_p)P(u_p)du_p. \tag{10}$$

Substituting (3) into (10) and taking into account the known properties of the δ-function, we arrive at the analytical expression for the absorption coefficient $\tilde{K}(\tilde{\omega}^{ph},\tilde{a},\tilde{c})$ of a system of semiconductor QDs with a distribution of random minor semiaxes:

$$\frac{\tilde{K}(\tilde{\omega}^{ph})}{\tilde{K}_0} = \sum_{\nu,\nu',s}\frac{\tilde{K}_{\nu,\nu'}(\tilde{\omega}^{ph})}{\tilde{K}_0},\quad \frac{\tilde{K}_{\nu,\nu'}(\tilde{\omega}^{ph})}{\tilde{K}_0} = \tilde{I}_{\nu,\nu'}(u_s)\left|\frac{df_{\nu,\nu'}(u)}{du}\Big|_{u=u_s}\right|^{-1}P(u_s), \tag{11}$$

where $\tilde{K}_0 = \tilde{A}^{-1}\tilde{E}_g$ is the normalization factor, u_s are the roots of the equation $f_{\nu,\nu'}(u_s) = 0$.

At $\gamma_F = 0$ for IPBM we have the interband overlap $\tilde{I}_{\nu,\nu'} = \delta_{n_{\rho o},n'_{\rho o}}\delta_{n_{zo},n'_{zo}}\delta_{m,-m'}$ for OSQD, or $\tilde{I}_{\nu,\nu'} = (J_{1+|m|}(\alpha_{n_{\rho p}+1,|m|})/J_{1-|m|}(\alpha_{n_{\rho p}+1,|m|}))^2\delta_{n_{zp},n'_{zp}}\delta_{n_{\rho p},n'_{\rho p}}\delta_{m,-m'}$ for PSQD, where $\alpha_{n_{\rho p}+1,|m|}$ is the positive root of the Bessel function, and the selection rules $m = -m'$, $n_{zo} = n'_{zo}$, $n_{\rho o} = n'_{\rho o}$, or $n_{\rho p} = n'_{\rho p}$, $n_{zp} = n'_{zp}$ [31], while at $\gamma_F \neq 0$ one should calculate the interband overlap (9) in accordance with the selection rules $m = -m'$, $n_{\rho o} = n'_{\rho o}$, or $n_{\rho p} = n'_{\rho p}$ [41], respectively. Note that the contributions of non-diagonal matrix elements to the energy values are about 1% for IPBM of OSQD and PSQD; then in the Born-Oppenheimer approximation of the order b_{max} for the AC we get [31, 41]

$$f_{\nu,\nu'}(u) = \lambda_1 - \sum_{j=0}^{b_{max}} \check{E}^{(j)} u^{j-2}. \tag{12}$$

For example, at $\gamma_F = 0$ the coefficients $\check{E}^{(j)}$ are defined by [31]

$$\begin{aligned}
&\check{E}^{(j)} = (2E_g)^{-1}E^{(j)}_{io}\,\omega^{2-j}_{\rho;n_o}(\bar{c})(1+\mu_e/\mu_h) \quad \text{or} \quad \check{E}^{(j)} = (2E_g)^{-1}E^{(j)}_{ip}\,\omega^{2-j}_{z;n_{\rho p}}(\bar{a})(1+\mu_e/\mu_h),\\
&\omega_{\rho;n_o}(\bar{c}) = \pi n_o/(a\bar{c}), \quad \omega_{z;n_{\rho p}}(\bar{a}) = \alpha_{n_{\rho p}+1,|m|}/(\bar{a}c); \qquad (13)\\
&E^{(0)}_{io} = a^2/4, \quad E^{(1)}_{io} = (2n_{\rho o}+|m|+1), \quad E^{(2)}_{io} = (6n_{\rho o}|m|+2+6n_{\rho o}+6n^2_{\rho o}+|m|^2+3|m|)a^{-2},\\
&E^{(3)}_{io} = 3(6n_{\rho o}+3|m|+2+|m|^2+6n^2_{\rho o}+6n_{\rho o}|m|+4n^3_{\rho o}+6|m|n^2_{\rho o}+2|m|^2 n_{\rho o})a^{-4}/2;\\
&E^{(0)}_{ip} = c^2, \quad E^{(1)}_{ip} = (2n_{zp}+1), \quad E^{(2)}_{ip} = +3(2n_{zp}+2n^2_{zp}+1)c^{-2}/4,\\
&E^{(3)}_{ip} = 3(3n^2_{zp}+7n_{zp}+2n^3_{zp}+3)c^{-4}/16.
\end{aligned}$$

For the Lifshits-Slezov distribution Figs. 2 and 3 display the total absorption coefficients $\tilde{K}(\tilde{\omega}^{ph})/\tilde{K}_0$ and the partial absorption coefficients $\tilde{K}_{\nu,\nu}(\tilde{\omega}^{ph})/\tilde{K}_0$, that form the corresponding partial sum (11) over a fixed set of quantum numbers ν, ν' at $m = -m' = 0$. As a result of averaging (10) a series of curves with finite with and height are observed instead of a series of δ-functions. One can see that, in the first place, the summation over the quantum numbers $n_o = n_{zo}+1 = 1, 2, 3, 4, 5$ (or $n_p = n_{\rho p}+1 = 1, 2, 3$) enumerating the nodes of the wave function with respect to the fast variable gives the corresponding principal maxima of the total AC for the ensemble of QDs with distributed dimensions of minor semiaxis. Secondly, the summation over the quantum number $n_{\rho o} = 0, 1, 2, 3, ..., 8$ (or $n_{zp} = 0, 1, 2, ..., 15$) that labels the nodes of the wave function with respect to the slow variable leads to the increase of amplitudes of these maxima and to secondary maxima arising in the case of sparer energy levels of IPBM of OSQDs (or PSQDs).

In the regime of strong dimensional quantization the frequencies $\Delta\tilde{\omega}^{ph}_{100}/(2\pi) = (2\pi\hbar)^{-1}(\tilde{W}_{100,100} - \tilde{E}_g)$ of the interband transitions $(h \to e)$ in GaAS between the levels $n_o = 1, n_{\rho o} = 0, m = 0$ for OSQD or $n_p = 1, n_{zp} = 0, m = 0$ for PSQD at the fixed values $\tilde{a} = 2.5a_e$ and $\tilde{c} = 0.5a_e$ for OSQD or $\tilde{a} = 0.5a_e$ and $\tilde{c} = 2.5a_e$ for PSQD, are equal to $\Delta\tilde{\omega}^{ph}_{100}/(2\pi) = 16.9$THz at $\gamma_F = 0$ and $\Delta\tilde{\omega}^{ph}_{100}/(2\pi) = 15.9$THz at $\gamma_F = 10$, or $\Delta\tilde{\omega}^{ph}_{100}/(2\pi) = 33.3$THz at $\gamma_F = 0$ and $\Delta\tilde{\omega}^{ph}_{100}/(2\pi) = 31.5$THz at $\gamma_F = 2$ correspond to the infra-red spectral region [9, 10], taking the band gap value $(2\pi\hbar)^{-1}\tilde{E}_g = 346$THz into account. In Fig. 3 one can see the quantum-confined Stark effect that consist in the reduction of the absorption energy (light frequency) at the expense of lowering the energy of both (e) and (h) bound states due to the electric field effect. The total ACs at $F \neq 0$, shown by solid lines, qualitatively correspond to the total AC at $F = 0$, shown by dashed lines, but have lower magnitudes and smooth behavior, in spite of the additional contribution to the partial ACs of

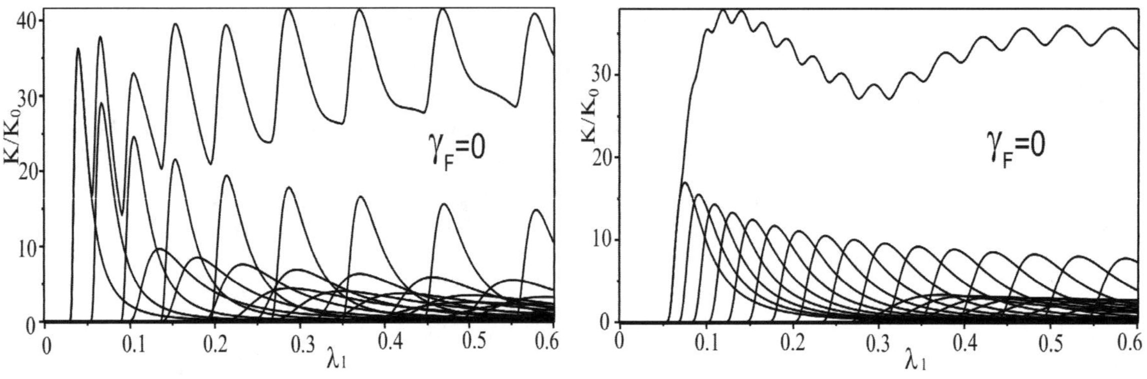

Figure 2. Absorption coefficient K/K_0, Eq. (11), consisting of a sum of the first partial contributions vs the energy $\lambda = \lambda_1$ of the optical interband transitions for the Lifshits-Slezov distribution, using the functions $f^{h\to e}_{\nu,\nu'}(u)$ for GaAs ($h \to e$) without electric field: for ensemble of OSQDs $\bar{c} = 0.5$, $a = 2.5$ (left panel) and for ensemble of PSQDs $\bar{a} = 0.5$, $c = 2.5$ (right panel).

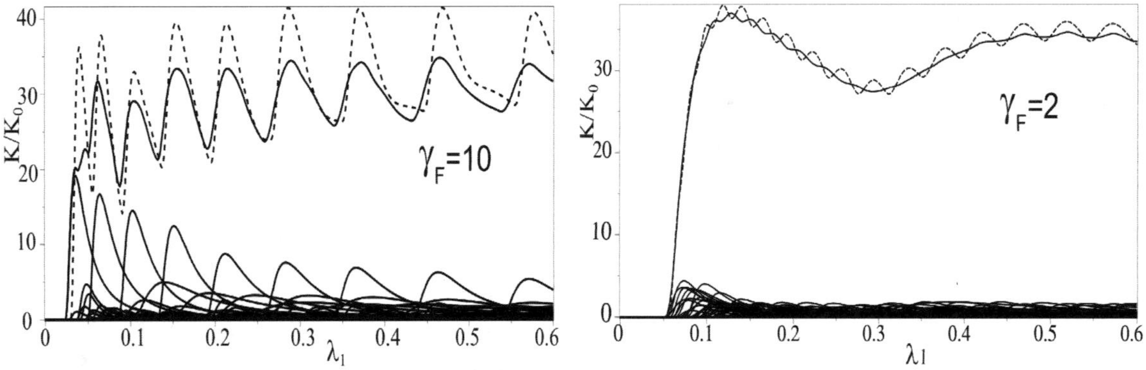

Figure 3. The same as in Fig. 2, but in the presence of electric field $2\gamma_F = F/F_0^*$. For comparison, the corresponding absorption coefficient without electric field is given by dashed line.

the overlap integral (9) from the interband transition $n_{zo} \neq n'_{zo}$ or $n_{zp} \neq n'_{zp}$ in OSQD or PSQD, also shown in Fig. 3.

At the same parameters of the QDs the frequencies of the interband transitions ($lh \to e$) in InSb are equal to $\Delta\tilde{\omega}^{ph}_{100}/(2\pi) = 68.5$THz for OSQD or $\Delta\tilde{\omega}^{ph}_{100}/(2\pi) = 87.2$THz for PSQD, while the frequencies of the interband transitions ($hh \to e$) in InSb are equal to $\Delta\tilde{\omega}^{ph}_{100}/(2\pi) = 78.6$THz for OSQD or $\Delta\tilde{\omega}^{ph}_{100}/(2\pi) = 102$THz for PSQD. These values correspond to the infrared spectral region with longer wavelength, similar to [24], with the band gap value $(2\pi\hbar)^{-1}\tilde{E}_g = 44$THz taken into account. One can see that the behavior of total ACs for parabolic dispersion law for IPBM of InSb, shown in Fig. 4, is similar to that for GaAs (Fig. 2), while the behavior of AC for nonparabolic dispersion law, shown in Fig. 5, is essentially different. In particular, for OSQDs it grows faster with increasing λ_1, while for PSQDs it goes to a plateau before starting to grow. Indeed, with increasing quantum numbers $n_{\rho o}$ or n_{zp} that characterize the excitation of slow motion, the maxima of partial ACs decrease for parabolic dispersion law, while for the nonparabolic one the maxima of partial ACs increase.

With decreasing semiaxis the threshold energy increases, because the "effective" band gap

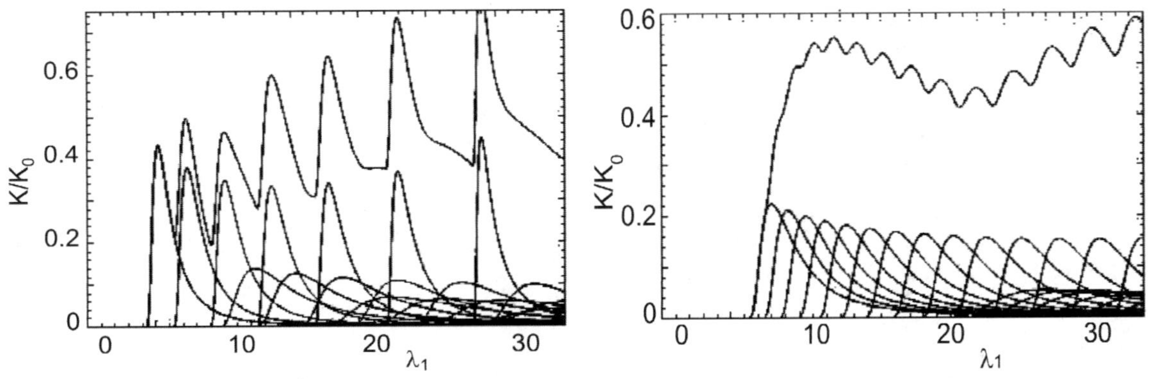

Figure 4. The same as in Fig. 2, but for InSb ($hh \rightarrow e$) interband transition.

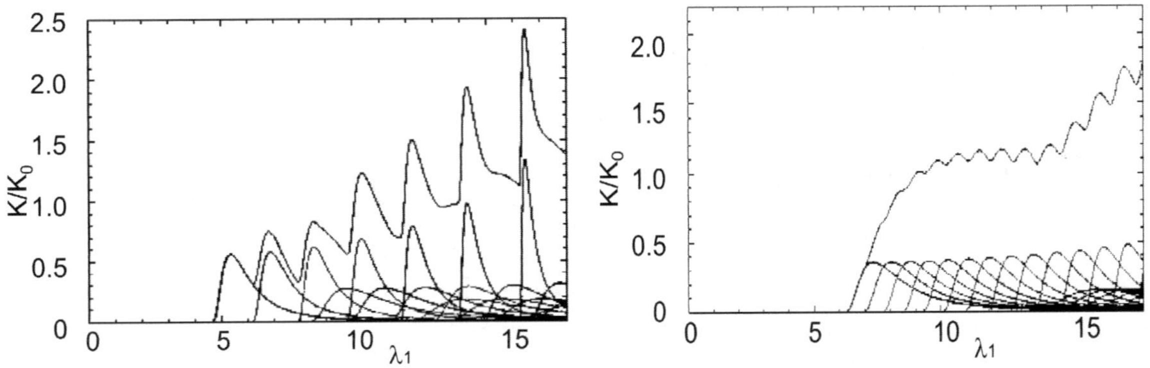

Figure 5. The same as in Fig. 2, but for InSb ($lh \rightarrow e$) interband transition.

width increases, which is a consequence of the dimensional quantization enhancement. Therefore, the above frequency is greater for PSQD than for OSQD, because the SQ implemented in two direction of the plane (x,y) is effectively larger than that in the direction of the z axis solely at similar values of semiaxes. Higher-accuracy calculations reveal an essential difference in the frequency behavior of the AC for interband transitions in systems of semiconductor OSQDs or PSQDs having a distribution of minor semiaxes, which can be used to verify the above models.

4. Conclusion

In this paper we demonstrate the efficient methods of calculating the electron and hole states in the presence of an electric field for spherical and spheroidal QDs with parabolic and non-parabolic dispersion laws. Our analysis shows that the adiabatic approach provides a useful theoretical tool for describing absorption coefficients in an ensemble of spheroidal QDs with random distribution of minor semiaxes. Further development and applications of our approach and SNAs are associated with the investigation of exciton effects, as well as electron-electron and hole-hole interactions in spheroidal QDs.

The work was supported partially by RFBR (grants No 10-02-00200 and 11-01-00523).

[1] Bimberg D, Grundman M, Ledentsov N 1999 *Quantum Dot Heterostructures* (Wiley, New-York)
[2] Alferov Zh 1998 *Semiconductors* **32** 1
[3] Li Bin et al 2007 *Phys. Lett. A* **367** 493
[4] Lamouche G, Lépine Y 1994 *Phys. Rev. B* **49** 13452
[5] Hawrylak P, Wojs A 1996 *Semicond. Sci. Technol.* **11** 1516

[6] Sarkisyan H A 2002 *Mod. Phys. Lett. B* **16** 835
[7] Dvoyan K G et al 2012 *Int. J. Mod. Phys. Conf. Ser.* **15** 40
[8] Nazmitdinov R G 2009 *Phys. Part. Nucl.* **40** 71
[9] Dvoyan K G et al 2009 *Nanoscale Res. Lett.* **4** 106; Dvoyan K G et al 2010 *Proc. SPIE* **7998** 79981F
[10] Dvoyan K G et al 2007 *Nanoscale Res. Lett.* **2** 601
[11] López S et al 2008 *Physica E* **40** 1383
[12] Barseghyan M, Kirakosyan A, Duque C 2009 *Eur. Phys. J. B* **72** 521
[13] Gharaati A, Khordad R 2010 *Superlattices Microstruct* **48** 276
[14] Filikhin I, Suslov V M, Vlahovic B 2006 *Phys. Rev. B* **73** 205332
[15] Filikhin I el al, 2009 *Physica E* **41** 1358
[16] Efros Al L, Efros A L 1982 *Sov. Phys. Semicond.* **16** 772
[17] Lifshits I M and Slezov V V 1958 *Sov. Phys. JETF.* **35** 479
[18] Moiseev K et al 2007 *Tech. Phys. Lett.* **33** 295
[19] Moiseev K et al 2009 *Semiconductors* **43** 1102
[20] Kane E O 1957 *J. Phys. Chem. Sol.* **1** 249
[21] Askerov B 1985 *Electronic Transport Phenomena in Semiconductors* (Nauka, Moscow)(in Russian)
[22] Kazaryan E, Meliksetyan A, Sarkisyan H 2007 *Tech. Phys. Lett.* **33** 964
[23] Kazaryan E, Meliksetyan A, Sarkisyan H 2010 *J. Comput. Theor. Nanosci.* **7** 486
[24] Atoyan M S et al 2011 *Physica E* **43** 1592
[25] Cantele G et al, 2000 *J. Phys.: Cond. Matt.* **12,** 9019
[26] Van den Broek M and Peeters F M, 2001 *Physica E* **11,** 345.
[27] Trani F et al, 2005 *Phys. Rev. B* **72,** 075423
[28] Lepadatu A-M et al, 2010 *J. Appl. Phys.* **107,** 033721
[29] Bagga A, Ghosh S, and Chattopadhyay P K 2005 *Nanotechnology* **16** 2726
[30] Gusev A A et al 2010 *Lect. Notes Comp. Sci.* **6244** 106
[31] Gusev A A et al 2012 *Phys. Atom. Nucl.* **75** 1210
[32] Gusev A A et al 2010 *J. Phys. Conf. Ser.* **248** 012047
[33] Gusev A A et al 2010 *Phys. Atom. Nucl.* **73** 352
[34] Chuluunbaatar O et al 2007 *Comput. Phys. Commun.* **177** 649
[35] Chuluunbaatar O et al 2009 *Comput. Phys. Commun.* **180** 1358
[36] Chuluunbaatar O et al 2008 *Comput. Phys. Commun.* **179** 685
[37] Gerdt V et al 2006 *Lect. Notes Comp. Sci.* **4194** 194
[38] Chuluunbaatar O et al 2008 *Comput. Phys. Commun.* **178** 301
[39] Chuluunbaatar O et al 2007 *Lect. Notes Comp. Sci.* **4770** 118
[40] Gusev A A et al 2012 *Lect. Notes Comp. Sci.* **7442** 155
[41] Gusev A A et al 2013 *Phys. Atom. Nucl.* **76** submitted
[42] Kazaryan E M, Petrosyan L S, Sarkisyan H A 2003 *Physica E* **16** 174
[43] Zoheir M, Manaselyan A Kh, Sarkisyan H A 2008 *Physica E* **40** 2945

Electron tunneling in double quantum dots and rings

I Filikhin, S G Matinyan and B Vlahovic

North Carolina Central University,
1801 Fayetteville St. Durham, NC 27707, USA,
E-mail: ifilikhin@nccu.edu

Abstract. The electron tunneling in double quantum dot (DQD) is studied by changing of inter-dot distance. Localization of an electron is calculated for each level of the whole spectrum of the confinement states of DQD. For identical QDs the tunneling goes consecutively from high energy levels to the ground state. It is changed essentially when the QDs of the system are non-identical. We show that violation of symmetry of the DQD geometry makes it difficult for the tunneling. For double quantum ring (DQR) we demonstrate that the electron states with the same type of rotation symmetry play an important role for tunneling.

1. Introduction

The dynamical tunneling in double quantum ring and dot (DQR [1] and DQD) is related to charge transport through the QD array. We study correlation between electron localization and energy spectrum of single electron confinement states in DQD and DQR with dependence on distance between QDs. When distance between the objects is large, the electron is localized in one object. When the distance decreases, the electron is tunneling; its wave function is spread over the whole double system. The cases of identical and non-identical QDs in DQD are considered. DQR may be considered as a variant of DQD with non-identical QDs. The tunneling may be changed essentially when the QDs of the system are non-identical. Generally, a type of the tunneling is defined by a dependence of confinement energy on quantum numbers due to the anti-crossing mechanism. Violation of symmetry of the DQD geometry leads to the change of such dependence for QDs of DQD and changes the type of the tunneling.

In the presented work we restrict the consideration by two dimensional quantum objects. This restriction allows us to use low computer resources for the numerical calculations of the whole spectrum of electron confine states (more than hundred states) with good accuracy. A 3D model has no principal difficulties in these studies. The 3D calculations are in progress.

2. Formalism

We study quantum dots and rings composed of InGaAs in a GaAs substrate utilizing a ***kp***-perturbation single sub-band approach with the effective potential approach [2,3]. Two dimensional (2D) objects are considered. To describe tunnelling of a single electron in doublet quantum objects we made some definition. Probability of localization of electron into region Ω_γ (γ =1, 2) is defined by $N_{n,\gamma} = \iint_{\Omega_\gamma} |\Phi_n(x,y)|^2\, dxdy$, where $\Phi_n(\rho,z)$ is wave function of electron, n is the number of the state. Ω_γ (γ=1, 2) are dictated by the QD shapes. Let us define as tunnelling measure parameter $\sigma = \dfrac{N_{n,1} - N_{n,2}}{N_{n,1} + N_{n,2}}$, with the range of [-1,1]. Obviously, when $\sigma = 0$, the electron will be in QD1 (Ω_1) or in QD2 (Ω_2) with equal probability (here we assume, that the QD1 and QD2 have the same shape). This happens when electron is tunneling between QD1 and QD2. The case $\sigma = 1$ ($\sigma = -1$) corresponds to the case when the election is strongly located in QD1 (QD2).

3. InAs/GaAs double quantum dots

3.1 Identical QDs

The circle shaped InAs/GaAs dots compose the double quantum system. The DQD is considered to be two dimensional. The radius of each QD is 40 nm. The inter-dot distance is changed in this study. Visualization for the tunnelling is given In Fig. 1 where the parameter σ is presented along the energies of single electron spectrum in the InAs/GaAs DQD for different inter-dot distances. When distance between the objects is large enough, an electron is localized in one of the objects $\sigma \approx 1$ ($\sigma \approx -1$); when the distance decreases, the electron is tunneling so that its wave function is spread over the whole double system and $\sigma \approx 0$. Tunnelling occurs in all the states for inter-dot distances less than 15 nm. There are no tunnelled states when a is larger 40 nm. One can see that the obtained picture (in Fig. 1) is symmetric relative axis of $\sigma = 0$. From Fig. 1, one can see that for each energy level (or very close energy levels) there are four or two values of σ. There are quasi quadruplets or quasi doublets of the DQD spectrum. The spectrum has degeneracy order of two for each level when the QDs are independent. This degeneracy due to the rotation symmetry of each QDs (which are identical). Energy of the states can be approximated by the relation: $E_{n,l} \sim \hbar^2/2m^*(n+|l|)/R^2$, where n and l are radial and orbital quantum numbers, R is radius of QD. The state with l and $-l$ are degenerate. There are quasi-doublet and quadruplet states. The states with σ between $|\sigma|=1$ and $\sigma \approx 0$ are tunnelled states. The symmetry of pictures in Fig. 1 reflects the symmetry between QDs in DQD due to their identity.

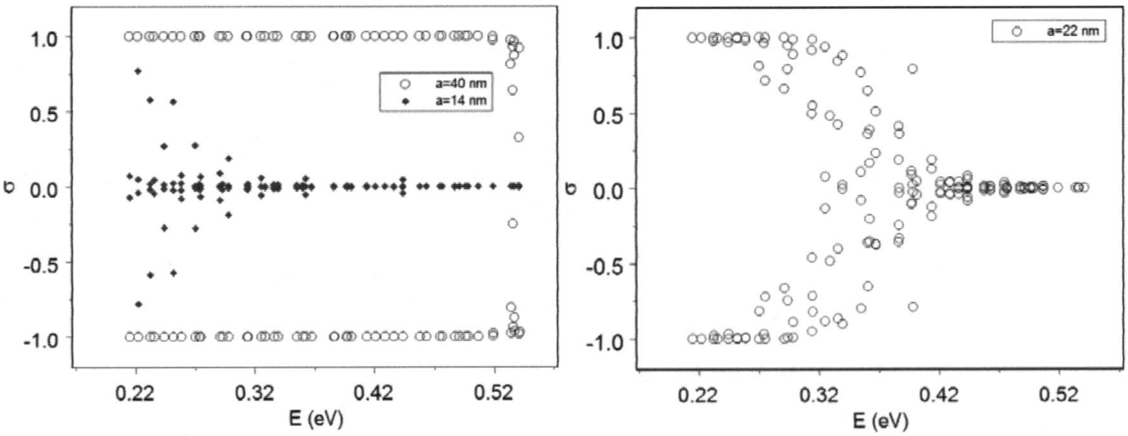

Figure 1. The parameter sigma for energies of the confinement single electron states in InAs/GaAs DQD. Inter-dot distances are left) 14 nm, 40 nm; right) 22 nm.

3.2 Non-identical QDs

Let us consider non-identical DQDs. Firstly we return to circle shaped DQDs discussed above (see Fig. 1). Initial radii of the QDs are 40 nm. The shape one of the QDs (right QD) was gradually decreased. The inter-dot distance was fixed with the value of 3 nm. The effect of non-identity on the tunnelling parameter σ be seen in Fig. 2. We used ratio $\xi = R_2 / R_1$ (R_1=40 nm) of the radii of the QDs for measure of the non-identity.

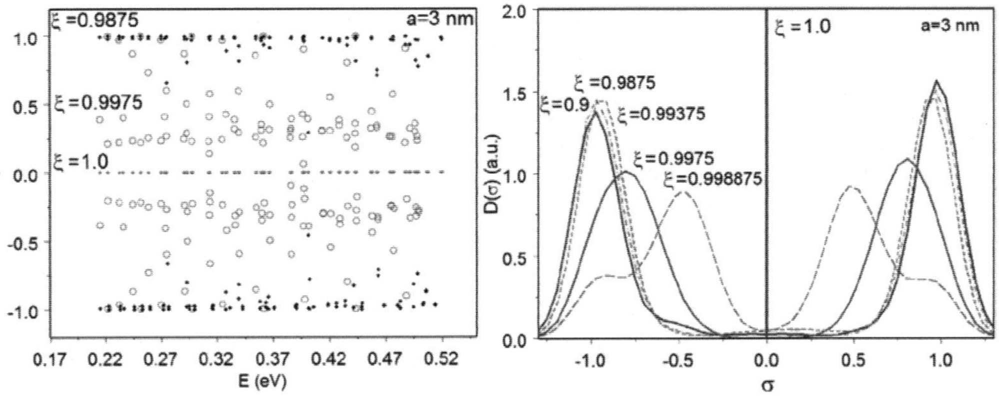

Figure 2. left) Parameter σ for different coefficient of asymmetry ξ in DQD with R_1=40 nm and $R_2 = \xi R_1$; right) Density function $D(\sigma)$ of σ for various ξ.

We are comparing the cases $\xi = 1$, $\xi = 0.9975$ and $\xi = 0.9875$ in Fig. 2 left). One can conclude that when $\xi = 1$ the wave function is distributed uniformed between QDs, when $\xi = 0.9975$ there are few states of single electron whose wave functions are well localized in the left or right QDs. For $\xi = 0.9875$ all states have strong localization in the left or right QDs. The efficiency of tunnelling depends on the ratio of asymmetry of QDs shapes. The larger asymmetry gives larger number of non-tunnelled states. It is important to note that the small violation of the QD symmetry drastically affects the tunnelling. In Fig. 2 right) the last result is presented using the density functions $D(\sigma)$ of σ shown for various ξ. The function $D(\sigma)$ is defined as a probability density for the variable σ to have a value σ with the normalization $\int_{-1}^{1} D(\sigma) d\sigma = 1$. We conclude that the dependence of the tunnelling on ξ is continuous, but the separation of the electron spectral states between two cases – "independent QDs" (no tunnelling states) to "full coupled QD" occurs for ξ in small vicinity of 1.

Note also that the inter-dot distance when all states of the DQD spectrum are tunnelled is depend essentially on value of ξ. For non-symmetric DQD (ξ <1) this inter-dot distance is decreased. Thus we can conclude that a violation of symmetry of the DQD geometry creates difficulties for the tunnelling.

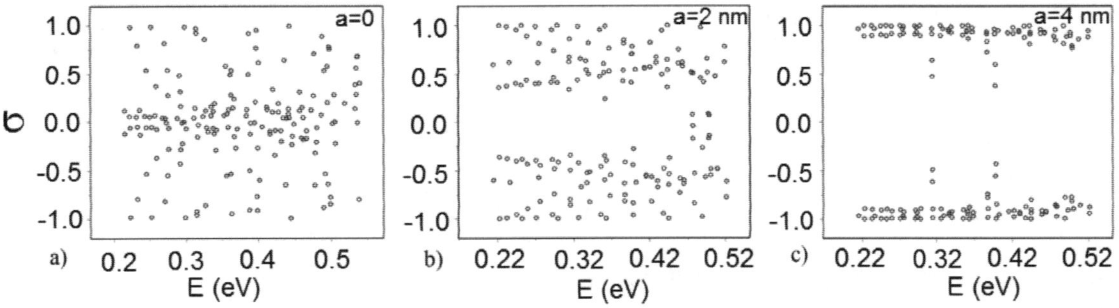

Figure 3. σ-parameter for the asymmetric DQD with R_1=40 nm and R_2=39.9 nm for different inter-dot distances: a) a =0, b) a =2nm, c) a =4 nm.

In Fig. 3 the dependence of the tunneling on inter-dot distance is shown. Violation of the symmetry of this DQD system, defined by ξ, is 0.9975. For the distances decreasing to a =0 there are non-tunnelled states in the spectrum. Comparing the cases of symmetric and non-symmetric DQD tunnelling, one can conclude that there is possibility for a =0 when the charge transport through non-symmetric DQD will be larger than through symmetric DQD due to the degeneracy (double) of the spectrum of symmetric DQD. In other words the number of electron level which allows the tunnelling for non-symmetric DQD may be larger than the number of such levels for symmetric DQD.

4. InAs/GaAs double quantum rings

The double concentric quantum ring is an example of the double quantum object with violation of the shape symmetry of constituent objects due to difference of median radii of the inner and outer rings. In Fig. 3 left) we show 2D DQR. Energy of a single electron in single QR of this DQR is related by the followed formula with radial and orbital quantum number: $E_{n,l} \sim \hbar^2/2m^*(n^2/W^2 + l^2/R^2)$, $D << R$, where W is the width of the QR, and R is median radius. The single electron levels in QR are separated by the $n = 1,2,3,...$ bands, with in-band states with $|l|$=0,1,2,... . The σ-parameter for spectrum of concentric DQR is shown in Fig. 3. We describe approximately the states using the "good" radial quantum numbers of the each ring. The description corresponds to the case of "independent rings". In Fig. 3, the calculated points (in the coordinates (E, σ)) are connected for $|l|$=0,1,2,... by a line. One can obtained some "traces" for each radial number n =1,2,3 as it shown in Fig. 3. We conclude that tunnelling in DQRs occurs through the levels having the same type of wave function symmetry defined by l for the rings (see [4]). The tunnelling occurs when $|\sigma| << 1$. For these states the tunnelling is possible by means the anti-crossing mechanism [4].

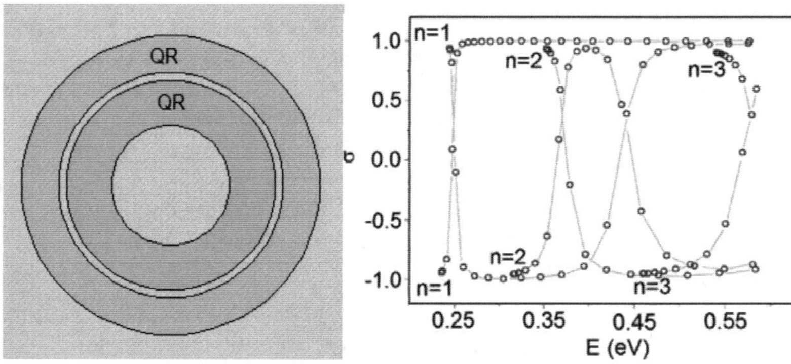

Figure 3. left) DQR shape; right) The σ-parameter for spectrum of concentric DQR. The traces are created by consequence increasing the orbital quantum number $|l|$=0,1,2,... for states with fixed n =1,2,3.

5. Conclusion

We studied dynamical tunnelling in the DQDs changing of inter-dot distance. Identical and non-identical QDs in DQD were considered. The two types of the tunnelling were found. We show that the electron states with the same type of rotation symmetry play important role for the tunnelling in double nano-sized systems. Generally, violation of symmetry of the DQD geometry diminishes the tunnelling.

This work was supported by the NSF (HRD-0833184) and NASA (NNX09AV07A).

References

[1] Kuroda T et al. 2005 *Phys. Rev.* **B72**, 205301
[2] Filikhin I, Suslov V M and Vlahovic B 2006 *Phys. Rev.* **B73**, 205332
[3] Filikhin I, Matinyan S G and Vlahovic B 2011 *Phys. Lett.* **A375**, 620
[4] Filikhin I, Matinyan S G and Vlahovic B 2012 *J. Comp. and Theor. Nanoscience* **9**, 1033

Electromagnetic Wave Emission from Intrinsic Josephson Junction Stacks: A Review of Theory and Simulation

M Machida[1], Y Ota[2], T Koyama[3], and H Matsumoto[3]

[1] CCSE, Japan Atomic Energy Agency, 6-9-3 Higashi-Ueno Taito-ku, Tokyo, 110-0015, Japan
[2] Riken
[3] Institute for Materials Research, Tohoku University, 2-1-1 Katahira Aoba-ku, Sendai, 980-8577, Japan

E-mail: machida.masahiko@jaea.go.jp, otayuki@riken.go.jp, tkoyama@imr.tohoku.ac.jp, matumoto@ldp.phys.tohoku.ac.jp

Abstract.
We review theoretical and numerical research developments in intrinsic Josephson junctions in terms of spontaneous THz electromagnetic-wave emission and report our recent multi-scale simulation scheme to further enhance the radiation power. Starting with theoretical treatments on the couplings among the stacked Josephson junctions, we reveal that solitonic excitation is essential for large spontaneous emission. Finally, we propose an experimental setup, in which an intrinsic Josephson junction is enwrapped by the dielectric materials and reveal that radiation power enhancement occurs by tunning the dielectricity.

1. Introduction

The discovery of clear THz emission from layered High-Tc cuprate $Bi_2Sr_2CaCu_2O_{8-\delta}$ (Bi-2212) [1] was a great mileston for the community which had intensively studied intrinsic Josephson junctions (IJJ's) and related issues. The experimental observation has triggered new types of theoretical and numerical studies on IJJ's. For example, the connection of junctions with environmental vacuum has been not a primary issue but just a concerning one in the previous studies, while the connection mismatch is now a central one [2] to understand the emission mechanism and enhance the radiation power. In addition, although the magnetic field application has been long regarded to be essential for the spontaneous emission, the cavity resonance together with AC Josephson effect is now recognized as an promising mechanism than the resonant vortex flow [3].

Before the discovery of the zero-field emission by Ozyuzer et al., [1] the IJJ community has concentrated on the coupling between neighboring junctions and established two kinds of couplings. One of them is called inductive coupling, which arises from incomplete screening of the penetrated magnetic field within an atomic-scale junction [4],[5]. It can be theoretically characterized by the ratio of the magnetic field screening length to the single junction thickness. Then, the coupling strength is found to be quite large, since the magnetic field screening length extends over thousands of junctions. The capacitive coupling is its counterpart for the electric field [6],[7]. It similarly comes from incomplete screening of the induced electric field within a superconducting layer [7]. The electric field screening length is comparable to the superconducting layer thickness, and the coupling is much weaker than the inductive coupling. However, the capacitive coupling is a unique one, which emerges only for atomic-scale junction stacks like IJJ's. It has an important role in small-scale IJJ's with no application of the magnetic field. In Section II, we present a theoretical review on both the couplings.

The theoretical framework studing the THz radiation from IJJ's is basically grounded on the superconducting phase dynamics. The dynamical action of the superconducting phase was derived from the microscopic treatment on the Cooper-pair Josephson tunneling in IJJ's [8]. Both the couplings are naturally involved in the dynamical action. The inductive coupling is known to bring about the cavity resonance. It occurs when the cavity mode frequency matches with the Josephson frequency given by the applied voltage. Theoretically, the resonance is described by the equation formulating the in-plane spatial variation of the superconducting phase together with the edge boundary condition. The correspondent equation is the coupled sine-Gordon equation, which has been intensively investigated by IJJ community. Since the equation is nonlinear and several junctions are nonlinearly coupled, numerical simulation has been regarded as a primary theoretical tool. In Section III, we briefly present the latest numerical simulation techniques [2]. The simulation area includes not only internal area of IJJ but also external one surrounding the junction. Such a technical extention is required to correctly study the emission mechanism and examine the radiation power enhancement. We emphasize that the technique offers a new tool to examine performance as an emitter in not only IJJ's but also the conventional Josephson junction.

The electromagnetic wave radiation occurs, when both the electric and the magnetic fields are excited inside the junction. Normally, the AC electric field is uniformly excited in the voltage state without magnetic field. Then, the Josephson relation gives the relation between the applied voltage and AC frequency. This implies that DC to AC conversion always occurs in the voltage state of Josephson junctions. However, it has been well-known that the radiation power caused by such a simple conversion is mostly quite weak. On the other hand, in the observed large THz emission, the frequency is uniquely given by the cavity resonance frequency [1],[9],[10]. This clearly indicates that the cavity resonance causes to enhance the Josephson AC oscillation and efficiently transport the excitation power into the emitted wave. Namely, the cavity resonance excites AC magnetic field component characterized by the spatial variation of the superconducting phase. It has been well-known that such a phase variation is generally brought about by vortices and other solitonic excitations, i.e., nonlinear localized modes. Thus, the interest is whether or not solitonic excitation really emerges at the cavity resonance frequency, and if it does, which type of solitonic mode is then excited. In Section IV, we reveal the solitonic excitation form at the cavity resonance frequencies from the numerical simulation result.

2. Theory of Couplings in IJJ's

In this section, we briefly review the theoretical developments on IJJ's. The most important development in this field is the establishement of the couplings between neighboring junctions, since the couplings cause collective features unique for IJJ's. We explain two coupling mechanisms from an effective action of the superconducting phase. It is interesting to say that both the two couplings can be naturally obtained from a theoretical requirment that the superconducting phase dynamics is governed by the gauge invariance.

2.1. Microscopic Model

We begin with a microscopic description for the stacked IJJ schematically depicted in Fig.1. In Bi-2212, we regard double CuO_2 planes as a single superconducting layer whose thickness $s \sim 3$ Å and other block layers as an insulating layer $D \sim 12$ Å. It is well-known in Bi-2212 that c-axis transport is given by the quasi-particle and Cooper pair tunneling between neighboring superconducting layers via the insulating one. The Cooper pair tunneling occurs only below the superconducting transition temperature.

In the present microscopic model, we examine a case, in which the current is biased only parallel to c-axis and the magnetic field is applied only parallel to the junction plane as shown in Fig.1. Namely, the presence of pancake types of vortices are not taken into account. Extending the BCS model for a single tunneling-junction [11] into IJJ stack, the system's partition function is given by the following imaginary-time functional integral over the Grassman fields, $\hat{\psi}_{\sigma\ell}$, $\psi_{\sigma\ell}$, for electrons, the gauge fields φ_ℓ, $\mathbf{A}_\ell$, $A^z_{\ell+1,\ell}$, and the auxiliary field corresponding to the superconducting order parameter, Δ_ℓ, where ℓ is the layer index and the z-component of the vector potential is defined on the link as $A^z_{\ell+1,\ell} = \int_\ell^{\ell+1} dz A^z$,

$$Z = \prod_\ell \int \mathcal{D}\bar{\psi}_{\sigma\ell}\mathcal{D}\psi_{\sigma\ell}\mathcal{D}\mathbf{A}_\ell\mathcal{D}A^z_{\ell+1,\ell}\mathcal{D}\varphi_\ell\mathcal{D}\Delta_\ell e^{-\frac{1}{\hbar}S[\bar{\psi}_{\sigma\ell},\psi_{\sigma\ell},\mathbf{A}_\ell,A^z_{\ell+1,\ell},\varphi_\ell,\Delta_\ell]}, \tag{1}$$

where, S is the Euclidian action defined as $S = S_{\text{matter}} + S_{\text{field}}$ whose expressions are given as

$$S_{\text{matter}} = \sum_\ell \int_0^{\beta\hbar} d\tau \int d\mathbf{r}\Big[\frac{|\Delta_\ell|^2}{g} + \bar{\psi}_{\sigma\ell}(\mathbf{r},\tau)(\hbar\partial_\tau + ie\varphi_\ell)\psi_{\sigma\ell}(\mathbf{r},\tau) + \bar{\Delta}_\ell\psi_{\downarrow\ell}\psi_{\uparrow\ell} + \Delta_\ell\bar{\psi}_{\uparrow\ell}\bar{\psi}_{\uparrow\ell}$$

$$-\bar{\psi}_{\sigma\ell}(\mathbf{r},\tau)\Big(\frac{\hbar^2}{2m}(\nabla - i\frac{e}{\hbar c}\mathbf{A}_\ell)^{\mathbf{2}} + \mu_\ell\Big)\psi_{\sigma\ell}(\mathbf{r},\tau) + (\mathbf{T}_{\ell+1,\ell}e^{\frac{\mathbf{e}}{\mathbf{hc}}\mathbf{A}^{\mathbf{z}}_{\ell+1,\ell}}\bar{\psi}_{\sigma\ell+1}\psi_{\sigma\ell} + \text{c.c})\Big], \tag{2}$$

and

$$S_{\text{field}} = \sum_\ell \int_0^{\beta\hbar} d\tau \int d\mathbf{r}\frac{D}{8\pi}\Big(\epsilon_c {E^z_{\ell+1,\ell}}^2 + {B^x_{\ell+1,\ell}}^2 + {B^y_{\ell+1,\ell}}^2\Big), \tag{3}$$

where, $T_{\ell+1,\ell}$ is the tunneling matrix element between ℓth and $(\ell+1)$th superconducting layers, $E^z_{\ell+1,\ell}$ is the electric field $(\equiv \frac{1}{cD}\frac{\partial A^z_{\ell+1,\ell}}{\partial\tau} - \frac{\varphi_{\ell+1}-\varphi_\ell}{D})$ along the z-axis, ϵ_c is the dielectric constant of the insulating barriers, and $B^x_{\ell+1,\ell}$ $(\equiv \frac{1}{D}\frac{A^z_{\ell+1,\ell}}{\partial y} - \frac{A^y_{\ell+1}-A^y_\ell}{D})$ and $B^y_{\ell+1,\ell}$ $(\equiv \frac{A^x_{\ell+1}-A^x_\ell}{D} - \frac{1}{D}\frac{A^z_{\ell+1,\ell}}{\partial x})$ are the magnetic fields parallel to the junction plane.

Following the standard technique to integrate out the fermion degree of freedom together with reasonable approximations, we have the following effective action,

$$S_{\text{eff}} = \sum_\ell \int_0^{\beta\hbar} d\tau \int d\mathbf{r}_{||}\Big[\frac{D}{8\pi}\Big(\epsilon_c {E^z_{\ell+1,\ell}}^2 + {B^x_{\ell+1,\ell}}^2 + {B^y_{\ell+1,\ell}}^2\Big) + msn_s {v^{||}_{s\ell}}^2 + \frac{s}{8\pi\mu_f^2}(\frac{\phi_0}{2\pi c}\frac{\partial\theta_\ell}{\partial\tau} + \varphi_\ell)^2$$

$$+J cos P_{\ell+1,\ell}(\tau,\mathbf{r}_{||})\Big] + \sum_\ell \int_0^{\beta\hbar} d\tau \int_0^{\beta\hbar} d\tau' \int d\mathbf{r}_{||}\Big[(-\alpha(\tau-\tau')\cos(\frac{P_{\ell+1,\ell}(\tau,\mathbf{r}_{||}) - P_{\ell+1,\ell}(\tau',\mathbf{r}_{||})}{2}\Big], \tag{4}$$

where $\mu_f \equiv \sqrt{\frac{\lambda^2_{\text{TF}}}{4\pi\epsilon}}$ is a constant related to the Thomas-Fermi screening length λ_{TF}, s is the thickness of the superconducting layers, n_s is the superfluid density, and $P_{\ell+1,\ell}$ and $v^{||}_{s\ell}$ are, respectively, the gauge-invariant phase difference and superfluid velocity defined as $P_{\ell+1,\ell} \equiv \theta_{\ell+1} - \theta_\ell - \frac{2\pi}{\phi_0}A^z_{\ell+1,\ell}$ and $v^{||}_{s\ell} \equiv \frac{\hbar}{2m}(\nabla_{||}\theta_\ell - \frac{2\pi}{\phi_0}\mathbf{A}^{||}_\ell)$. In obtaining Eq.(4), we use a local approximation for the integral kernels associated with the intra-layer current, charge fluctuation, and Cooper pair tunneling current. Consequently, msn_s, $\frac{s}{8\pi\mu_f^2}$, J are local and constant coefficients [8]. These approximations are valid for the low energy dynamics, with which the tunneling process is relevant. On the other hand, the integral kernels, $\alpha(\tau-\tau')$ in Eq.(4) describes the frequency-dependent quasi-particle tunneling, which brings about the dissipation current proportional to the time derivative of the gauge invariant phase difference with a dissipation parameter $\beta \equiv \frac{4\pi\sigma\lambda_c}{\sqrt{\epsilon_c}c}$. The form is equivalent with the Caldeira-Leggett's action [8]. See Ref. [8] for the details of the discussion.

2.2. Dynamical Equations and Couplings

From the action Eq.(4), we can derive the following coupled equations for the gauge-invariant phase difference $P_{\ell+1,\ell}$ and the charge density ρ_ℓ,

$$\frac{\epsilon_c\phi_0}{2\pi c^2 D}\frac{\partial^2 P_{\ell+1,\ell}}{\partial t^2}+\frac{4\pi}{cD}(\epsilon_c\mu_f^2-\lambda_{ab}^2)(\frac{\partial\rho_{\ell+1}}{\partial t}-\frac{\partial\rho_\ell}{\partial t})-\frac{4\pi}{c}\frac{\lambda_{ab}^2}{sD}\beta(\frac{dP_{\ell+2,\ell+1}}{dt}+\frac{dP_{\ell,\ell-1}}{dt}-2\frac{dP_{\ell+1,\ell}}{dt})$$
$$-\frac{\phi_0}{2\pi D}\frac{\partial^2 P_{\ell+1,\ell}}{\partial x^2}+\frac{4\pi}{c}(sinP_{\ell+1,\ell}+\beta\frac{dP_{\ell+1,\ell}}{dt})-\frac{4\pi}{c}\frac{\lambda_{ab}^2}{sD}(sinP_{\ell+2,\ell+1}+sinP_{\ell,\ell-1}-2sinP_{\ell+1,\ell})=0, \tag{5}$$

$$\frac{\phi_0\epsilon_c}{2\pi cD}(\frac{\partial P_{\ell+1,\ell}}{\partial t}-\frac{\partial P_{\ell,\ell-1}}{\partial t})+\frac{4\pi\epsilon_c\mu_f^2}{D}(\rho_{\ell+1}+\rho_{\ell-1}-2\rho_\ell)=4\pi s\rho_\ell \tag{6}$$

We notice that Eq.(5) and (6) contains two parameters, which are the electric field screening length, μ_f and the magnetic field screening length in the in-plane direction, λ_{ab}. These parameters differ extremely, i.e., $\mu_f \ll \lambda_{ab}$. The terms including the coefficient λ_{ab} describe the inductive coupling between junctions [4],[5], while the counterparts for μ_f do the capacitive one [6],[7].

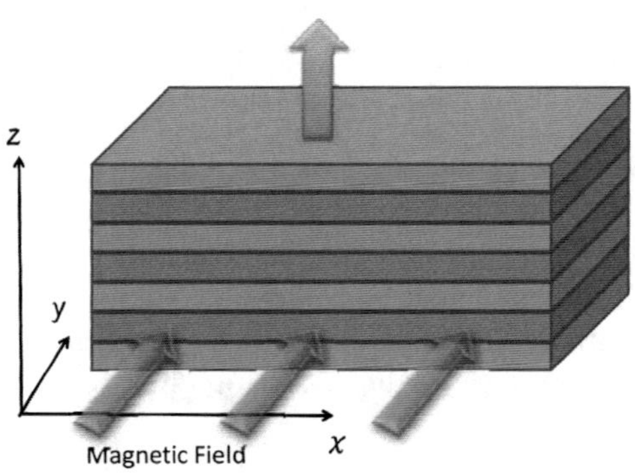

Figure 1. A schematic view for intrinsic Josephson junctions, in which the superconducting CuO_2 planes and insulating layers are alternatively stacked along z(c)-axis. The magnetic field and external current are supposed to be applied only in the layer parallel and perpendicular directions, respectively.

2.3. Dynamical Equations for Inductive and Capacitive Couplings

By elminating the terms including μ_f, i.e., $\mu_f \to 0$ in Eq.(5) and (6), and using the normalized units $t = t'/\omega_p, \omega_p = \frac{c}{\sqrt{\epsilon_c}\lambda_c}$ and $x = x'/\lambda_c$ we have the coupled sine-Gordon equation taking account of only the inductive coupling,

$$\frac{\lambda_{ab}^2}{sD}(\frac{\partial^2 P_{\ell+2,\ell+1}}{\partial t'^2}+\frac{\partial^2 P_{\ell,\ell-1}}{\partial t'^2}-2\frac{\partial^2 P_{\ell+1,\ell}}{\partial t'^2})+\frac{{\lambda_{ab}}^2}{sD}\beta(\frac{\partial P_{\ell+2,\ell+1}}{\partial t'}+\frac{\partial P_{\ell,\ell-1}}{\partial t'}-2\frac{\partial P_{\ell+1,\ell}}{\partial t'})$$
$$-\frac{\partial^2 P_{\ell+1,\ell}}{\partial x'^2}+\frac{{\lambda_{ab}}^2}{sD}(\sin P_{\ell+1,\ell}+\sin P_{\ell,\ell-1}-2\sin P_{\ell+1,\ell})-(\sin P_{\ell+1,\ell}+\beta\frac{\partial P_{\ell+1,\ell}}{\partial t'}+\frac{\partial^2 P_{\ell+1,\ell}}{\partial t'^2})=0. \tag{7}$$

This equation was initially examined by Sakai et al., in their analytical way [4], while a pioneering simulation on the Josephson vortex dynamics was made by Kleiner et al. [12] The main issue of the equation was the existence of the multiple plasma excitation modes propagating mainly along the in-plane direction, called the transverse plasma, and the resonant vortex dynamics with the multiple plasma modes [3]. The resonant vortex flow has been intensively investigated by several authors in the community. To our knowledge, the most surprising physics was the emergence of the in-phase locked vortex flow due to the resonance between the in-phase like propagating mode and flowing vortex lattice. The in-phase locked flux flow promises high-power radiation of THz wave [3]. However, there has been no clear evidence of such high-power radiation in the presence of the layer parallel magnetic field.

By assuming uniformity of the superconducting phase along the junction plane, all the terms originated from the inductive coupling vanish in Eq.(5) and Eq.(6), and a simple equation appears as follows.

$$\partial_{t'}^2 P_{\ell+1,\ell} + \beta\partial_{t'} P_{\ell+1,\ell} + \sin P_{\ell+1,\ell} = I/j_c$$

$$+\frac{\epsilon_c\mu_f^2}{sD}\Big[\sin P_{\ell+2,\ell+1} - 2\sin P_{\ell+1,\ell} + \sin P_{\ell,\ell-1}\Big] + \frac{\epsilon_c\mu_f^2}{sD}\beta\Big[\frac{dP_{l+2,l+1}}{dt'} - 2\frac{dP_{l+1,l}}{dt'} + \frac{dP_{l,l-1}}{dt'}\Big]. \quad (8)$$

This equation was firstly derived by Machida et al., through the micrscopic treatment [8]. We immediately notice that if one drops the third term in the right-hand side, then Eq.(8) coincides with Koyama-Tachiki model derived by their phenomenological treatment [6]. The model has been investigated initially by Machida et al., [7] and recently by Shukrinov et al., [13],[14]. The main physics of the model is the existence of the localized rotating mode, which has been intensively studied in nonlinear physics community motivated by a counter-intuitive concept, in which fully stable localized excitaions emerge even on non-integrable discrte lattice models like solitonic excitations [7],[13],[14].

3. Numerical Simulation Techniques for THz Radiation

In the previous section, we derived the general equations describing both the inductive and capacitive couplings. In this section, we present the latest numerical techniques to simulate the electromagnetic wave emission from IJJ's. The research target is further radiation power enhancement.

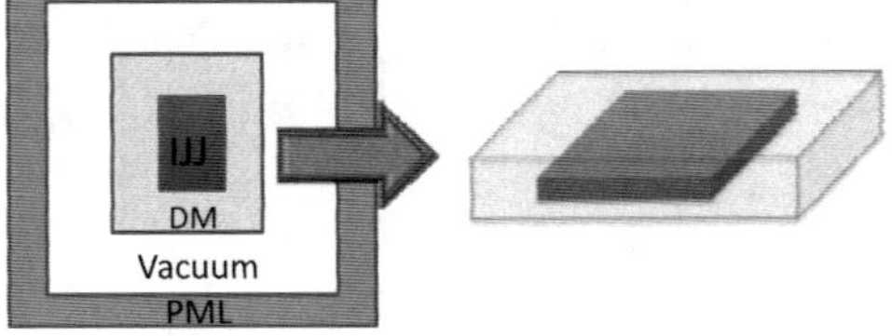

Figure 2. A schematic view for the simulation region to examine the THz radiation mechanism and the radiation power enhancement. The region are composed of IJJ, the surrounding vacuum, and the perfectly matched layer (PML). In the present case, it is noted that IJJ is enwrapped by a dielectric medium (DM) to improve the mismatch in the external radiation.

4. Multiscale Modeling for Electromagnetic Wave Emission

In addition to the coupled sine-Gordon Eq.(5) and (6), the EM fields in the dielectric medium or the vacuum outside the IJJ's device as schematically shown in Fig.2 is described by the Maxwell equations [2],

$$\frac{\epsilon(x,z)}{c}\partial_t E_x(x,z,t) = -\nabla_z B_y(x,z,t), \tag{9}$$

$$\frac{1}{c}\partial_t B_y(x,z,t) = \nabla_x E_z(x,z,t) - \nabla_z E_x(x,z,t). \tag{10}$$

where the dielectric constant $\epsilon(x,z)$ is chosen as $\epsilon(x,z) = \epsilon_D$ inside the dielectric medium or $\epsilon(x,z) = 1$ in the vacuum. An IJJ device is located in the region $-L_x/2 < x < L_x/2$, where L_x is the in-plane length in the x-direction. In this case, we apply the external current given as, $j_{\text{ext}}(x,z) = j_{\text{ext}}\theta(x + L_x/2)\theta(x - L_x/2)[\theta(-W_z/2 - z) + \theta(z - W_z/2)]$, where $\theta(x)$ is the step function and the position $z = \pm W_z$ is chosen inside the electrodes.

In the present simulation, we solve Eqs.(5),(6),(9), and (10) in a finite box as shown in Fig.2, though the surrounding space is infinitely large. To avoid unphysical reflection of the EM waves at the artificial box boundary, we introduce the PML (perfectly matched layer) enwrapping the simulation box [2]. The PML is a fictitious layer that absorbs the outgoing EM waves without any reflection. In PML, the EM fields are decomposed, depending on the propagating directions. For example, B_y is decomposed into two components as $B_y = B_{yx} + B_{yz}$, where B_{yx} (B_{yz}) is the component propagating in the x (z) direction. The decomposed fields are assumed to obey the following quasi-Maxwell equations which bring about the rapid decrease for the decomposed fields,

$$\frac{1}{c}\partial_t E_{xz} = -\nabla_z B_y - \sigma_{xz}E_{xz}, \quad \frac{1}{c}\partial_t E_{zx} = \nabla_x B_y - \sigma_{zx}E_{zx}, \tag{11}$$

$$\frac{1}{c}\partial_t B_{yx} = \nabla_x E_z - \sigma_{yx}B_{yx}, \quad \frac{1}{c}\partial_t B_{yz} = -\nabla_z E_x - \sigma_{yz}B_{yz}, \tag{12}$$

where $E_{xz} = E_x$ and $E_{zx} = E_z$. The explicit expressions for the damping factors $\sigma_{\alpha\beta}$ used in this paper, which damp the EM waves propagating inside the PML, are given in Ref.[2].

Outside the IJJ device, the grid pitch is given as $\Delta_x = \Delta_z = 0.01\lambda_c \sim 2\mu$m (Bi-2212 IJJ's), since the wavelength of emitted EM waves is on the order of λ_c. On the other hand, the grid pitch in the interanl junction area is in a scale of λ_J, which is quite small compared to λ_c. In addition, it is noted that a mesa sample in which THz emission occurs has a thickness $D \sim 1\mu$m, that is, $D < \Delta_z$. To perform the calculations in such a real multiscale system, one needs an approximations in solving Eqs.(5),(6), and Eqs.(9),(10) simultaneously. Thus, we impose a coarse-grained condition at the junction edges that the EM field averaged over all the junctions along the z-direction is continuously connected to the EM field outside the IJJ's. This assumption was verified by a comparison with experimental observations. Moreover, it is noted that we treat IJJ's whose $N = 20$ junctions, applying the periodic boundary condition along the z-direction, though real mesas emitting EM waves contain thousands of junctions. The reason is that the system containing several hundreds of junctions is too large for the present numerical resource. One needs parallelization techniques to divide the box region as Fig.2 with keeping each PE's calculation cost equivalent. The parallelization is the future task. More details of the numerical scheme to solve the coupled dynamical equations (5),(6),(9), and (10) is given in Ref.[2].

5. Numerical results and discussions

Now, let us present the numerical simulation results. We perform various simulations for an IJJ device covered by a dielectric film of thickness $0.02\lambda_c$ and in-plane size $0.5\lambda_c$ as

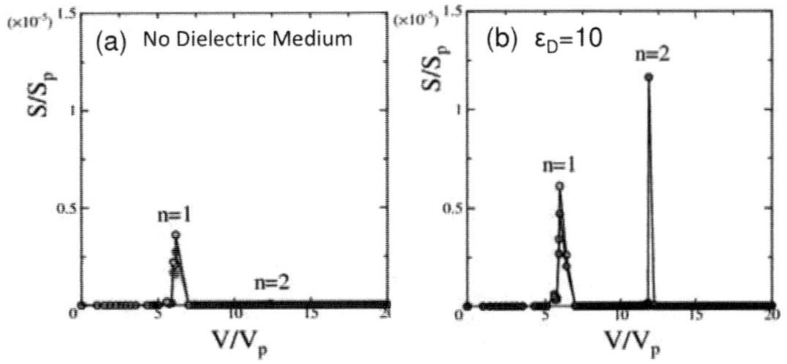

Figure 3. The simulation results for the applied voltage dependences of the radiation power in (a) no dielectric medium covered and (b) covered cases, in which the dielectric constant $\epsilon_D = 10$. The red and green circles denote the power in the directions [100] and [001], respectively.

schematically shown in Fig.2. The device parameter values are taken as $\frac{\epsilon_c \mu_f}{sD} = 0.1$, $\frac{\lambda_{ab}^2}{sD} = 10^4$, $\beta \equiv \sigma(\hbar\omega_p/2eDj_c) = 0.05$, and $\beta_L \equiv \sigma_L(\hbar\omega_p/2eDj_c) = 100$ with ω_p being the Josephson plasma frequency $\omega_p = c/\sqrt{\epsilon_c}\lambda_c$.

Figure 3 shows the voltage dependences of the power of the EM waves (Poynting vector) emitted in the two directions [100] and [001]. In these simulations, we take the value $\epsilon_D = 10$ for the dielectric constant of the covered dielectric medium, since a good impedance matching between the IJJ's and the outside space occurs, i.e., $\epsilon_D = \epsilon_c$. In the case of no dielectric medium, the emission peak can be observed only at $n = 1$ cavity mode excitation in an agreement with the observation by Ozyuzer et al., [1]. On the other hand, in the case covered by the dielectric medium, two peaks $n = 1$ and $n = 2$ cavity mode resonances are observed. Here, it is noted that $n = 1$ and $n = 2$ cavity resonances are, respectively, characterized by the relations as $V/V_p \sim \nu_1/\omega_p = \pi(\lambda_c/L) = 2\pi$ and $V/V_p \sim \nu_2/\omega_p = 2\pi(\lambda_c/L) = 4\pi$ with $V_p = \hbar\omega_p/2e$. Comparing between these figures, the peak intensity of the resonant modes is found to become larger in the presence of the dielectric cover. Then, we also check that AC EM fields excited by the Josephson oscillations penetrates well into the dielectric medium, that is, the impedance matching is improved in the system. It is also noticed that the enhancement of the $n = 2$ resonant peak is large compared with that of the $n = 1$ resonance peak and the emission from the $n = 2$ mode is directed mainly in the [100] direction, because the power in the [001] direction is negligibly small, which is expected in the dipole type of emission. To understand this unexpected increase in the peak intensity of the $n = 2$ mode, we plot a snapshot of the phase differences $P_{\ell+1,\ell}(x)$ in Fig.4(a). As seen in this figure, the phase differences exhibit an alternate structure along the c-direction as $\cdots = P_{2\ell+1,2\ell}(x,t) = -P_{2\ell,2\ell-1}(x,t)$. Such an alternate one has been also suggested at $n = 1$ cavity resonance in Refs. [15] and [16]. As seen in Fig.4(a), the spatial variation along the in-plane direction exhibits localized solitonic structures with π-phase shift, i.e., $0 \to \pi$ or $\pi \to 0$. In contrast, such solitonic variations are not at all observed except for the resonances at $n = 1$ and 2. From these results, one understands that the strong radiation comes from the excitation of solitonic modes.

In order to seek whether an optimum value of ϵ_D whose suitable choice leads to stronger emission or not, we examine the emission power variation by changing its value. Fig.4(b) shows the power of EM waves emitted from $n = 2$ resonant mode as a function of ϵ_D. From these results, one may expect that the optimum value of ϵ_D which leads to the strongest emission.

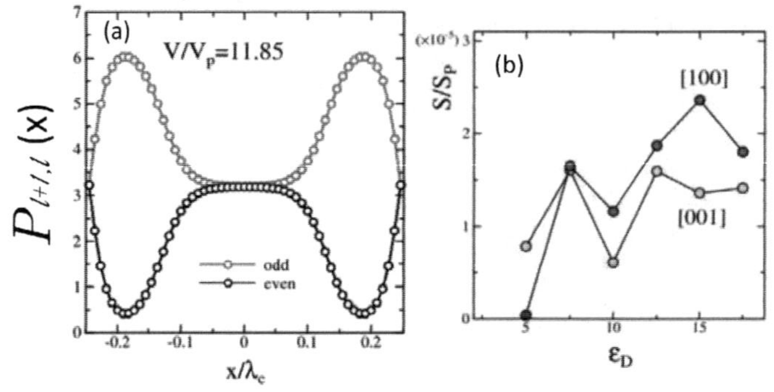

Figure 4. The simulation results for (a) the spatial profile snapshot of $P_{\ell+1,\ell}$ in the radiation state assigned as $n = 2$ and (b) the dielectric constant dependences of the radiation power.

6. Summary

We reviewed the theoretical framework for the intrinsic Josephson junction systems. Starting with the microscopic tunneling model, we reach the general equations of the superconducting phase and the charge density and derive approximate equations describing the inductive or capacitive coupling. Finally, we introduced the latest numerical scheme to examine the electromagnetic wave radiation in realistic situations and presented typical simulation results. An idea to enhance the radiation power was suggested together with the numerical confirmations.

Acknowledgments

The authors wish to acknowledge valuable discussions with K. Kadowaki, R. Kleiner, H-B. Wang, A. E. Koshelev, and Y. Shukrinov. M.M. was partially supported by Grant-in-Aid for Scientific Research from the Ministry of Education, Culture, Sports, Science and Technology of Japan (MEXT).

References

[1] Ozyuzer L, Koshelev A E, Kurter C, Gopalsami N, Li Q, Tachiki M, Kadowaki K, Yamamoto T, Minami H, Yamaguchi H, Tachiki T, Gray K E, Kwok W K, and Welp U, 2007 *Science* **318** 1291
[2] Koyama T, Matsumoto H, Machida M, Ota Y, 2011 *Supercond. Sci. Technol.* **24** 085007
[3] Machida M, Koyama T, and Tachiki M, 2000 *Physica* C **330** 85
[4] Sakai S, Bodin P, and Pedersen N F, 1993 *J. Appl. Phys.* **73** 2411
[5] Bulaevskii L N, Zamora M, Baeriswyl D, Beck H, and Clem J R, 1994 *Phys. Rev.* B **50** 12831
[6] Koyama T and Tachiki M, 1996 *Phys. Rev.* **B** 16183
[7] Machida M, Koyama T, and Tachiki M, 1999 *Phys. Rev. Lett.* **83** 4618
[8] Machida M, Koyama T, Tanaka A, and Tachiki M, 2000 *Physica* C **331** 85
[9] Tsujimoto M, Yamaki K, Deguchi K, Yamamoto T, Kashiwagi T, Minami H, Tachiki M, Kadowaki K, and Klemm R A, 2010 *Phys. Rev. Lett.* **105** 037005
[10] Wang H B, Gue'non S, Gross B, Yuan J, Jiang Z G, Zhong Y Y, Grünzweig M, Iishi A, Wu P H, Hatano T, Koelle D, and Kleiner R, 2010 *Phys. Rev. Lett.* **105** 057002
[11] Eckern U, Schön G, and Ambegaokar V, 1984 *Phys. Rev.* B **30** 6419
[12] Kleiner R, 1994 *Phys. Rev.* B **50** 6919
[13] Shukrinov Yu M and Mahfouzi F, 2007 *Phys. Rev. Lett.* **98** 157001
[14] Shukrinov Yu M, Hamdipour M, and Kolahchi M R, 2009 *Phys. Rev.* B **80** 014512
[15] Lin S and Hu X, 2008 *Phys. Rev. Lett.* **100** 247006.
[16] Koshelev A E, 2008 *Phys. Rev. B* **78** 174509

Distributed Josephson Junction between Multiband Superconductors

Vladislav V. Kurin

Institute for Physics of Microstructure of the Russian Academy of Science, GSP-105, 603950 Nizhny Novgorod, Russia

E-mail: kurin@ipm.sci-nnov.ru

Abstract. A simple hydrodynamic model of multiband superconductors describing Leggett interband collective excitations and their interactions with electromagnetic waves is applied to description of dynamics of distributed Josephson Junction between multiband superconductors. A modified Sin-Gordon equation is obtained and excitation of Leggett modes by moving Josephson vortices is investigated.

1. Introduction

It is well known that multifolded Fermi surface superconductors [1, 2] may feature the so-called Leggett modes [3], that essentially are collective excitations of phase difference of order parameters in different bands. Multiband superconductors, in accordance with the modern view, include the conventional ones such as Nb, and the recently discovered novel superconductors as MgB_2 and the FeAs based ones. Excitation of the Leggett modes in lumped Josephson junctions was earlier considered by some authors theoretically, see for example, [4] and possibly observed experimentally [5]. Attempts have been made to generalize the theory for the case of distributed contacts [6], but the approach of tunnelling Hamiltonian used in this paper is seemed to be unsatisfactory for correct description of field distribution in the vicinity of Josephson Junction and, as a consequence, of Josephson Dynamics .

In this work we propose a hydrodynamic self-consistent model describing the dynamics of distributed Josephson junctions between multiband superconductors.

2. Hydrodynamic description of Multiband Superconductors

The dynamics of superconducting condensates within electrodes of Josephson junction can be described by hydrodynamical equations for superconducting velocities $\mathbf{v}_j$ and concentrations n_j. These equations simply express general laws of number of particles and momentum conservation. We will limit yourself by the simplest case of sufficiently low temperatures so that the dynamics of quasiparticles will not be relevant. In such a case hydrodynamic equations look like

$$\dot{N}_j + \operatorname{div} N_j \mathbf{v}_j = \frac{2}{\hbar} \sum_k J_{j,k} \sqrt{N_j N_k} \sin(\theta_j - \theta_k), \tag{1}$$

$$\dot{\mathbf{v}}_j + (\mathbf{v}_j \nabla)\mathbf{v}_j + \frac{\nabla p_j(N_j)}{m_j N_j} = -\frac{e}{m_j} \left\{ \mathbf{E} + \frac{1}{c}[\mathbf{vB}] \right\} \tag{2}$$

where $N_j, \mathbf{v}_j$ are concentrations and velocities of superconducting electrons in j-th band, $J_{j,k}$ energies of interband couplings, p_j are the pressure of electrons in j-th band for which we will use a simple expression for degenerate Fermi gas

$$p_j = \frac{(3\pi^2)^{2/3}}{5}\frac{\hbar^2}{m_j}N^{5/3}, \tag{3}$$

implying that corrections to compressibility due to superconductivity which are of order Δ/ε_F are very weak. It corresponds to well known Thomas-Fermi approximation. Calculating the derivatives $\frac{\partial p_j}{m_j \partial N_j}$ which are squared velocities of Bogolubov-Anderson modes in corresponding bands s_j^2 we have well known relations $s_j^2 = v_{F,j}^2/3$ where Fermi velocities are defined by standard relations $m_j v_{F,j} = (3\pi^2 N_j)^{1/3}\hbar$.

Concentrations are related to module of order parameters $N_j = 2|\Psi_j|^2$ and velocities are expressed via their phase by standard gauge invariant relation

$$2m_j\mathbf{v}_j = \hbar\nabla\theta_j + \frac{2e}{c}\mathbf{A}, \tag{4}$$

where we explicitly took into account negative sign of electronic charge. Under Coulomb gauge condition div$\mathbf{A} = 0$ the second term in RHS of Eq. 4 describes pure vortex motion while the first term is pure potential only if there are no Abrikosov vortices.

Expressing electromagnetic fields via vector and scalar potentials, using the vector identity $(\mathbf{v}\nabla)\mathbf{v} = \nabla v^2/2 - \mathbf{v}\times\nabla\times\mathbf{v}$ and substituting the expression 4 for velocity via phase into equations of motion 2 and integrating it we are able to find the Bernoulli-like equations

$$\hbar\dot{\theta} + m_j v_j^2 - 2e\varphi + 2\mu_j(N_j) - 2\mu_j(N_j^0) = 0, \tag{5}$$

where $\mu_j = m_j v_{F,j}^2/2$ are chemical potentials in corresponding bands, N_j^0 are equilibrium concentrations, which are, in such kind of description, phenomenological parameters which should be defined from spatially homogeneous many bands Bardeen-Cooper-Schriffer theory. These relations define an expression for concentration via phase and electromagnetic, vector and scalar, potentials. It can be called as extended London equation since it does not include derivatives of concentrations. It is naturally gauge invariant since phase and scalar potentials enters only in combination $\hbar\dot{\theta} - 2e\varphi$. With a good accuracy neglecting terms of order the relation 5 can be inverted and represented in the form

$$n = N_j - N_j^0 = \frac{\rho_j}{2}(2e\varphi - \hbar\dot{\theta} - m_j v_j^2), \tag{6}$$

where

$$\rho_j = \frac{N_j}{m_j s_j^2} = \left(\frac{\partial\mu}{\partial N}\right)^{-1} = \frac{p_{F,j}^2}{\pi^2\hbar^3 v_{F,j}}$$

are the densities of states per unit volume per two spins. These quantities determine inertia of the phase difference oscillations of order parameters, $v_{F,j}, p_{F,j}-$ are Fermi velocities and momenta. Taking into account quadratic terms give rise to somewhere more complicated expression

$$n_j = \frac{1}{2}\left(\frac{\partial\mu_j}{\partial N_j}\right)^{-1}\left[2e\varphi - \hbar\dot{\theta}_j - m_j v_j^2 - \frac{\partial^2\mu_j}{2\partial N_j^2}\left(\frac{\partial\mu}{\partial n}\right)^{-2}(2e\varphi - \hbar\dot{\theta})^2\right], \tag{7}$$

Let us notice here that that potential and vortex motions turns out to be connected via term $m_j v_j^2$.

Electromagnetic fields obey Maxwell equations which being rewritten in terms of electromagnetic potentials in Coulomb gauge div$\mathbf{A} = 0$ are

$$\Delta\mathbf{A} - \frac{1}{c^2}\frac{\partial^2 \mathbf{A}}{\partial t^2} + \frac{4\pi}{c}\mathbf{j} = \frac{1}{c}\frac{\partial \nabla\varphi}{\partial t}; \quad \Delta\varphi = 0 \tag{8}$$

Hydrodynamic equations 1 with state equation 3 and Maxwell equations form a closed system describing dynamics of multiband superconductor.

2.1. Linear approximation and Leggett modes

Now let us consider the theory of linear waves using hydrodynamic approach suggested. Introducing the perturbation of concentrations $n_j = N_j - N_j^0$ and linearizing LHS of continuity equation and Eq. 5 provided velocities and perturbation concentrations to be small we find expressions for concentrations via phases $2n_j = \rho_j(2e\varphi - \hbar\dot{\theta})$ and, then, assuming Coulomb gauge div$\mathbf{A} = 0$, from continuity equations we find nonlinear wave equations for phases

$$\ddot{\theta}_j - 2e\hbar^{-1}\dot{\varphi} - s_j^2\Delta\theta_j + \rho_j^{-1}\sum_k \beta\sin(\theta_j - \theta_k) = 0, \tag{9}$$

which describes potential collective excitations in superconductor. Electric potential here should be found from Poisson equation $\Delta\varphi = 2\pi e\sum_l \rho_l(2e\varphi - \hbar\dot{\theta}_l)$. Here $n_j, \mathbf{v}_j$ are perturbations of the electron densities and velocities in the j-band, the coefficient $\hbar\beta/2 = 2J(\Psi_1^0\Psi_2^0)$ defines the coupling between the phases of order parameters in different bands. We used the notations of Leggett [3], by which J is the interband coupling energy,

$$\rho_j = \frac{N_j}{m_j s_j^2} = \left(\frac{\partial\mu}{\partial N}\right)^{-1} = \frac{p_{F,j}^2}{\pi^2\hbar^3 v_{F,j}}$$

are the densities of states per unit volume per two spins, they determine inertia of the phase difference oscillations of order parameters so that the frequency of spatially homogeneous Leggett oscillations is $\omega_0^2 = 4(\rho_1^{-1} + \rho_2^{-1})\beta$; $s_j^2 = v_{F,j}^2/3$ are velocities of the Bogolubov-Anderson modes in the respective bands; $(v,p)_{F,j}$ are, respectively, the Fermi velocities and momenta.

Let us consider a simple two-band case. It is possible two equilibrium homogeneous phase distributions depending on sign of interband interaction β. If $\beta > 0$ the stable configuration will be $\theta_1^0 = \theta_2^0$ but in the case of repulsive interband interaction $\beta < 0$ stable phase difference $\theta_1^0 = \theta_2^0 + \pi$. Linearizing the sine term in equations in the vicinities of corresponding steady state and excluding scalar potentials by using Poisson equation $-\Delta\varphi = -4\pi e\sum_l n_l$ we come to two equations for perturbations of electronic densities

$$\ddot{n}_j - s_j^2\Delta n_j + \omega_j^2\sum_l n_l + |\beta|\sum_k\left(\frac{n_j}{\rho_j} - \frac{n_k}{\rho_k}\right) = 0 \quad ,$$

where we introduce the notation ω_j for plasma frequencies in the j-th band. Plasma frequencies and densities of state are not independent since the values $q_j^2 = 4\pi e^2\rho_j$ representing squared inverse screening length in the j-th band are related to plasma frequencies by $\omega_j = q_j s_j$. Searching for solution in the form $\sim \exp(-i\omega t + i\mathbf{kr})$ we find the dispersion relation for all potential waves, plasmon and Leggett ones

$$\prod_{j=1,2}\left(\omega^2 - s_j^2k^2 - \omega_j^2 - \frac{\beta}{\rho_j}\right) = \left(\omega_1^2 - \frac{\beta}{\rho_2}\right)\left(\omega_2^2 - \frac{\beta}{\rho_1}\right).$$

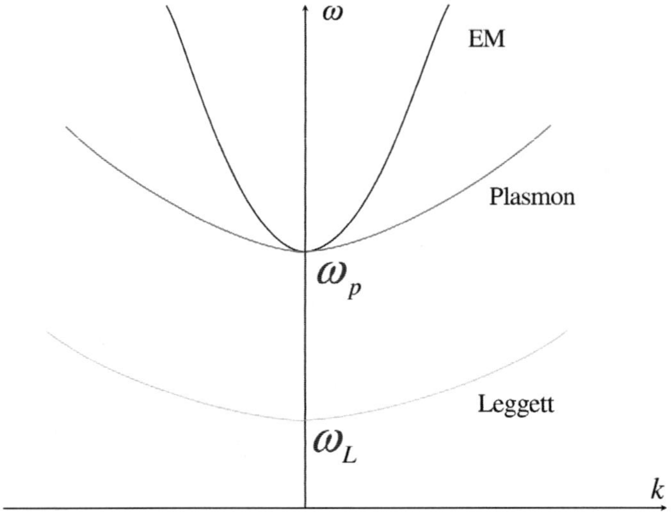

Figure 1. Dispersion of electromagnetic, plasma and Leggett waves in two-band superconductor

This equation reproduce the dispersion of potential waves obtained in [3, 7] using different approach. Of course it does not take into account dissipation of potential waves due to Landau damping and due to pair breaking. In a such approach they should be taken into account phenomenologically. Dispersion of electromagnetic waves follow from equation for vector potential and is given by $\omega^2\varepsilon_t - c^2k^2 = 0$ where $\varepsilon_t = 1 - \omega_p^2/\omega^2$ and plasma frequency is defined as $\omega_p^2 = \sum_j \omega_j^2$. Typical dispersion of all modes, electromagnetic, plasma and Leggett ones are qualitatively shown on fig.1 In the next section we consider different mechanisms of Leggett mode excitation by Josephson vortices moving in distributed Josephson junction between multiband superconductors.

3. Modified Sine-Gordon Equation

In this section we consider dynamics of distributed Josephson Junction between multiband superconductors and will obtain a modified Sine-Gordon equation taking into account Leggett modes outside the junction, both localized and leaky ones. For simplicity we consider the symmetric case of Josephson junction between identical two band superconductor and limit yourself only by linear approximation. Josephson Junction represents a thin tunnelly transparent dielectric layer and from point of view of external problem it should be described as boundary conditions both for hydrodynamic and Maxwell equations. Let us begin with hydrodynamic boundary conditions. It can be obtained by trivial generalization of one-band boundary condition and can written in the form

$$\left.\frac{\partial\Psi_j}{\partial y} + \frac{2ie}{\hbar c}A_y\Psi_j\right|_u = \left.\sum_k \delta_{j,k}\Psi_k \exp\left\{\frac{-2ie}{\hbar c}\int_{-d/2}^{d/2} A_y dy\right\}\right|_d .$$

Two electrodes of Josephson junction we denote by indices u for $y > d/2$ and d for $y < -d/2$, $\delta_{j,k}$ are the coefficients giving tunneling rate from k-th band of the lower superconductor to j-th band of the upper one. We can rewrite it via modules and phases of order parameters as

$$-\frac{e\hbar|\Psi_j^u|^2}{m_j}\left(\frac{\partial\theta_j}{\partial y}+\frac{2e}{\hbar c}A_y\right)=\frac{e\hbar}{m_j}\sum_k \delta_{j,k}|\Psi_j^u||\Psi_k^d|\sin\left(\theta_j^u-\theta_j^d+\frac{2e}{\hbar c}\int_{-d/2}^{d/2}A_y dy\right)$$

from there we just have the standard expression for Josephson current

$$j_{i,j}=j_{i,j}^c\sin\left(\theta_i^u-\theta_j^d+\frac{2e}{\hbar c}\int_{-d/2}^{d/2}A_y dy\right),\; j_{j,k}^c=\frac{e\hbar}{m_j}\delta_{j,k}|\Psi_j^u||\Psi_k^d|, \tag{10}$$

where $j_{j,k}^c$ are critical current densities defining tunneling between different bands. This boundary condition defines tunnel current inside Josephson junction and it can be rewritten in the terms of velocities as

$$v_{y,j}^{u,d}=\sum_l \gamma_{j,k}^{u,d}\sin(\theta_j^d-\theta_k^u),\; \gamma_{k,l}^{u,d}=\delta_{j,k}\frac{\hbar|\Psi_j^u||\Psi_k^d|}{2m_j|\Psi_j^{u,d}|} \tag{11}$$

where we introduce the new coupling coefficients $\gamma_{j,k}$ To obtain electrodynamic boundary conditions we write down Maxwell equation for interior of the junction and take y component of equation 8 for vector potential assuming $\varphi = 0$ what is possible due to fulfilment of equation $\Delta\varphi = 0$. Assuming a Josephson junction being sufficiently thin and searching for solution in the form, (common factor $\exp(-i\omega t + i\chi x)$ is omitted)

$$A_x=i\chi a_0\frac{2\Lambda y}{(d+2\Lambda)}+i\chi f_0\sinh\chi y,\; A_y=a_0\left(1+\frac{\chi^2\Lambda y^2}{(d+2\Lambda)}\right)+\chi f_0\cosh\chi y,\; \varphi=i\frac{\omega}{c}f_0\sinh\chi y,$$

where the terms with f_0 are gauge functions, for amplitude of y-component of vector potential a_0 we find an equation

$$c\left[-\frac{\omega^2}{c^2}\varepsilon+\frac{\chi^2}{1+{}^{2\Lambda}\!/_{d}}\right]a_0+4\pi j_y^t=4\pi j_{ext} \tag{12}$$

where Λ is an effective penetration depth related to impedance of Josephson interface $Z = E_x(d/2)/B_z(d/2)$ as $\Lambda = cZ/i\omega$, j_t is a tunneling current, j_{ext}-external current flowing into the junction. Tunnel current and effective length To make this equation closed we should find from solution of external problem unknown quantities, effective length Λ and tunnel current j_y^t To find external field we will use linearized in the vicinity of corresponding steady state the equation 9 for phases which is equivalent to equations 10 for concentrations. Let us write down the solutions of linearized equations for external with respect Josephson junction region

$$A_x=-i\frac{\kappa_t}{\chi}A_t e^{-\kappa_t\tilde{y}}+i\chi f_1 e^{-\chi\tilde{y}},\; A_y=A_t e^{-\kappa_t\tilde{y}}-|\chi|f_1 e^{-\chi\tilde{y}},$$

$$\varphi=i\frac{\omega}{c}\left(-A_t\sum_{i,l}\frac{b_l}{k_l^2\kappa_l}n_{i,l}e^{-\kappa_l\tilde{y}}+f_1 e^{-\chi\tilde{y}}\right),$$

$$\delta\theta_j=-\frac{2e}{\hbar c}\left\{-\frac{A_t}{q_j^2}\sum_l\frac{b_l}{\kappa_l}e^{-\kappa_l\tilde{y}}\left[n_{j,l}+\frac{q_j^2}{k_l^2}\sum_i n_{i,l}\right]+f_1 e^{-\chi\tilde{y}}\right\},$$

$$n_j = \frac{i\omega}{c}\frac{A_t}{4\pi e}\sum_l \frac{b_l}{\kappa_l} n_{j,l} e^{-\kappa_l \tilde{y}},\; v_{yj} = \frac{eA_t}{m_j c}\left\{-\frac{1}{q_j^2}\sum_l b_l\, e^{-\kappa_l \tilde{y}}\left[n_{j,l} + \frac{q_j^2}{k_l^2}\sum_i n_{i,l}\right] + e^{-\kappa_t \tilde{y}}\right\}. \quad (13)$$

Here we introduced phase perturbations as $\theta_j = \theta_j^0 + \delta\theta_j$, $k_l = k_l(\omega)$, $l = 1,2$ are solutions of dispersion equation for potential waves Eq.10, $l = 1$ denotes plasma wave, $l = 2$-Leggett one, $n_{j,l}$ -density perturbations eigenvectors what are found from 10, the indices j,l denote the band and sort of potential wave respectively, κ_t, κ_l transversal wave numbers of electromagnetic and potential waves defined as $\kappa_t = \sqrt{\chi^2 - \varepsilon_t\omega^2 c^{-2}}$ and $\kappa_l = \sqrt{\chi^2 - k_l^2(\omega)}$, A_t-coefficient of excitation of electromagnetic wave, b_l-relative, respect to A_t coefficients of potential mode excitation, the terms with f_1 is a gauge functions not affecting gauge conditions. Finding expressions for coefficients A_t, f_0, f_1, Λ via a_0 from electromagnetic boundary conditions expressing continuity of potentials and tangential fields we will be able to find jumps of phase differences $\delta\theta_j^u - \delta\theta_k^d$ and tunnel currents. Having this procedure carried out we come to the modified Sine-Gordon equation for phase variable ψ defined as $2e\,(d+2\lambda)\,B_z = \hbar c\partial_x\psi$ where λ is a London penetration length.

$$\frac{\hbar}{2e}\frac{1}{4\pi}\left[\frac{c^2\chi^2}{d+2\lambda} - \frac{\varepsilon\omega^2}{d}\frac{d+2\Lambda}{(d+2\lambda)}\right]\psi + \sum_{j,k} j_{i,k}^c \sin\left[\left(\hat{H}_i + \hat{H}_k\right)\psi\right] = j_{ext} \quad (14)$$

In such form this equation valid for positive interband coupling $J > 0$ when stable steady state is $\theta_1^0 = \theta_2^0$. For the case of negative $J < 0$ when stable unperturbed phase difference between bands is a π the critical currents $j_{j,k}^c$ in Eq.14 should be replaced by $(-1)^{j+k} j_{j,k}^c$. Expressions for effective length Λ and for operators $\hat{H}_i$ defining phase jumps via ψ are nonlocal and in Fourier representation take the form

$$\Lambda \approx \lambda - \lambda^2\chi^2 \sum_l \frac{F_l(\omega)}{\kappa_l},\; F_l(\omega) = \frac{b_l}{k_l^2}\sum_i n_{i,l},$$

$$\hat{H}_j \approx \left\{\frac{1}{2} - \frac{\chi^2\lambda^2}{d+2\lambda}\sum_l \frac{G_{j,l}(\omega)}{\kappa_l}\right\},\; G_{j,l}(\omega) = b_l\left[\frac{1}{q_j^2}n_{j,l} + \frac{1}{k_l^2}\sum_i n_{i,l}\right],$$

where coefficients of potential wave excitation b_l are still left undefined. They should be found using hydrodynamic boundary condition. For sufficiently low transparency of Josephson junction a superconducting phase distribution in exterior can be found from approximate conditions $v_{y,j} = 0$ what yields a simple equation for coefficients b_l

$$\sum_l b_l\left[n_{j,l} + \frac{q_j^2}{k_l^2}\sum_i n_{i,l}\right] = q_j^2 \quad (15)$$

To accomplish all calculations we need to solve dispersion relation and find $k_l(\omega)$, then find eigenvectors for electronic density perturbation $n_{j,l}(\omega)$, and, in conclusion, to solve equations Eqs.15 for excitation coefficients. Then we have expression for functions $F_l(\omega), G_{j,l}(\omega)$ To carry out such a procedure analytically we will use natural assumption that working frequency ω and frequency of Leggett mode $\omega_L^2 \approx\sim \beta(\rho_1^{-1} + \rho_2^{-1})$ is much smaller than plasma frequencies ω_j and consider the case of strongly asymmetric bands so that $\omega_1 \gg \omega_2$, $s_1 \gg s_2$, $\rho_1 \gg \rho_2$. For such a case we have a simple formulas for transversal wave numbers for electromagnetic, plasma and Leggett modes

$$\kappa_t \approx \lambda^{-1},\; \kappa_1 \approx q_1,\; \kappa_2 \approx \sqrt{\frac{\omega_L^2 - \omega^2}{s_2^2} + \chi^2}$$

and explicit expressions for coefficients

$$F_1 \approx \frac{\omega_1^2}{\omega^2},\ F_2 \approx -\frac{\omega_2^2 s_1^4}{\omega_1^2 s_2^4},\ \tilde{G}_{1,1} \approx 1,\ \tilde{G}_{1,2} \approx \frac{q_2^2}{q_1^2},\ \tilde{G}_{2,1} \approx \frac{s_1^2}{s_2^2},\ \tilde{G}_{2,2} \approx -\frac{s_1^2}{s_2^2} \tag{16}$$

and are able to write down explicitly a modified Sine-Gordon equation taking into account Leggett interband collective excitation. So far we assumed applicability of Fourier transform to phase distributions. But if we would understand ω, χ and their functions as operators $\chi = -i\partial_x, \omega = i\partial_t$ we will be able to apply our approach to more general functions. It appear to be strongly nonlocal both in time and space due to presence of transverse wavenumber of Legett mode $\kappa_2(\omega, \chi)$ in denominators in expressions for effective length and for Josephson phase difference. Now let us apply this modified Sine-Gordon equation to some concrete problems

3.0.1. Swihart-Leggett modes in Josephson Junction. Here we consider distortion of dispersion of Josephson plasma mode sometimes called Swihart mode due of coupling to Leggett waves. Linearizing the term $\sin\left(\hat{H}_i + \hat{H}_k\right)\psi$ we find dispersion of modified plasma waves

$$\omega^2 - \Omega_J^2 = v_s^2\chi^2\left(\frac{2\varepsilon}{q_1 d} + 1 + \frac{2\varepsilon\omega_J^2}{s_2 q_1^2 d\sqrt{\omega_L^2 - \omega^2 + s_2^2\chi^2}}\right),$$

where we introduced two effective frequencies $\Omega_J^2 = (8\pi ed)(\hbar\varepsilon)^{-1}(j_{1,1}^c \pm 2j_{1,2}^c + j_{2,2}^c,\ \omega_J^2 = (8\pi ed)(\hbar\varepsilon)(\pm j_{1,2}^c + j_{2,2}^c,$ the first one plays role of Josephson plasma frequency and Swihart velocity $v_s^2 = c^2\, d[\varepsilon(d + 2\lambda)]^{-1}$. The first term in the parentis

3.0.2. Excitation of Swihart-Leggett modes by Josephson vortices. Modified Sine-Gordon equation as and the old one in high frequency limit has an approximate solution corresponding to dense vortex chain wave $\psi^0 = \Omega t - hx$ where h, Ω are dimensionless magnetic field and Josephson frequency $h = 2e(\hbar c)^{-1}(d + 2\lambda)B_z, \Omega = 2edE_y\hbar^{-1}$ where B_z, E_y are magnetic and electric field inside the junctions. For such a wave all nonlocal corrections in Josephson current disappear since $\partial_x^2\psi = \partial_t^2\psi = 0$ and the only small nonzero terms defined by contributions from imaginary part of dielectric permittivity resulted from quasiparticle tunneling $\varepsilon = 4i\pi\sigma\omega^{-1}$ and external current in RHS of Eq.14 give unperturbed IV characteristic $\Omega \sim \sigma^{-1}E_y$. In the equation for the next approximation the term $\sum_{i,k} j_{i,k}^c \sin\psi^0$ will be the source for perturbation ψ^1 for which we have an expression via complex amplitude $\psi^1 = \mathrm{Im}(\Psi^1 e^{ihx - i\Omega t})$ where Ψ^1 is

$$\Psi^1 = -\frac{8\pi edj^c}{\hbar\varepsilon}\frac{1}{\Omega^2 + i\gamma\Omega - v_s^2h^2\left(1 + \frac{2\varepsilon\omega_J^2}{s_2q_1^2d\sqrt{\omega_L^2 - \Omega^2 + s_2^2h^2}}\right)}. \tag{17}$$

and we introduced $j^c = \sum_{i,k} j_{i,k}^c$ and $\gamma \sim \sigma/d$ as a dimensionless conductance of the junction. Dissipation of the Leggett mode can be taken into account by changing $\Omega \to \Omega + i\nu$ under the square root in the denominator, where ν is a damping rate of Leggett mode. Calculating the next approximation we find the correction for I-V curve $j_c\overline{\psi^1\cos\psi^0} = 0.5j_c\,\mathrm{Im}(\Psi^1)$ and we can see that position and width of resonance peak on I-V curve will be changed due to Leggett modes excitation.

3.0.3. Excitation of Leggett modes by vortices due to Raman mechanism. Instead of linearized relation 6 let us use more precise equation 18 and neglect by small term $\sim \frac{\partial^2\mu}{2\partial n^2}\left(\frac{\partial\mu}{\partial n}\right)^{-2}$. We come to the following relation between concentrations and phases of order parameters

$$n = \frac{\rho_j}{2}\left[2e\varphi - \hbar\dot{\theta} - m_j v_j^2\right], \tag{18}$$

what result in nonlinear equation

$$\ddot{\theta}_j - 2e\hbar^{-1}\dot{\varphi} - s_j^2\Delta\theta_j + \rho_j^{-1}\sum_k \beta\sin(\theta_j - \theta_k) = -\frac{1}{4\hbar m_j}\frac{\partial}{\partial t}\left(\hbar\nabla\theta_j + \frac{2e}{c}\mathbf{A}\right)^2, \tag{19}$$

which differs from linear equation 9 by the term in RHS giving coupling of vortex and potential motions. If we consider the Josephson junction with external static magnetic field applied then this last term will bring to excitation of Leggett mode with Josephson frequency Ω just as it was considered in previous section. Let us consider this additional with respect to considered process in more detail. Earlier considered mechanism was resulted from matching boundary conditions on Josephson junction interface and how it follows from Eq14 amplitude of Leggett mode is proportional to χ^2. The new one can be more effective since excitation force $-m_j\hbar^{-1}\partial_t v_j^2$ can excite Leggett mode even for $\chi = 0$ due to imbalance of chemical potentials in different bands when this mode is pure neutral . Excitation of Leggett mode now can be estimated using perturbation theory. We can calculate the superconducting velocity staying in RHS of Eq.19 using the modified Sine-Gordon equation 14 derived above and expressions for fields found in linear approximation 13. Let us consider the case of strong magnetic field $B = e^{-y/\lambda}(B^0 + B^1 e^{ihx - i\Omega_J t}$ and the term in RHS of Eq.19 is resulted from their combination. Solving linearized equation Eq.19 we find the amplitude of phase perturbation in exciting Leggett mode

$$\sim \frac{i\Omega_J}{s_2\Omega_L}\frac{e^2\lambda^2 B^0 B^1}{c^2\hbar}\left[\frac{1}{m_1} - \frac{1}{m_2}\right]\frac{\lambda}{2 - iq_y\lambda}, \tag{20}$$

here q_y is a wavenumber of Leggett mode, and B^0, B^1 are defined by solution of Sine-Gordon equation Eq.14. Here we used an approximation of strong asymmetric band as we did earlier for derivation of relations 16. So the mechanism considered will result in resonance at $\Omega_J \sim \Omega_L$ just as it was for pure linear mechanism. For low magnetic field the leading term for Leggett wave exitation due to such mechanism will be $\sim (v^1)^2 e^{2ihx - 2i\Omega_J t}$ and results in resonance peculiarity at $2\Omega_J \sim \Omega_L$ on IV curve.

Thus we have developed a hydrodynamical model of multiband superconductors and derived a modified Sine-Gordon equation describing the dynamics of a distributed Josephson junction between multiband superconductors and studied the excitation of Leggett modes by dense chain of moving Josephson vortices. Though a hydrodynamical model is unable to describe kinetic effects resulting in decay of collective excitation such as Landau damping and Cooper pair breaking but qualitatively such effects can be taken into account in hydrodynamical scheme as some effective damping rates which should be calculated using kinetic equations [8].

References

[1] V.A. Moskalenko, Physics of Metals and physical metallurgy, ,**8**, 503 (1959) in Russian.
[2] H. Suhl, B. T. Matthias, L. R. Walker, Phys.Rev. Lett.,**3**, 552 (1959).
[3] A. J. Leggett, Progress of Theoretical Physics **36**, 901 (1966).
[4] D.F. Arterberg, E. Demler, B. Yanko, Phys. Rev. B **66**, 214507 (2002)
[5] Ya. G. Ponomarev et al, Pis'ma v ZhETF, **85** 52 (2007)
[6] Y. Ota, M. Machida, T. Koyama, H. Matsumoto, Phys. Rev. Lett. **102**, 237003 (2009)
[7] S.G. Sharapov, V.P. Gusynin, H. Beck, Eur. Phys. J. B **30**, 45 (2002).
[8] N. Kopnin, Theory of nonequilibrium superconductivity, Clarendon Press, Oxford, (2001)

Dubna-Nano 2012 IOP Publishing
Journal of Physics: Conference Series **393** (2012) 012015 doi:10.1088/1742-6596/393/1/012015

Zero-bias conductance quantization in a normal / superconducting junction of nano wire

Yasuhiro Asano[1] **and Yukio Tanaka**[2]

[1] Department of Applied Physics, Hokkaido University, Sapporo 060-8628, Japan
Center for Topological Science & Technology, Hokkaido University, Sapporo 060-8628, Japan
[2] Department of Applied Physics, Nagoya University, Nagoya 464-8603, Japan

E-mail: asano@eng.hokudai.ac.jp

Abstract. We discuss a strong relationship between Majorana fermions and odd-frequency Cooper pairs which appear at a disordered normal nano wire attached to a topologically nontrivial superconducting one. The zero-bias differential conductance in a normal / superconducting nano wire junctions is quantized at $2e^2/h$ irrespective of degree of disorder, length of disordered segment, and random realization of disordered potential. Such behaviors are exactly the same as those in the anomalous proximity effect of p_x-wave spin-triplet superconductors. We show that odd-frequency Cooper pairs assist the unusual transport properties.

1. Introduction

Finding of Majorana fermions (MFs) and controlling of Majorana bound states (MBSs) are hot research issues in condensed matter physics [1] from the view of potential application of MBS to the topological quantum computation [2, 3]. Although we have already known a number of promising systems hosting MFs, the semiconductor nano wire fabricated on top of a superconductor has attracted much attention [4, 5, 6, 7, 8, 9]. Actually, a plenty of theoretical studies have discussed MFs or MBSs in such nano wires [10, 11, 12, 13, 14, 15, 16]. These researches have stimulated a number of theoretical investigation on unusual charge transport phenomena through the MBS in normal-metal / superconductor (NS) and superconductor /normal-metal/ superconductor (SNS) junctions on nano wires [17, 18, 19, 20, 21, 22, 23]. The zero-bias conductance peak reported very recently would be considered as an experimental evidence of MFs (MBS) [24, 25].

Odd-frequency Cooper pairing was originally proposed to understand nature of unconventional superfluidity and superconductivity[26]. Ubiquitous appearance of the odd-frequency pairs at the surface of superconductors and near the interface of superconducting junctions has been established and widely accepted in recent years [27, 28, 29, 30, 31]. In particular, the odd-frequency Cooper pairs make the background of the anomalous proximity effect in a diffusive normal metal attached to a spin-triplet superconductor: (i) the large zero-energy quasiparticle density of states in a normal metal [32], (ii) the quantized zero-bias conductance at twice of the Sharvin's value in diffusive NS junctions [33], (iii) the fractional current(J)-phase(φ) relationship of $J \propto \sin(\varphi/2)$ in diffusive SNS junctions [34], (iv) the zero-bias anomaly in nonlocal conductance spectra [35] and (v) the anomalous surface impedance in NS bilayers [36].

In a previous paper [37], we have shown that disordered NS and SNS junctions of nano wire indicate the properties of (i)-(iii) when the superconducting nano wire is topologically nontrivial by tuning the Zeeman field and the spin-orbit coupling. We have also suggested a strong relationship between the Majorana fermions and the odd-frequency Cooper pairs. In this paper, we theoretically study details of the differential conductance, the local density of states, and the pairing function in disordered NS nano wires.

2. Model

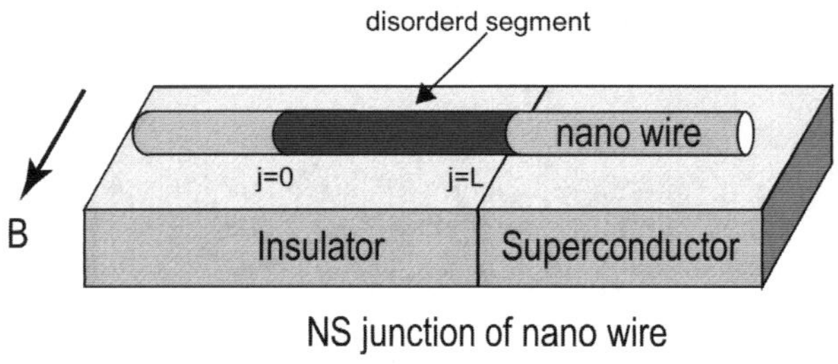

Figure 1. A Schematic figure of a normal / superconducting junction of a nano wire.

Let us consider a nano wire with strong spin-orbit coupling fabricated on a junction of an insulator and a metallic superconductor as shown in Fig. 1. A segment on the insulator and that on the superconductor are in the normal and the superconducting states, respectively. The diameter of the nano wire is sufficiently small so that the number of propagating channel is unity for each spin degree of freedom. We describe the present nano wire by using the tight-binding model in one-dimension,

$$\begin{aligned} H_0 &= -t\sum_{j,\alpha}\left(c^\dagger_{j+1,\alpha}c_{j,\alpha}+c^\dagger_{j,\alpha}c_{j+1,\alpha}\right)+i\frac{\lambda}{2}\sum_{j,\alpha,\beta}\left[c^\dagger_{j+1,\alpha}(\hat{\sigma}_2)_{\alpha,\beta}c_{j,\beta}-c^\dagger_{j,\alpha}(\hat{\sigma}_2)_{\alpha,\beta}c_{j+1,\beta}\right] \\ &+ \sum_{j,\alpha,\beta}c^\dagger_{j,\alpha}\left\{(2t-\mu)\hat{\sigma}_0-V_{ex}\hat{\sigma}_3\right\}_{\alpha,\beta}c_{j,\beta} \end{aligned} \tag{1}$$

$$H_d = \sum_{1\le j\le L,\alpha} V_j c^\dagger_{j,\alpha}c_{j,\alpha}, \tag{2}$$

$$H_s = \sum_{j\ge L+1}\left[\Delta e^{i\varphi}c^\dagger_{j,\uparrow}c^\dagger_{j,\downarrow}+\Delta e^{-i\varphi}c_{j,\downarrow}c_{j,\uparrow}\right], \tag{3}$$

where $c^\dagger_{j,\alpha}(c_{j,\alpha})$ is the creation (annihilation) operator of an electron at the lattice site j with spin $\alpha = (\uparrow \text{ or } \downarrow)$, t denotes the hopping integral, μ is the chemical potential, and Δ is the pair potential in the superconducting segment. The Pauli matrices in spin space are denoted by $\hat{\sigma}_j$ for $j = 1-3$ and the unit matrix of 2×2 is $\hat{\sigma}_0$. The on-site potential in the normal segment is given randomly in the range of $-W/2 \le V_j \le W/2$ for $1 \le j \le L$. We measure the energy and the length in units of t and the lattice constant, respectively. Throughout this paper, we fix several parameters as $\mu = t$, $W = 2t$, the superconducting macroscopic phase $\varphi = 0$, and the pair potential at the zero temperature $\Delta_0 = 0.01t$. The number of samples used for the random ensemble averaging is typically 1000. By tuning the external magnetic field B as shown in Fig. 1, it is possible to introduce the exchange potential V_{ex}. For $V_{ex} > V_c \equiv \sqrt{\Delta_0^2+\mu^2}$, the number of propagating channels in the nano wire is unity and the superconducting segment becomes topologically nontrivial. Therefore MBS is expected at the boundary between the normal and the superconducting segments.

In the following, we study the transport properties of two nano wires. One is the nano wire with $V_{ex} = 1.5t$ and $\lambda = 0.5t$, which we call topological nano wire hosting MF. The other is the non-topological nano wire with $V_{ex} = \lambda = 0$ in which MF is absent. The latter one is studied as a reference. By comparing the results of the two nano wires, we discuss anomalous charge transport properties of the topological nano wire.

3. Results

At first, we study the normal conductance g_N of disordered nanowires based on the linear response theory with using the recursive Green function method [38]. The Hamiltonian is given by $H_0 + H_d$. In Fig. 2, the normal conductance g_N in units of $2e^2/h$ is plotted as a function of the length of the disordered segment L. The results show $\ln(g_N) \propto -L/\xi_{AL}$ with ξ_{AL} being the localization length because one-dimensional disordered wires are in the localization regime. ¿From Fig. 2, we estimate that ξ_{AL} for the topological and the non-topological nano wires are about 11 and 8.5 lattice constants, respectively. The localization length of the topological nano wire is slightly larger than that of the reference nanowire. This may be due to the breaking down the time-reversal and the spin-rotation symmetries. The normal nano wires are in the Anderson insulators. Namely there is no propagating channels at the fermi level.

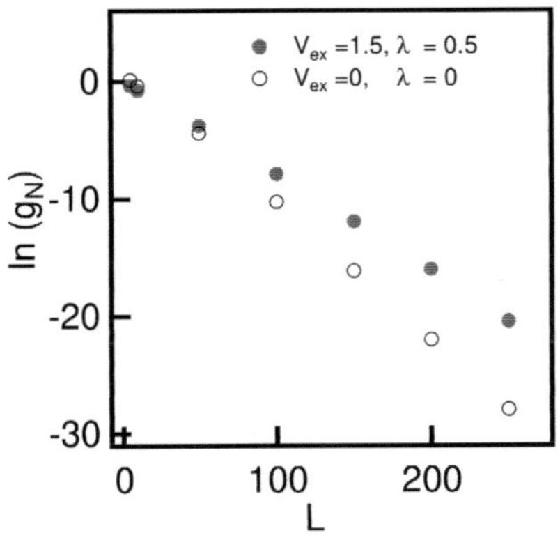

Figure 2. The normal conductance in a disordered nano wires. The results for $V_{ex} = 1.5t$ and $\lambda = 0.5t$ are plotted with filled symbols. Those for $V_{ex} = \lambda = 0$ are shown with open symbols.

Secondly, we study the differential conductance g_{NS} of NS junctions based on the Blonder-Tinkham-Klapwijk formula [39],

$$g_{NS} = \frac{e^2}{h} \sum_{\alpha,\beta} \left[\delta_{\alpha,\beta} - |r^{ee}_{\alpha,\beta}|^2 + |r^{he}_{\alpha,\beta}|^2 \right]_{E=eV} \tag{4}$$

where we consider the Hamiltonian $H_0 + H_d + H_s$. In Eq. (4), $r^{ee}_{\alpha,\beta}$ and $r^{he}_{\alpha,\beta}$ are the normal and Andreev reflection coefficients of the junction at energy E. By solving the Bogoliubov-de Gennes equation numerically, we obtain the Green function

$$\check{\mathcal{G}}_E(j,j') = \left[\begin{array}{cc} \hat{G}_E(j,j') & \hat{F}_E(j,j') \\ -\hat{F}^*_{-E}(j,j') & -\hat{G}^*_{-E}(j,j') \end{array} \right], \tag{5}$$

where the symbols $\cdot\hat{\cdot}\cdot$ and $\cdot\check{\cdot}\cdot$ indicate 2×2 matrices in spin space and 4×4 matrices in spin plus Nambu space. The Andreev reflection coefficients can be calculated from the asymptotic behavior of the Green function [40]. We show the differential conductance g_{NS} in units of $2e^2/h$ as a function of the bias-voltage eV for the non-topological nano wire in Fig. 3(a) and the topological nano wires in Fig. 3(b). The conductance for the non-topological nano wires in (a) decreases with increasing the length of disordered segment L for all eV. The similar tendency can be seen also in the conductance in the topological nano wires in (b) for finite

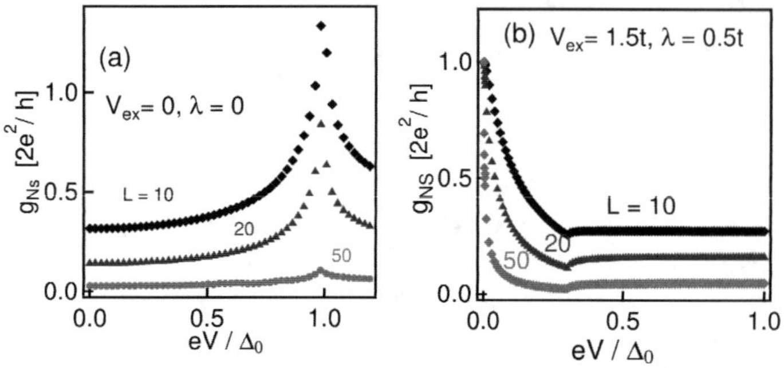

Figure 3. The differential conductance of NS nano wires for several choices of length of disordered segment L. The results for $V_{ex} = \lambda = 0$ and those for $V_{ex} = 1.5t$ and $\lambda = 0.5t$ are shown in (a) and (b), respectively.

eV. However the zero-bias conductance of the topological nano wires is quantized at $2e^2/h$ irrespective of L, which is an intrinsic phenomenon of Majorana fermion. The results suggest a perfect transmission channel in the disordered nanowire even in the localization regime.

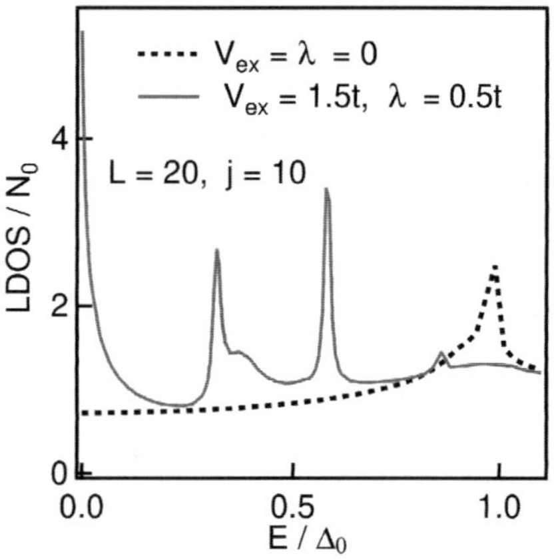

Figure 4. The local density of states at the center of the disordered segment. The results for $V_{ex} = \lambda = 0$ and those for $V_{ex} = 1.5t$ and $\lambda = 0.5t$ are shown with broken and solid lines, respectively.

Next we discuss the local density of states (LDOS) defined by

$$N(E,j) = -\frac{\mathrm{Im}}{\pi}\mathrm{Tr}\left[\check{\mathcal{G}}_E(j,j)\right]. \tag{6}$$

The LDOS in the disordered nano wire is shown in Fig. 4, where we plot the LDOS at $j = 10$ as a function of E for $L = 20$. The LDOS for the non-topological nano wire (broken line) is almost flat around the zero energy and is slightly smaller than the density of states at the fermi level in clean normal nano wire N_0. On the other hand, the LDOS for the topological nano wire (solid line) has large zero-energy peak, which reflects the presence of the resonant state in the normal disordered wire. Strictly speaking, the resonant state is no longer a bound state because a quasiparticle can escape from the disordered normal segment to a semi-infinitely long clean normal lead wire ($j \leq 0$). This is responsible for the finite width of the zero-energy peak. The LDOS also shows a peak around $E = 0.3\Delta_0$, where the number of the propagating channels in the superconducting segment changes from 0 to 2. The dip structures in the differential conductance around $eV = 0.3\Delta_0$ in Fig. 3(b) reflect this fact. The LDOS shows another peak structures around $E = 0.6\Delta_0$ and $E = 0.85\Delta_0$ as shown in Fig. 4. At present, we do not know well the reason of the peaks. In the conductance, there are neither peak nor dip structures in the conductance at corresponding bias voltage as shown in Fig. 3(b).

Finally we analyze the pairing function in the normal segment. The anomalous Green function can be decomposed into

$$\hat{F}_E(j,j) = i\,[f_0(j,E) + \mathbf{f}(j,E)\cdot\hat{\sigma}]\,\hat{\sigma}_2. \qquad (7)$$

The spin-singlet pairing function is denoted by f_0 and the spin-triplet pairing functions are f_i for $i = 1-3$. According to the definition, the spin configuration of f_3, $f_{\uparrow\uparrow} = -f_1 + if_2$ and $f_{\downarrow\downarrow} = f_1 + if_2$ are $(|\uparrow\downarrow\rangle + |\downarrow\uparrow\rangle)/\sqrt{2}$, $|\uparrow\uparrow\rangle$ and $|\downarrow\downarrow\rangle$, respectively. In Fig. 5, we show the pairing function in non-topological and topological nano wire in (a) and (b), respectively. In the non-topological nano wire, the spin-triplet components are zero. Only the spin-singlet component has finite amplitude. The real part of f_0 is the even function of E, whereas the imaginary part of f_0 is the odd-function of E, (i.e., $f_0(-E) = f_0^*(E)$). This behavior means the even-frequency spin-singlet Cooper pairs penetrate into the normal segment. On the other hand in the topological nano wire, $f_{\uparrow\uparrow}$ is the most dominant component. In fact, Re(f_0) takes its maximum value -0.14 at $E = 0$. Thus the amplitudes of the even-frequency spin-singlet component is much smaller than $f_{\uparrow\uparrow}$. The results in Fig. 5(b) show that the real part of $f_{\uparrow\uparrow}$

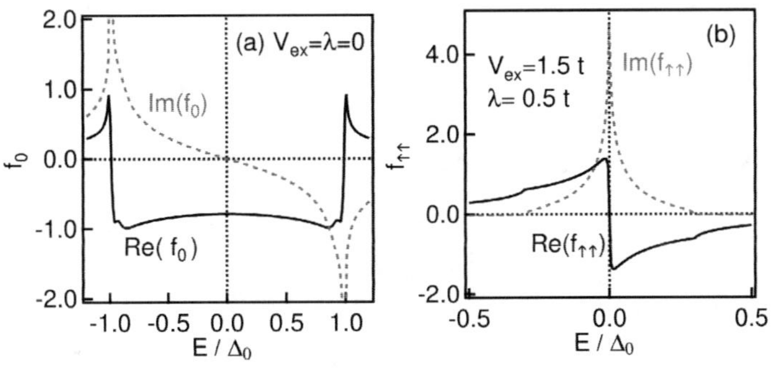

Figure 5. The pairing function at the center of the disordered segment. The spin-singlet component f_0 for $V_{ex} = \lambda = 0$ are shown in (a). The equal-spin triplet component $f_{\uparrow\uparrow}$ for $V_{ex} = 1.5t$ and $\lambda = 0.5t$ are shown in (b). We choose $L = 20$. The results calculated at $j = L/2 = 10$.

is the odd function of E, whereas the imaginary part of $f_{\uparrow\uparrow}$ is the even function of E, (i.e., $f_{\uparrow\uparrow}(-E) = -f_{\uparrow\uparrow}^*(E)$). This fact means the penetration of odd-frequency Cooper pairs into the normal segment of the topological nano wire [31, 36]. The odd-frequency Cooper pairs assist the conductance quantization at the zero-bias in Fig. 3(b). To make this statement clearer, we show so called pair density defined by

$$n_i(j) = 2\int_0^\infty dE\,\mathrm{Re}[f_i(j,E)]\mathrm{Im}[f_i(j,E)], \qquad (8)$$

at the center of disordered segment $j = L/2 = 10$ as a function of V_{ex} in Fig. 6. The pair density for even-frequency spin-singlet component n_0 decreases with decreasing V_{ex}. The pair density for the odd-frequency equal spin-triplet component $n_{\uparrow\uparrow}$ is negative, which is also found in anomalous proximity effect of spin-triplet superconductors and is responsible for unusual response to the electromagnetic field [36]. The amplitude of $n_{\uparrow\uparrow}$ slightly increases with V_{ex} in non-topological wire for $V_{ex} < V_c \approx 1$. For $V_{ex} > V_c$, the amplitude of $n_{\uparrow\uparrow}$ suddenly grows. In the topological wire $V_{ex} > V_c$, the fraction of the odd-frequency pairs becomes almost unity. Thus we conclude that the odd-frequency Cooper pairs assists the conductance quantization at the zero-bias in Fig. 3(b). We note that the pair density in Eq. (8) is obtained average over a number of samples with different random potential configurations. In our simulation, we find very large sample-to-sample fluctuations of $n_{\uparrow\uparrow}$ for $V_{ex} > V_c$. In fact, the results of $n_{\uparrow\uparrow}$ for $V_{ex} > V_c$ seems to be not a smooth function of V_{ex}. This may be a characteristic feature in the localization regime.

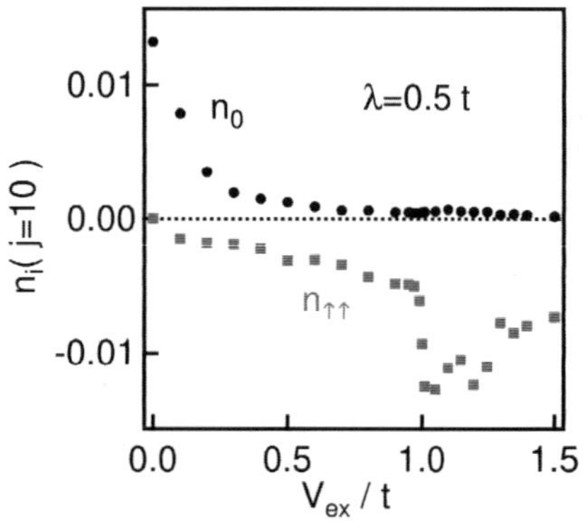

Figure 6. The pair density is plotted as a function of V_{ex}, where λ is fixed at $0.5t$. We choose $L = 20$. The results calculated at $j = L/2 = 10$.

4. Conclusion

We have studied transport properties of disordered normal / superconducting (NS) junctions of nano wires which have strong spin-orbit coupling. When the superconducting pair potential is absent, the normal nano wires are in the Anderson localization regime. By introducing the exchange field V_{ex}, superconducting nano wires become topologically nontrivial when the exchange potential is larger than a critical value of V_c. For $V_{ex} > V_c$, the superconducting nano wires host the Majorana fermions at their edges. In such NS junction, the zero-bias conductance is quantized at the universal value of $2e^2/h$ irrespective of the length of disordered normal segment, degree of disorder, and the random potential configurations. The penetration of the Majorana fermions into the normal segment is responsible for the universal behavior. We have also studied the pairing function in normal segment and found that the amplitude of the odd-frequency Cooper pairs suddenly grows for $V_{ex} > V_c$. We conclude that the odd-frequency Cooper pairs and the Majorana fermions always coexist with each other in solids.

5. acknowledgement

This work was supported by KAKENHI(No. 22540355) and the "Topological Quantum Phenomena" (No. 22103002) Grant-in Aid for Scientific Research on Innovative Areas from the Ministry of Education, Culture, Sports, Science and Technology (MEXT) of Japan.

References

[1] Wilczek F 2009 *Nat. Phys.* **5** 614; Franz M 2010 *Physics* **3** 24
[2] Ivanov D A 2001 *Phys. Rev. Lett.* **86** 268
[3] Das Sarma S, Freedman M and Nayak C 2005 *Phys. Rev. Lett.* **94** 166802
[4] Sato M, Takahashi Y and Fujimoto S 2009 *Phys. Rev. Lett.* **103** 020401
[5] Sau J D, Lutchyn R M, Tewari S and Das Sarma S 2010 *Phys. Rev. Lett.* **104** 040502
[6] Alicea J 2010 *Phys. Rev. B* **81** 125318
[7] Potter A C and Lee P A 2010 *Phys. Rev. Lett.* **105** 227003
[8] Lutchyn R M, Sau J D and Das Sarma S 2010 *Phys. Rev. Lett.* **105** 077001
[9] Oreg Y, Refael G and von Oppen F 2010 *Phys. Rev. Lett.* **105** 177002
[10] Tewari S, Stanescu T D, Sau J D and Das Sarma S 2011 *New Journal of Physics* **13** 065004
[11] Lutchyn R M, Stanescu T D and Das Sarma S 2011 *Phys. Rev. Lett.* **106** 127001
[12] Golub A, Kuzmenko I and Avishai Y 2011 *Phys. Rev. Lett.* **107** 176802
[13] Flensberg K 2010 *Phys. Rev. B* **82** 180516
[14] Romito A, Alicea J, Refael G and von Oppen F 2011 *arXiv*:1110.6193; Alicea J 2012 *arXiv*:1202.1293
[15] Bena C, Sticlet D and Simon P 2012 *Phys. Rev. Lett.* **108** 096802
[16] Potter A C and Lee P A 2011 *Phys. Rev. B* **83** 094525
[17] Akhmerov A R, Dahlhaus J P, Hassler F, Wimmer M and Beenakker C W L 2011 *Phys. Rev. Lett.* **106** 057001 Brouwer P W, Duckheim M, Romito A and von Oppen F 2011 *Phys. Rev. B* **84** 144526

[18] Qu C, Zhang Y, Mao L and Zhang C 2011 *arXiv*:1109.4108
[19] Gibertini M, Taddei F, Polini M and Fazio R 2011 *arXiv*:1111.4656.
[20] Roy D, Bolech C J and Shah N 2012 *arXiv*:1202.6222
[21] Wimmer M, Akhmerov A R, Dahlhaus J P and Beenakker C W J 2011 *New Journal of Physics* **13** 053016
[22] Liu X-J 2012 *arXiv*:1204.0570
[23] Lin C H, Sau J D and Das Sarma S 2012 *arXiv*:1204.3085
[24] Mourik V, Zuo K, Frolov S M, Plissard S R, Bakkers E P A M and Kouwenhoven L P 2012, *Science* **336** 1003
[25] Das A, Ronen Y, Most Y, Oreg Y, Heiblum M and Shtrikman H 2012 *arXiv*:1205.7073; Rokhinson L P, Liu X, Furdyna J K 2012 *arXiv*:1204.4212; Deng M T, Yu C L, Huang G Y, Larsson M, Caroff P and Xu H Q 2012 *arXiv*:1204.4130.
[26] Berezinskii V L 1974 *JETP Lett.* **20** 287
[27] Bergeret F S, Volkov A F and Efetov K B 2001 *Phys. Rev. Lett.* **86** 4096
[28] Asano Y, Tanaka Y and Golubov A A 2007 *Phys. Rev. Lett.* **98** 107002; Braude V and Nazarov Yu V 2007 *Phys. Rev. Lett.* **98** 077003; Eschrig M and Löfwander T 2008 *Nature Phys.* **4** 138.
[29] Tanaka Y, Golubov A A, Kashiwaya S and Ueda M 2007 *Phys. Rev. Lett.* **99** 037005 (2007); Tanaka Y, Tanuma Y and Golubov A A 2007 *Phys. Rev. B* **76** 054522
[30] Tanaka Y and Kashiwaya S 1995 *Phys. Rev. Lett.* **74** 3451; Tanaka Y and Kashiwaya S 1997 *Phys. Rev. B* **56** 892; Asano Y 2001 *Phys. Rev.B* **64** 014511
[31] Tanaka Y and Golubov A A 2007 *Phys. Rev. Lett.* **98** 037003.
[32] Tanaka Y, Asano Y, Golubov A A and Kashiwaya S 2005 *Phys. Rev. B* **72** 140503(R)
[33] Tanaka Y and Kashiwaya S 2004 *Phys. Rev.B* **70** 012507
[34] Asano Y, Tanaka Y and Kashiwaya S 2006 *Phys. Rev. Lett.* **96** 097007
[35] Asano Y, Tanaka Y, Golubov A A and Kashiwaya S 2007 *Phys. Rev. Lett.* **99** 067005
[36] Asano Y, Golubov A A, Fominov Y and Tanaka Y 2011 *Phys. Rev. Lett.* **107** 087001
[37] Asano Y and Tanaka Y 2012 *arXiv*:1204.4226
[38] Lee P A and Fisher D S 1981 *Phys. Rev. Lett.* **47** 882
[39] Blonder G E, Tinkham M and Klapwijk T M 1982 *Phys. Rev. B* **25** 4515
[40] Asano Y, Tanaka Y and Kashiwaya S 2004 *Phys. Rev. B* **69** 214509; Asano Y 2001 *Phys. Rev. B* **64** 224515

Journal of Physics: Conference Series **393** (2012) 012016 doi:10.1088/1742-6596/393/1/012016

Collective excitonic and plasmonic excitations on a surface of 3D topological insulator

D K Efimkin[1], **Yu E Lozovik**[1,2]

[1] Institute of Spectroscopy RAS , 142190, Troitsk, Moscow Region, Russian Federation.
[2] Moscow institute of Physics and Technology, 141700, Dolgoprudniy, Moscow Region, Russian Federation.

E-mail: lozovik@isan.troitsk.ru

Abstract. Properties of collective excitonic and plasmonic excitations on the surface of three-dimensional topological insulator are reviewed. Due to spin-momentum locking for electrons populating the surface states plasmonic excitation manifests itself as coupled spin- and density-wave and can be called "spin-plasmon". Its excitation induces spin polarization of topological insulator surface that is perpendicular to the spin-plasmon momentum. Excitons appear on the surface when the gap in the surface spectrum is magnetically induced. Excitons correspond to the set of subgap levels with unusual dependence of the energy on orbital angular momentum quantum number and can be called "chiral". Contrary to conventional excitons chiral ones resonantly contribute to Hall component of optical conductivity tensor and play significant role in magneto-optical Faraday and Kerr effects. They resonantly enhance Faraday effect and reduce Kerr effect.

1. Introduction

Topological insulator (TI) is new class of solids that has nontrivial topology of the band structure (see [1] and references therein). TIs as conventional insulators have forbidden band in the bulk but also contain exotic topologically protected conducting states on their surface (3D) or border (2D). Recently a "new generation" of three-dimensional topological insulators (the binary compounds Bi_2Se_3, Bi_2Te_3 and others), retaining topologically protected behavior at room temperatures, were investigated experimentally [2]. Surface states of these materials are described by the Dirac-like Hamiltonian for two-dimensional massless particles that is given by

$$H_0 = v_{\rm F}(p_x\sigma_y - p_y\sigma_x), \tag{1}$$

where $v_{\rm F}$ is the Fermi velocity of electron; the Pauli matrices σ_x and σ_y act in the spaces of its spin projections. Spectrum is formed by conduction ($\gamma = 1$) and valence ($\gamma = -1$) bands that have linear dispersion low $\epsilon_{p\gamma} = \gamma v_{\rm F} p$ and touch each other. This Hamiltonian is similar to one that describes electrons in graphene but electron spin is substituted by pseudospin which is connected with lattice degree of freedom [3].

The direction of spin projection of electrons populating the surface states of TI is determined by their momentum. Such spin-momentum locking manifests itself in different transport phenomena that include coupled diffusion of charge and spin [4], inverse galvanomagnetic effect [5] and contact effects [6]. Collective plasmonic excitation in degenerate electron gas populating

TI surface states manifests itself as coupled charge- and spin- density waves and can be called spin-plasmon [7, 8]. Latter it was proposed to apply spin-plasmons in spin accumulator [9]. Comprehensive analysis of the physical properties of spin-plasmons was made in the paper [8].

The surface spectrum is protected from external perturbations that do not break the time reversal symmetry, for example, from nonmagnetic disorder [10]. External exchange field that can be created by ordered magnetic impurities introduced to a TI bulk or surface [11] acts only on magnetic moment of the surface states and leads to the gap formation in the spectrum. The Hamiltonian describing electrons on the surface of TI in the presence of external exchange field is given by

$$H_0 = v_{\mathrm{F}}(p_x\sigma_y - p_y\sigma_x) + m\sigma_z, \tag{2}$$

where value m parameterizes coupling strength of the z-component of exchange field with electron's spin. Other components of exchange field can be excluded by the gauge transformation that shifts Dirac point in momentum space. Exchange field leads to the formation of the gap in the spectrum $\epsilon_{\mathbf{p}\gamma} = \gamma\sqrt{v_{\mathrm{F}}^2 + m^2}$. Excitonic states that correspond to bound states of electron from conduction band and hole from valence were considered in [12]. It interesting that the excitonic state with minimal energy has finite value of orbital momentum quantum number and can be called chiral.

Breaking of the time reversal symmetry leads to half-integer quantum Hall effect on the TI surface that causes topological magnetoelectric effect in the TI bulk [13]. The topological magnetoelectric effect leads to quantized magneto-optical Kerr and Faraday effect [14] without external magnetic field. In thin film of topological insulator which width is less then length of incident electromagnetic wave Faraday angle θ_F and Kerr angle θ_K are universal: $\tan\theta_F = \alpha$ and $\tan\theta_K = 1/\alpha$, where $\alpha \approx 1/137$ is fine structure constant. The role of substrate [15], interference effects in thick TI film [14], oblique incidence of electromagnetic wave [16] and frequency dispersion of conductivity [14] was investigated. Important role of the chiral excitons in magneto-optical Faraday and Kerr effects was outlined in our paper [17].

Here we review our theoretical study on spin-plasmons [8] in degenerate electron gas populating the surface surface of TI and chiral excitons [17] on the magnetically gapped surface of TI. Both of them can be effectively described within equations of motion approach that is also discussed.

Review is organized in following way. In section 2 the mathematical approach that was used for description of excitons and plasmons is descussed. Section 3 and 4 are devoted to physical properties of plasmons and excitons, respectively. Section 5 is devoted to brief summary.

2. Equation of motion approach

Excitations are called collective ones if they can not be described within single-particle picture. Starting point of their investigation is the Hamiltonian that includes interaction between particles. For electrons populating surface states of TI the Hamiltonian is given by

$$H = \sum_{\mathbf{p}\gamma} \xi_{p\gamma} a^+_{\mathbf{p}\gamma} a_{\mathbf{p}\gamma} + \frac{1}{2}\sum_{\mathbf{q}\mathbf{p}\mathbf{p}'}\sum_{\gamma_1\gamma_2\gamma_1'\gamma_2'} V_c(\mathbf{q})\langle f_{\mathbf{p}+\mathbf{q}\gamma_1'}|f_{\mathbf{p}\gamma_1}\rangle\langle f_{\mathbf{p}'-\mathbf{q}\gamma_2'}|f_{\mathbf{p}'\gamma_2}\rangle a^+_{\mathbf{p}+\mathbf{q}\gamma_1'} a_{\mathbf{p}'-\mathbf{q}\gamma_2'} a_{\mathbf{p}'\gamma_2} a_{\mathbf{p}\gamma_1}, \tag{3}$$

where $a_{\mathbf{p}\gamma}$ is the destruction operator for the electron with momentum $\mathbf{p}$ from the band γ; $|f_{p,\gamma}\rangle$ denotes the corresponding wave function in spinor space that is eigenvectors of Hamiltonians (1) or (2), respectively; $\xi_{\mathbf{k}\gamma} = \epsilon_{\mathbf{k}\gamma} - E_F$ is the dispersion law of electrons that includes chemical potential E_F; $V_c(\mathbf{q}) = 2\pi e^2/\varepsilon q$ is the two-dimensional Fourier transform of Coulomb interaction potential and ε is effective dielectric permittivity of half-space formed by topological insulator surface.

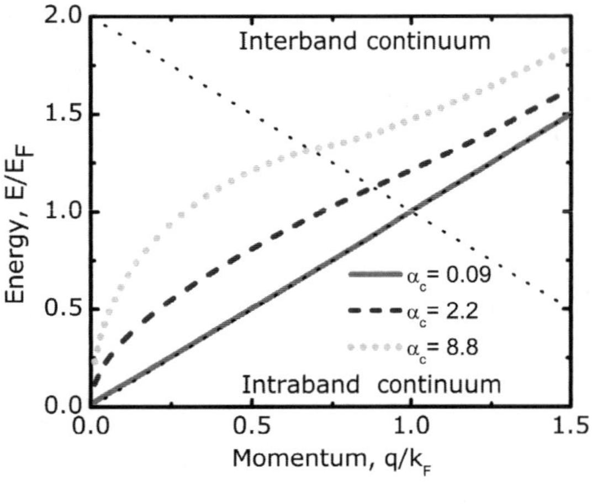

Figure 1. Plasmon dispersion at different values of parameter α_c. Dotted lines denote borders of intraband ($w < v_F q$) and interband ($w > 2E_F - v_F q$) single particle continuums

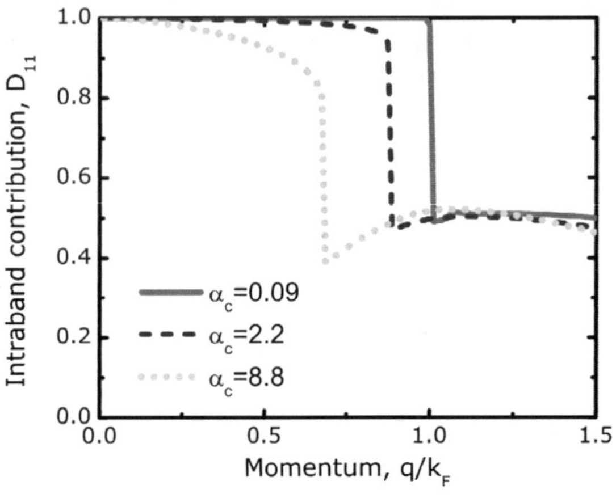

Figure 2. Dependence of total weight of intraband transitions that contribute to wave function of plasmon at different values of dimensionless parameter α_c

Creation operators for plasmons and excitons have the *same* form and are given by

$$Q_{\mathbf{q}}^{+} = \sum_{\mathbf{p}\gamma\gamma'} C_{\mathbf{pq}}^{\gamma'\gamma} a_{\mathbf{p+q}\gamma'}^{+} a_{p\gamma}, \tag{4}$$

where $C_{\mathbf{pq}}^{\gamma'\gamma}$ are weights of inraband ($\gamma = \gamma'$) and interband ($\gamma \neq \gamma'$) single-particle transitions. Plasmonic excitation contains both interband and intraband contributions while exciton contains only interband ones. Both of them can be represented as composite bosons and their creation operators satisfy corresponding commutation relations $[Q_q, Q_{q'}^{+}] = \delta_{qq'}$. Creation operator of collective excitation obeys equation of motion $[H, Q_{\mathbf{q}}^{+}] = E_q Q_{\mathbf{q}}^{+}$, where E_q is its energy and H is Hamiltonian of interacting electrons (3).

There is only single dimensionless parameter that determines physical properties of the collective excitations on the surface of TI that is given by $\alpha_c = e^2/\epsilon\hbar v_F$. In case of spin-plasmons it corresponds to ratio between Coulomb interaction energy of degenerate electron gas to its kinetic energy and is usually denoted as parameter r_s. In case of excitons it corresponds to the ratio between electron-hole interaction energy to their kinetic energies. Weak-coupling regime $\alpha_c \ll 1$ corresponds to experimental relevant conditions [2]. In this case Random phase approximation (RPA) [18] can be employed for solution of equation of motion for creation operation of spin-plasmon and equation of motion of excitons has the simple approximate analytical solution [17]. Also we considered the case of zero temperature.

3. Spin-plasmons

3.1. Internal structure of plasmons

We find that in RPA wave function of spin-plasmon is given by

$$C_{\mathbf{pq}}^{\gamma'\gamma} = \frac{\langle f_{\mathbf{p+q},\gamma'} | f_{\mathbf{p}\gamma} \rangle N_{\mathbf{q}}}{E_{\mathbf{q}} + \xi_{\mathbf{p}\gamma} - \xi_{\mathbf{p+q},\gamma'} + i\delta}, \tag{5}$$

where $N_{\mathbf{q}}$ is normalization factor and E_q is the dispersion law of plasmons that can be obtained from the following equation

$$1 - V_c(\mathbf{q}) \sum_{\mathbf{p}\gamma\gamma'} \frac{\left|\langle f_{\mathbf{p+q},\gamma'} | f_{\mathbf{p}\gamma}\rangle\right|^2 (n_{\mathbf{p}\gamma} - n_{\mathbf{p+q},\gamma'})}{E_{\mathbf{q}} + \xi_{\mathbf{p}\gamma} - \xi_{\mathbf{p+q},\gamma'} + i\delta} = 0. \tag{6}$$

Here $n_{p\gamma}$ is occupation number of electronic states $|p,\gamma\rangle$. The same equation was obtained within diagrammatic approach based on RPA to plasmonic excitations in Dirac electronic gas on a surface of topological insulator [7] and graphene [19].

For numerical calculations we used the following set of parameters $v_{\mathrm{F}} = 0.62 \cdot 10^6 \mathrm{m/s}$ and $\epsilon_{\mathrm{TI}} = 80$ that corresponds to $\alpha_{\mathrm{c}} = 0.09$. Dispersion law of spin-plasmons on the surface of TI is presented in Fig.1. The dispersion laws for plasmons in Dirac electronic gas in suspended graphene ($\alpha_{\mathrm{c}} = 8.8$) and graphene on SiO_2 substrate ($\alpha_{\mathrm{c}} = 2.2$) are presented for comparison. (For graphene parameter α_{c} is given by $ge^2/\hbar v_{\mathrm{F}}\varepsilon$ where $g = 4$ is degeneracy factor of Dirac cones). For $\alpha_{\mathrm{c}} = 0.09$ dispersion law of plasmon tends to the border of continuum of intraband single-particle transitions ($w < v_{\mathrm{F}}q$). At $q \simeq k_F$ it intersects continuum of iterband single-particle transitions ($w > 2E_F - v_{\mathrm{F}}q$) and spin-plasmon become damped since conservation laws of energy and momentum allow transfer energy between spin-plasmon and electron-hole pairs.

Plasmonic excitation consists of intraband and interband single-particle transitions. Their total contribution can be characterized by their weight $D_{\gamma'\gamma}$ that is given by

$$D_{\gamma'\gamma} = \sum_{\mathbf{p}} |C^{\gamma'\gamma}_{\mathbf{pq}}|^2 (n_{\mathbf{p}\gamma} - n_{\mathbf{p+q},\gamma'}). \tag{7}$$

Dependence of weight of the intraband contribution D_{11} on momentum of the spin-plasmon for different values of dimensionless coupling α_{c} is presented in Fig.2. The undamped plasmon for all values of parameter α_{c} consists mainly of intraband transitions. In single-particle continuum contributions of the interband and the interband transitions approximately coincide.

3.2. Spin- and density- waves

In the spin-plasmon state distribution of the electrons is shifted along its momentum $\mathbf{q}$ in the momentum space and spin-polarization of the TI surface appears due to the locking between spin and momentum of single electron. By the same reason spin polarization of the surface of TI appears in the current-carrying state [5]. Value of the spin polarization of the TI surface in the spin-plasmon state versus its momentum is presented in Fig.3.

Due to spin-momentum locking plasmonic excitations manifests themselves as coupled spin- and density- waves. Direction of spin polarization is perpendicular to plasmon momentum. Dependence of amplitudes of spin- and density- waves (A_s and A_ρ) on momentum of spin-plasmon in dimensionless units $\sqrt{\rho \cdot n_q}$ (ρ is the concentration of electrons and $n_{\mathbf{q}}$ is the concentration of plasmons with momentum $\mathbf{q}$) is presented in Fig.3. The values of their amplitudes have the same order. Similar effect appears in electron gas in the presence of spin-orbital interaction in which plasmonic excitation is also accompanied by the spin-density wave. But for the experimental relevant parameters its amplitude is considerable less then amplitude of the charge wave.

3.3. Scattering of spin-plasmons

We considered scattering of spin-plasmons on the external potential and magnetic field. Fermi's Golden rule was used for the calculation of the probability of the spin-plasmon scattering. The angle probability distribution of plasmon scattering can be presented in the form:

$$w_V(q,\phi) = |V(2q \sin\frac{\phi}{2})|^2 |\Phi_V(q,\phi)|^2, \tag{8}$$

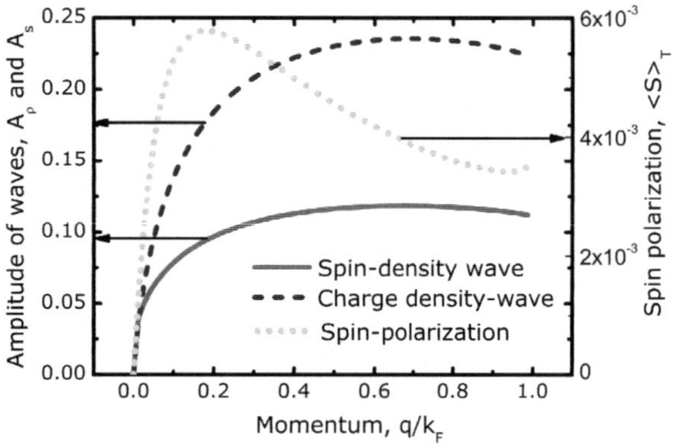

Figure 3. Amplitudes of spin- and charge-density waves and value of spin polarization versus momentum of spin-plasmon.

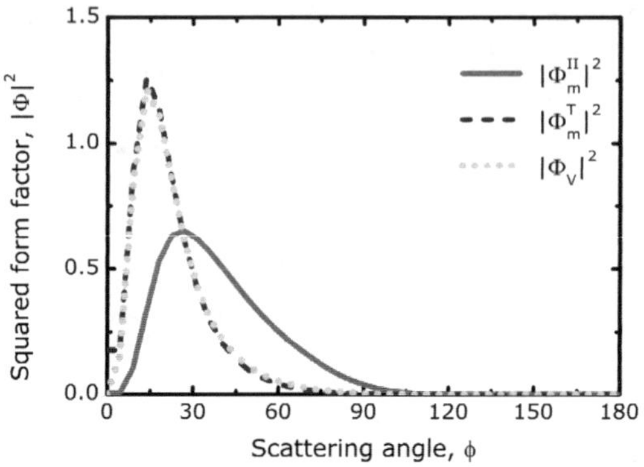

Figure 4. Dependence of squared electric and magnetic form-factors on scattering angle of spin-plasmon ϕ.

$$w_{\mathbf{M}}(q,\phi) = |\mathbf{M}(2q\sin\frac{\phi}{2}) \cdot \mathbf{\Phi}_{\mathbf{M}}(q,\phi)|^2, \tag{9}$$

where ϕ is the scattering angle; q is the absolute value of momentum of spin-plasmon; $V(p)$ and $M(p)$ are the Fourier transform of spatial distribution of the external potential and magnetic field that possess rotational symmetry. Here we introduced electric form-factor $\Phi_V(q,\theta)$ and vector of magnetic form-factors $\mathbf{\Phi}_{\mathbf{M}}(q,\theta)$ of spin-plasmon. Dependence of spin-plasmon form-factors on scattering angle is presented in Fig.4 for spin-plasmon momentum $q = 0.6$. For other values of the momentum the dependence is qualitatively the same. Due to complicated internal structure of spin-plasmon its forward and backward scattering are suppressed. Form-factors achieve their maximal value at finite scattering angle.

4. Chiral excitons

4.1. Structure of the set of excitonic states

Only excitons with zero momentum contribute to the optical conductivity tensor, hence we will suppose below that their momentum is zero and omit momentum index. If the renormalization of single-particle spectrum can be omitted [17] the equation of motion for exciton with zero total momentum is given by

$$2\epsilon_{\mathbf{k}} C_{\mathbf{k}} + \sum_{k'} V_{eff}(\mathbf{k},\mathbf{k}')C_{\mathbf{k}'} = EC_{\mathbf{k}}. \tag{10}$$

Here $V_{eff}(\mathbf{k},\mathbf{k}')$ is the efficient electron-hole interaction that is given by

$$V_{eff}(\mathbf{k},\mathbf{k}') = V_c(\mathbf{k}-\mathbf{k}')\frac{1}{2}\{\frac{v_{\mathrm{F}}^2 k^2}{\epsilon_{\mathbf{k}}\epsilon_{\mathbf{k}'}} + (1+\frac{m^2}{\epsilon_k \epsilon_{\mathbf{k}'}})\cos(\phi_{\mathbf{k}}-\phi_{\mathbf{k}'}) + i(\frac{m}{\epsilon_{\mathbf{k}}}+\frac{m}{\epsilon'_{\mathbf{k}}})\sin(\phi_{\mathbf{k}}-\phi_{\mathbf{k}'})\} \tag{11}$$

Equation (10) have been numerically solved in [12]. Excitonic state with minimal energy has with finite value of orbital angular momentum and there is no symmetry between the states with opposite values of orbital angular momentum. Hence excitons can be called "chiral". We find that in weak coupling regime $\alpha_c \ll 1$ the energy spectrum and effective interaction can be approximated in following way: $\epsilon_{\mathbf{k}} \approx |m| + v_{\mathrm{F}}^2 k^2/2|m|$ and $V_{eff}(\mathbf{k},\mathbf{k}') \approx V_c(\mathbf{k}-\mathbf{k}')[\cos(\phi_{\mathbf{k}}-\phi'_{\mathbf{k}}) + i\sin(\phi_{\mathbf{k}}-\phi'_{\mathbf{k}})]$. In this case the equation (10) coincides with Schrodinger equation for 2D hydrogen atom in which: 1) multipole momenta of Coulomb potential Fourier transform are *shifted* by $\delta l = 1$; 2) effective reduced mass of electron in 2D hydrogen atom problem is $\mu^\star = |m|/2v_{\mathrm{F}}^2$. The chiral excitons can be characterized by radial $n = 0, 1, ...$ and

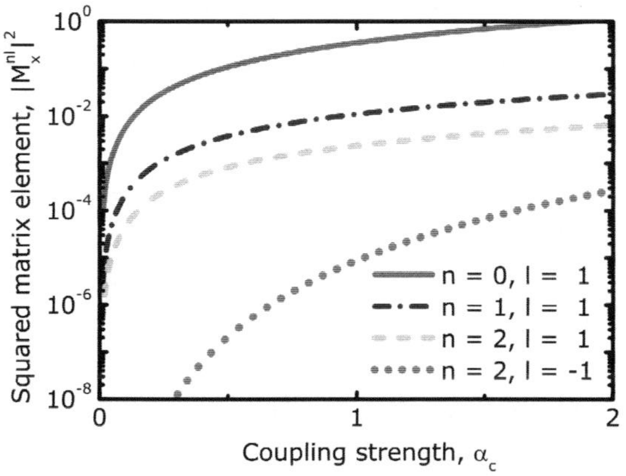

Figure 5. Dependence of squared dimensionless matrix element $|M_x^{nl}|^2$ for the four lowest-energy optical active excitonic states on dimesionless coupling constant α_c.

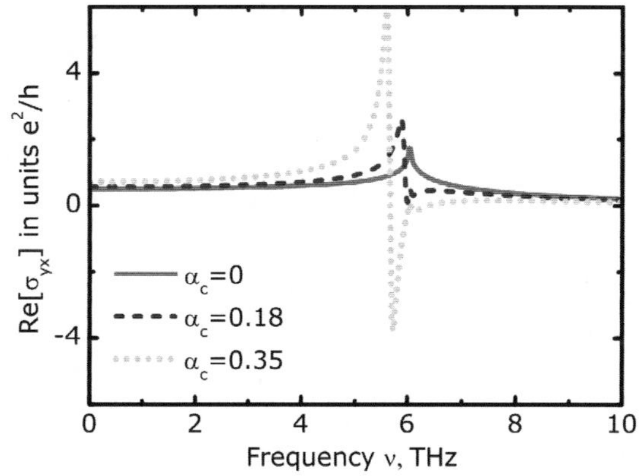

Figure 6. Frequency dependence of real part of Hall optical conductivity $\mathrm{Re}[\sigma_{yx}]$.

orbital angular $l = 0, \pm1, ...$ quantum numbers with $|l - 1| \leq n$. Their energy spectrum Ω_{nl} and wave functions in momentum space C_k^{nl} can be obtained by *shift* of well-known ones for a 2D hydrogen atom. State $|0, 1\rangle$ with minimal energy has orbital quantum number $l = 1$. Sign of the orbital quantum number that corresponds to the lowest-energy state depends on sign of m and therefore on the direction of exchange field.

4.2. Excitonic contribution to conductivity tensor

We used the linear response theory and the equation of motion approach for calculation of the excitonic contribution of optical conductivity tensor. Its components are given by

$$\sigma_{xx}^{ex}(\omega) = i\frac{e^2}{h}2\pi\sum_{nl}|M_x^{nl}|^2\frac{\omega}{E_{nl}}\frac{2m^2}{(\omega+i\gamma)^2 - E_{nl}^2}, \tag{12}$$

$$\sigma_{yx}^{ex}(\omega) = -\frac{e^2}{h}2\pi\sum_{nl}l|M_x^{nl}|^2\frac{2m^2}{(\omega+i\gamma)^2 - E_{nl}^2}, \tag{13}$$

where the summation is performed over all exciton quantum numbers; γ is phenomenologically introduced exciton decay rate; $\mathbf{M}^{nl}$ is dimensionless matrix element that depends only on dimensional coupling strength α_c and is given by

$$\mathbf{M}^{nl} = \frac{\hbar}{m}\sum_{\mathbf{k}} C_{\mathbf{k}}^{nl}\langle k, v|\mathbf{J}|k, c\rangle, \tag{14}$$

where vector $\mathbf{J} = v_\mathrm{F}(\sigma_y, -\sigma_x)$ is the single-particle current operator. Value $\mathbf{M}^{nl}$ characterizes coupling strength between exciton and external electromagnetic field. It has nonzero value only for levels with $l = \pm1$ and all other states are optically inactive. Dependence of squared absolute value of matrix element $|M_x^{nl}|^2$ on dimensional coupling strength for states for four optical active states that has the lowest energy is presented in Fig.5. Matrix elements are decreasing with increasing of quantum number n due to destructive interference of interband transitions in (14) since wave functions $C_{\mathbf{k}}^{nl}$ for high energy levels oscillate in momentum space. Also optical activity of levels with angular momentum $l = -1$ is considerable less then $l = 1$.

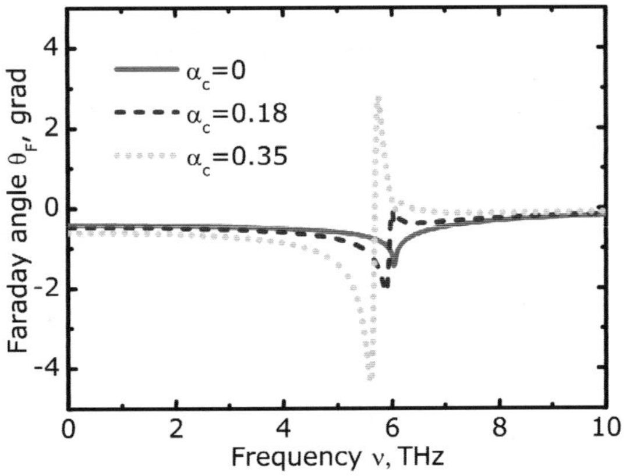

Figure 7. Frequency dependence of Faraday angle θ_F.

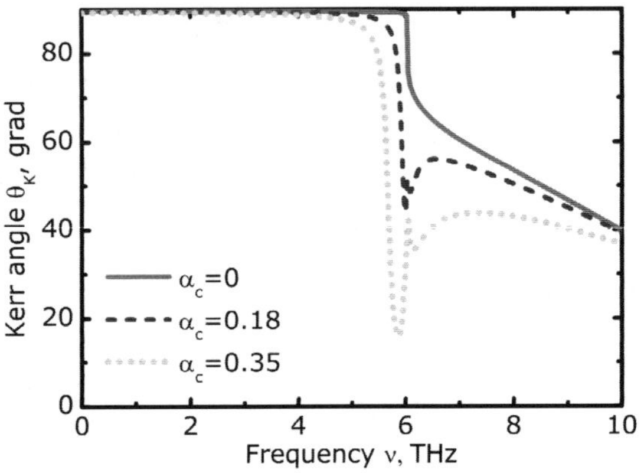

Figure 8. Frequency dependence of Kerr angle θ_K.

For all numerical calculations we used the following parameters: $m = 12.5$ meV, $\gamma = 0.25$ meV and $v_\mathrm{F} = 0.62 \times 10^6$ m/s. Also we used three values of dimensionless parameter α_c. Value $\alpha_\mathrm{c} = 0$ corresponds to the case of noninteractiong electrons on TI surface. Values $\alpha_\mathrm{c} = 0.18$ and $\alpha_c = 0.35$ correspond to values of effective permittivity $\epsilon = 20.5$ and $\epsilon = 10.5$, respectively (If the TI surface is subjected to the air with $\epsilon' = 1$ the corresponding values dielectric permittivity of TI are $\epsilon_\mathrm{TI} = 40$ and $\epsilon_\mathrm{TI} = 20$, respectively).

Contribution of the excitonic level $|n, l\rangle$ to Hall conductivity has the same sign as it orbital quantum number l. States with opposite orbital quantum numbers l and $-l$ are connected by the time reversal transformation. Due to the time reversal symmetry breaking by external exchange field chiral excitons on the surface of TI *do* resonantly contribute to optical Hall conductivity. Frequency dependence of real part of Hall conductivity is presented in Fig.6.

4.3. Magneto-optical effects in thin film of TI

We theoretically investigated magneto-optical effects at normal incidence in thin film of topological insulator which width is considerably less then wave length of electromagnetic field. The results can be easily generalized for more complicated geometry and oblique incidence.

Dependencies of Faraday angle θ_F and Kerr angle θ_K on frequency of incident electromagnetic wave are presented in Fig.7 and Fig.8 respectively. Values of Faraday and Kerr angles in low frequency limit where excitonic contribution is negligent tend to their universal values: $\tan\theta_F = \alpha$ and $\tan\theta_K = 1/\alpha$, where $\alpha \approx 1/137$ is fine structure constant. Resonance contribution of chiral excitons to Hall conductivity of the TI surface lead to resonant enhancement of Kerr effect. At resonant condition longitudinal part of the conductivity has sharp maximum that leads to resonant absorbtion of energy of electromagnetic field and resonant reduction of Kerr effect.

Conventional excitons in two- and three- dimensional systems in presence of time reversal symmetry do not contribute to Hall conductivity and, hence, do not manifests in magneto-optical effect in absence of magnetic field.

5. Summary

We reviewed properties of spin-plasmons in degenerate electronic gas populating the surface states and chiral excitons in presence of the magnetically induced gap in the surface spectrum. They both can be represented as composite bosons that are superposition of single-particle electron-hole transitions and can be effectively described within equation of motion based approach. But their physical properties are different. Spin-plasmons manifest themselves as coupled spin- and density- waves and their excitation also induces spin polarization of topological insulator that is perpendicular to their momentum. Chiral exciton that manifest itself as bound state of electron and hole and has unusual dependence of energy on its angular quantum number. Contrary to conventional ones chiral excitons resonantly contribute to Hall component of conductivity tensor. Chiral excitons resonantly enhance Faraday effect and resonantly reduce Kerr effect.

Acknowledgments
The work was supported by RFBR. D.K.E acknowledge support from Grant of the President of Russian Federation MK-5288.2011.2 and by scholarship from Dynasty Foundation.

[1] Hasan M Z, Kane C L 20101 *Rev. Mod. Phys*, **82**, 3045. Qi X L, Zhang S C 2010 *Rev. Mod. Phys.*, **83**, 1057.
[2] Chen Y L, Analytis J G, Chu Z H and et al. 2009 *Science*, **325**, 178. Hsieh D, Xia Y, Qian D et al. 2009 *Nature*, **460**, 1101. Xia Y, Wray L, Qian D 2009 *Nature Phys*, **5**, 398.
[3] Castro Neto A H, Guinea F, Peres N M R et al., 2009, *Rev. Mod. Phys.* **81**, 109.
[4] Burkov A A, Hawthorn D G 2010 *Phys. Rev. Lett.*, **105**, 066802.
[5] Culcer D, Hwang E H, Stanescu T D and et al. 2010, *Phys. Rev.* B, **82**, 155457.
[6] Modak S, Sengupta K and Sen D 2011 *arXiv:1203.4266.*
[7] Raghu S, Chung S B, Qi X L, Zhang S C 2010*Phys Rev Lett*, **104**, 116401.
[8] Efimkin D K, Lozovik Yu E, Sokolik A A 2012 *Nanoscale Research Letters*,**7**, 163.
[9] Appelbaum I, Drew H D, and Fuhrer M S 2011 *Appl. Phys. Lett. 98*, 023103.
[10] Nomura K, Koshino M, Ryu S 2007 *Phys. Rev. Lett.*, **99**, 146806.
[11] Chen Y L, Chu J H, Analytis J G et al. 2010, *Science* **329**, 659. Wray L A, Xu S-Y, Xia Y et al. 2011 *Nature Phys.* **7**, 32.
[12] Garate I, Franz M 2011, *Phys. Rev.* B **84**, 054403.
[13] Qi X L, Hughes T L, Zhang S C 2009 *Phys. Rev* .B, **78**, 195424. Essin A M, Moore J E, Vanderbilt D 2009 *Phys. Rev. Lett.* **102**, 146805.
[14] Tse W K and MacDonald A H 2010, *Phys. Rev.* B, **82**, 161104, Tse W K and MacDonald A H 2010 *Phys. Rev. Lett.*, **105**, 057401.
[15] Maciejko J, Qi X L, Drew H D et al., 2010 *Phys. Rev. Lett.*, **105**, 166803.
[16] Lan Y, Wan S, Zhang S C 2011 *Phys Rev.* B, **83**, 205109.
[17] Efimkin D K, Lozovik Yu E, in press.
[18] Sawada K, Brueckner K A, Fukuda N et al. 1957, *Phys. Rev.*, **108**, 507.
[19] Hwang E H, Das Sarma S 2007 *Phys. Rev.* B,**75**, 205418. Wunsch B, Stauber T, Sols F et al. 2006 . *New J Phys*, **8**, 318.

Dubna-Nano 2012 IOP Publishing
Journal of Physics: Conference Series **393** (2012) 012017 doi:10.1088/1742-6596/393/1/012017

Andreev Bound-states of vortices and surfaces in topological superconductors

Yuki Nagai, Hiroki Nakamura and Masahiko Machida
CCSE, Japan Atomic Energy Agency, 5-1-5 Kashiwanoha, Kashiwa, Chiba, 277-8587, Japan
E-mail: nagai.yuki@jaea.go.jp

Abstract. We propose an experimental way to determine the pairing states for the unconventional superconductor $Cu_xBi_2Se_3$. Using the spectral polynomial expansion method, we numerically calculate the local density of states (LDOS) in the multi-orbital tight-binding model and analytically obtain the zero-energy solution around a vortex. Both calculations reveal that a quantized vortex is spin-polarized for the gap function categorized by "pseudo scalar" which is the isotropic odd-parity order parameter. On the other hand, no clear spin-polarization around a vortex can observed for the gap function categorized by "polar vector" which breaks the rotational isotropy in the normal states.

1. Introduction

The topological insulators and superconductors have attracted much attention of theorists and experimentalists, because of their topologically protected nature of gapless surface states originated by non-trivial topology of the bulk wave function. So far, numerous works on topological insulators have been published [1, 2, 3, 4, 5, 6, 7, 8, 9, 10, 11], while topological superconductors have been not much studied due to relatively little variety.

Cu-intercalated Bi_2Se_3 ($Cu_xBi_2Se_3$), which is an electron-doped superconductor whose superconducting transition temperature $T_c \sim 0.3$K, has been regarded as a key compound to investigate the unconventional superconductivity with the non-trivial topology (topological superconductivity)[12, 13, 14, 15]. The mother compound Bi_2Se_3 is well-known as a 3-dimensional topological insulator by the non-trivial Z_2 topology. Recently, zero-bias conductance peaks (ZBCP's) were observed using the point contact spectroscopy[16, 17], which are well-known as a signature of unconventional superconductivity such as d-wave, sign-reversing s-wave and p-wave Cooper pairs[18, 19, 20]. The observed ZBCP is generally caused by the Andreev bound states formed at surfaces or in vortex cores where the phase variation of the order parameters occurs. On the other hand, it is also known that the ZBCPs are observable owing to the topological origin in the chiral p-wave superconductor such as Sr_2RuO_4, and its topologically-protected gapless states are regarded as Majorana fermions[20]. Then, the ZBCP observed in $Cu_xBi_2Se_3$ may be also a signature of a topological superconductivity. Therefore, it is important to obtain a clear evidence of the topological superconductivity.

In the superconductor, $Cu_xBi_2Se_3$, Sasaki *et al.* theoretically demonstrated how the Majorana fermion brings about ZBCP's [16]. Their theoretical calculations have revealed that the odd-parity spin-triplet pairing is the most-likely pairing to explain the observed ZBCP's. However, they could not select only a possible pairing among four possible types of odd-parity pairing in $Cu_xBi_2Se_3$, since the ZBCP can occur even in other pairing states responding to a variation of parameters in the original tight-binding model. Therefore, it is hard to identify the paring symmetry only through ZBCP, i.e. the energy dependence of the density of states.

In this paper, we propose two methods to determine the pairing state in $Cu_xBi_2Se_3$. In the first one, the strong anisotropic thermal conductivity can be employed as an evidence of the spin-polarized Cooper pairs referred as the gap function Δ_4[25], one of four possible odd-parity parings. In the second one, the spin-polarized vortex-core bound-states can be done as that of the isotropic spin-triplet Cooper pairs referred as the gap function Δ_2. Thus, there are two ways to distinguish between the gap function Δ_2 and Δ_4. We also discuss that some of the gap functions break the rotational isotropy in the normal states, regardless of the material parameters, in terms of the massive Dirac formulation. The structure of this paper is as follows. In Sec. 2, we discuss general properties of the tight-binding model Hamiltonian equivalent with the massive Dirac Hamiltonian which is a low-energy effective model and show that the gap function Δ_4 usually has point-nodes. In Sec. 3, two theoretical treatments to examine a single vortex numerically and analytically are presented. Section 4 shows details of the numerical calculation results. In Sec. 5, it is demonstrated with the solution of the zero-energy and the spin-polarized vortex usually occurs for the gap function Δ_4. Section 6 gives the conclusions.

2. General Background

An effective theory for the topological superconductor $Cu_xBi_2Se_3$ can be described by the Nambu representation of the two-orbital model with spin-orbit coupling, whose form is expressed by the massive Dirac Hamiltonian[25]:

$$H = \int d\boldsymbol{r} \begin{pmatrix} \bar{\psi}(\boldsymbol{r}) & \bar{\psi}_c(\boldsymbol{r}) \end{pmatrix} \begin{pmatrix} \hat{H}^-_{\text{eff}}(\boldsymbol{r}) & \Delta^-(\boldsymbol{r}) \\ \Delta^+(\boldsymbol{r}) & \hat{H}^+_{\text{eff}}(\boldsymbol{r}) \end{pmatrix} \begin{pmatrix} \psi(\boldsymbol{r}) \\ \psi_c(\boldsymbol{r}) \end{pmatrix}, \tag{1}$$

where

$$\hat{H}^{\pm}_{\text{eff}}(\boldsymbol{r}) = M(-i\boldsymbol{\nabla}) + A_x(-i\boldsymbol{\nabla})\gamma^1 + A_y(-i\boldsymbol{\nabla})\gamma^2 + A_z(-i\boldsymbol{\nabla})\gamma^3 \pm \epsilon(-i\boldsymbol{\nabla})\gamma^0. \tag{2}$$

Here, γ^i is 4×4 Dirac gamma matrix. The gamma matrices in the Dirac representation is expressed as $\gamma^0 = \hat{\sigma}_z \otimes 1$, $\gamma^{i=1,2,3} = i\hat{\sigma}_y \otimes \hat{s}_i$ and $\gamma^5 = \hat{\sigma}_x \otimes 1$ with 2×2 Pauli matrices $\hat{\sigma}_i$ in the orbital space and $\hat{s}_i$ in the spin space, $\psi(\boldsymbol{r})$ is the Dirac spinor, $\bar{\psi}(x) \equiv \psi^\dagger(x)\gamma^0$, $\bar{\psi}_c(x) \equiv \psi_c^\dagger\gamma^0$, and $\psi_c \equiv \mathcal{C}\bar{\psi}^T$, where $\mathcal{C}(\equiv i\gamma^2\gamma^0)$ is a representative matrix of charge-conjugation. The gap functions are defined as $\Delta^- \equiv \gamma^0\hat{\Delta}_i i\gamma^2$ and $\Delta^+ \equiv \gamma^0(\Delta^-)^\dagger\gamma^0$. M, A_ν and ϵ are the parameters depending on the material. It is noted that the order parameter Δ^- is determined by the on-site pairing interaction. According to Table I in Ref. [25], possible six gap functions are characterized by a pseudo-scalar, scalar, and polar vector (four-vector):

$$\Delta^- = \Delta_0,\ \Delta_0\gamma^5,\ \Delta_0\not{\alpha}\gamma^5, \tag{3}$$

where, Δ_0 is a scalar parameter, the Feynman slash $\not{\alpha}$ is defined by $\sum_\mu \gamma^\mu\alpha_\mu$, and the gap function is characterized by the four-vector α_μ. In the superconducting state, the gap functions represented by α_μ with finite α_i-components ($i = 1, 2, 3$) break the rotational symmetry and bring about the anisotropic electronic state, even if the normal state is isotropic. This induced anisotropy results in the angle-dependent thermal conductivity. Thus, the anisotropic thermal conductivity can be a clear evidence of the gap function represented by the four-vector α_μ [25].

Next, we discuss point-nodes of the gap function Δ_4. Here, we define the gap function Δ_{4a} with the four-vector $\alpha_\mu = (0, 0, 1, 0)$ ($\Delta^{12}_{\uparrow\uparrow} = \Delta^{12}_{\downarrow\downarrow} = -\Delta^{21}_{\uparrow\uparrow} = -\Delta^{21}_{\downarrow\downarrow}$). In this case, the gap function has the rotational symmetry around y-axis, and therefore the point-nodes can exist only on the k_y-axis (See, Fig. 1). In order to find the point-nodes, we must solve the above

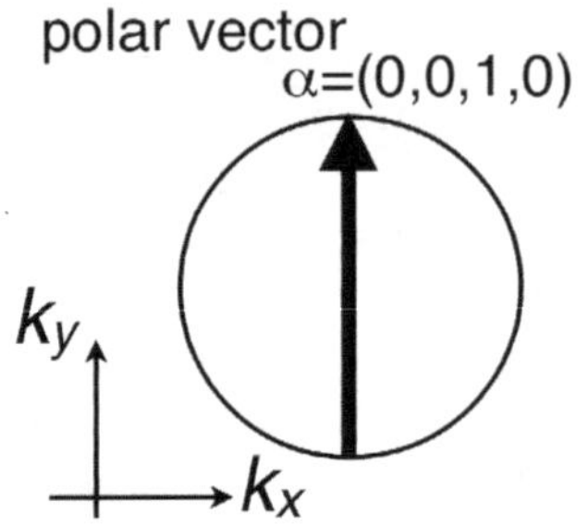

Figure 1. Schematic figure of the polar vector for the gap function Δ_{4a}.

Hamiltonian. The eigenvalues of the Hamiltonian can be obtained through the BdG equations described as

$$\begin{pmatrix} \gamma^0 \hat{H}_{\rm eff}^-(\boldsymbol{r}) & \gamma^0 \Delta^-(\boldsymbol{r}) \\ \gamma^0 \Delta^+(\boldsymbol{r}) & \gamma^0 \hat{H}_{\rm eff}^+(\boldsymbol{r}) \end{pmatrix} \begin{pmatrix} u(\boldsymbol{r}) \\ u_{\rm c}(\boldsymbol{r}) \end{pmatrix} = E \begin{pmatrix} u(\boldsymbol{r}) \\ u_{\rm c}(\boldsymbol{r}) \end{pmatrix}. \quad (4)$$

With the use of the relation $\lim_{k_x \to 0} A_x(\boldsymbol{k}) = \lim_{k_z \to 0} A_z(\boldsymbol{k}) = 0$ in this model, one can obtain the condition that the point-nodes exist:

$$A_y(k_x = 0, k_y, k_z = 0)^2 = \Delta_0^2 + \epsilon(k_x = 0, k_y, k_z = 0)^2 - M(k_x = 0, k_y, k_z = 0)^2. \quad (5)$$

In the normal state, the Fermi surface is determined by $\boldsymbol{A}(\mathrm{k})^2 = \epsilon(\boldsymbol{k})^2 - M(\boldsymbol{k})^2$, because of the eigenvalues $E = \sqrt{\boldsymbol{A}(\mathrm{k})^2 + M(\boldsymbol{k})^2} - \epsilon(\boldsymbol{k})$. Therefore, in the case of the small order-parameter $\Delta_0 \ll M, \epsilon$, the point-nodes usually exist near the Fermi level. As the above discussion, the gap functions represented by a polar-vector such as Δ_{4a} can be identified experimentally in terms of the anisotropy of the physical quantities such as an angle-dependent thermal transport measurements[25]. One can conclude that the gap function is Δ_4 if the thermal conductivity becomes anisotropic below T_c. On the other hand, we need to have another ways to determine the pairing function if the measured thermal conductivity is isotropic below T_c. Therefore, in this paper, we show that one can conclude that the gap function is Δ_2 if spin-polarized vortices are observed.

3. Model and Method

In this section, we introduce the models for both of the numerical and analytical calculations to determine the gap function Δ_2 represented as a pseudo-scalar.

In the numerical calculation, we use the mean-field tight-binding Hamiltonian in the Bogoliubov-de Gennes scheme,

$$H = \sum_{k_z} \sum_{i,j} \begin{pmatrix} c_i^\dagger & c_i^T \end{pmatrix} \begin{pmatrix} \hat{H}_{ij}(k_z) & \hat{\Delta} f(\boldsymbol{R}_i) \delta_{ij} \\ \hat{\Delta}^\dagger \delta_{ij} f(\boldsymbol{R}_i)^* & -\hat{H}_{ij}^*(-k_z) \end{pmatrix} \begin{pmatrix} c_j \\ c_j^* \end{pmatrix}, \quad (6)$$

where $c_i^\dagger$ is the 4-component creation operator at the i-th site on the two-dimensional triangle lattice and k_z denotes the momentum in the crystal c-axis. $\hat{\Delta}$ is 4×4 matrix whose elements are given as $\Delta_{\sigma\sigma'}^{lm}$ with orbital $l(m)$ and spin $\sigma(\sigma')$ indices. The normal state matrix $\hat{H}_{ij}(k_z)$ is given by

$$\hat{H}_{ij}(k_z) = \int d\boldsymbol{k}_\perp e^{i\boldsymbol{k}_\perp \cdot (\boldsymbol{R}_i - \boldsymbol{R}_j)} \hat{H}(\boldsymbol{k}_\perp, k_z), \quad (7)$$

with ab-plane momentum $\boldsymbol{k}_\perp$. The 4×4 matrix $\hat{H}(\boldsymbol{k}_\perp, k_z)$ is given as

$$\hat{H}(\boldsymbol{k}_\perp, k_z) = M(\boldsymbol{k}_\perp, k_z)\gamma^0 + \gamma^0 A_x(\boldsymbol{k}_\perp)\gamma^1 + \gamma^0 A_y(\boldsymbol{k}_\perp)\gamma^2 + \gamma^0 A_z(k_z)\gamma^3 + \epsilon(\boldsymbol{k}_\perp, k_z), \tag{8}$$

where, $M(\boldsymbol{k}_\perp, k_z) \equiv M_0 - 2\bar{B}_1(1-\cos(k_z)) - \bar{B}_2\eta(\boldsymbol{k}_\perp)$, $A_x(\boldsymbol{k}_\perp) \equiv (2/3)\bar{A}_2\sqrt{3}\sin(\sqrt{3}k_x/2)\cos(k_y/2)$, $A_y(\boldsymbol{k}_\perp) \equiv (2/3)\bar{A}_2(\cos(\sqrt{3}k_x/2)\sin(k_y/2) + \sin(k_y))$, $A_z(k_z) \equiv \bar{A}_1\sin(k_z)$, $\epsilon(\boldsymbol{k}_\perp, k_z) \equiv 2\bar{D}_1(1 - \cos(k_z)) + \bar{D}_2\eta(\boldsymbol{k}_\perp) - \mu$, and $\eta(\boldsymbol{k}_\perp) \equiv (3 - 2\cos(\sqrt{3}k_x/2)\cos(k_y/2) - \cos(k_y))$. We set $M_0 = 0.28$ eV, $\bar{A}_1 = 0.32$ eV, $\bar{A}_2 = 4.1/a$ eV, $\bar{B}_1 = 0.216$ eV, $\bar{B}_2 = 56.6/a^2$ eV, $\bar{D}_1 = 0.024$ eV, $\bar{D}_2 = 19.6/a^2$ and $a = 4.076$ Å as the material parameters for $Cu_xBi_2Si_3$[16]. For simplicity, we do not solve the gap-equation but use a spatial distribution form of the order parameter around a single vortex $f(\boldsymbol{R}_i)$ written as

$$f(\boldsymbol{R}_i) = e^{i\theta}\Delta_0 \frac{|\boldsymbol{R}_i|}{\sqrt{|\boldsymbol{R}_i|^2 + \xi^2}}, \tag{9}$$

where θ denotes the polar angle around c-axis, Δ_0 is the amplitude of the order-parameter and ξ is the coherence length. We solve this $8N \times 8N$ matrix BdG equations without self-consistent calculations. Here, N is the total number of lattice sites. To access large scale systems, we adopt the polynomial expansion scheme to calculate the local density of states (LDOS)$n(\omega, i)$[21, 22, 23]. In this scheme with the Chebyshev polynomials, LDOS $n(\omega, i)$ is expressed as

$$n(\omega, i) \sim \frac{1}{a}\sum_{n=0}^{n_c} \frac{\cos(n\arccos(\tilde{\omega}))}{\sqrt{1-\tilde{\omega}^2}} \frac{2g_n^{\mathrm{L}}}{\pi(1+\delta_{n,0})} \boldsymbol{e}(i)^T \boldsymbol{e}_n(i), \tag{10}$$

where $2N$-component unit-vector $\boldsymbol{e}(i)$ is determined by $[\boldsymbol{e}(i)]_\gamma = \delta_{i,\gamma}$ and $\boldsymbol{e}_n(i)$ is recursively generated by

$$\boldsymbol{e}_{n+1}(i) = 2\hat{K}\boldsymbol{e}_n(i) - \boldsymbol{e}_{n-1}(i), \ (n > 1), \tag{11}$$

$$\boldsymbol{e}_1(i) = \hat{K}\boldsymbol{e}(i), \quad \boldsymbol{e}_0(i) = \boldsymbol{e}(i). \tag{12}$$

Here, $g_n^{\mathrm{L}} = \sinh[n_c\eta(1 - n/n_c)]/\sinh[n_c\eta]$ is a coefficient in the Lorentz kernel[23, 24] and $\tilde{\omega} = (\omega - b)/a$ and $\hat{K} = (\hat{H} - b\hat{I})/a$ denotes renormalized energy and Hamiltonian matrix, respectively. n_c is a cut-off parameter, η denotes a smearing factor, and a and b are renormalization factors.

In the analytical calculation, we adopt the Dirac Hamiltonian which is the low-energy linearized model of the Hamiltonian (2):

$$\hat{H}^{\pm}_{\mathrm{eff}}(\boldsymbol{r}) = M_0 - i\partial_x\gamma^1 - i\partial_y\gamma^2 - i\partial_z\gamma^3 \pm \mu\gamma^0, \tag{13}$$

with the renormalized axes $(x, y, z) \to (\bar{A}_2 x, \bar{A}_2 y, \bar{A}_1 z)$. In this case, the order-parameter is determined by $\Delta^-(\boldsymbol{r}) \to \Delta^- f(\boldsymbol{r})$.

In the numerical and analytical calculations, we consider only a gap function $\hat{\Delta}_2 = \gamma^2\gamma^0$, since this does not break the rotational symmetry so that the solution can be obtained analytically.

4. Numerical Results

We consider $L_x \times L_y$ triangle lattice sites with the closed boundary. The single vortex is located at $(i_x, i_y) = (L_x/2, L_y/2)$. We adopt $L_x = 96$ and $L_y = 96$, where the matrix dimension size is 73728. To obtain the LDOS $n(\omega, i)$, the 40 k_z-points meshes are considered. We consider the gap-amplitude $\Delta_0 = 0.3$ eV and the chemical potential $\mu = 0.4$ eV for convenience of calculation. We use $a = 30$ eV, $b = -\mu$, $\eta = 1 \times 10^{-3}$ eV and $n_c = 8000$ as the parameters in the polynomial

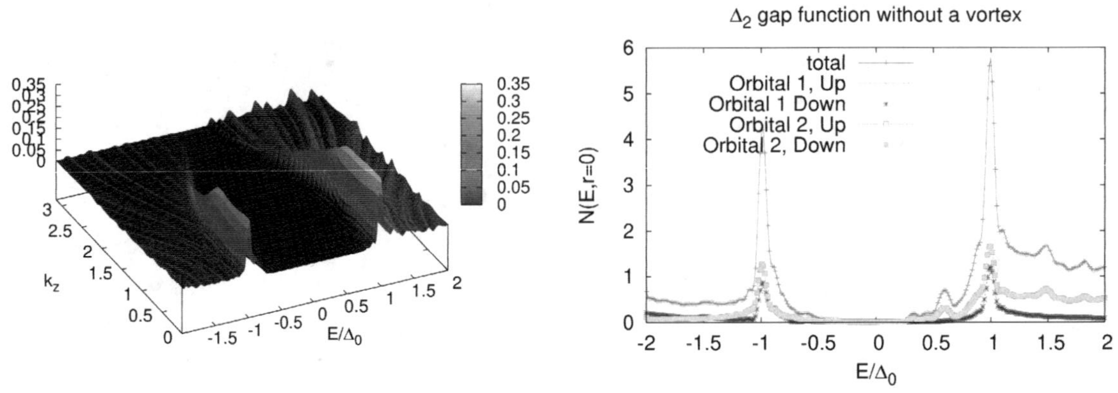

Figure 2. The energy dependence of the LDOS without a vortex in the superconductor with the gap function Δ_2 at the center $(i_x, i_y) = (L_x/2, L_y/2)$. (Left panel) the k_z-resolved LDOS. (Right panel) the total and partial LDOS.

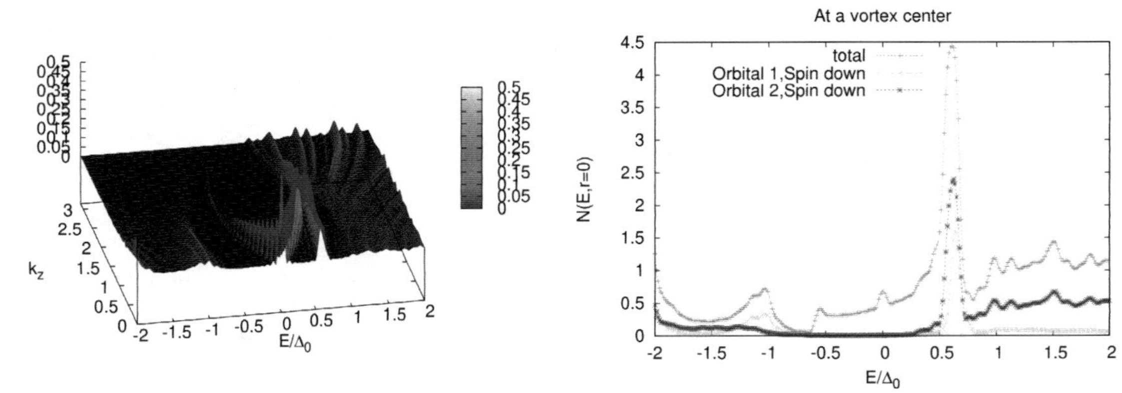

Figure 3. The energy dependence of the LDOS with a vortex in the superconductor with the gap function Δ_2 at the center $(i_x, i_y) = (L_x/2, L_y/2)$. (Left panel) the k_z-resolved LDOS. (Right panel) the total and partial LDOS.

expansion scheme. Without a vortex ($f(\boldsymbol{R}_i) = \Delta_0$), the gap function $\hat{\Delta}_2$ is fully-gapped, and the spin-degeneracies of each orbitals are found as shown in Fig. 2. With a vortex, one can find the zero-energy bound states at a vortex core. We find two bound states in the k_z-resolved LDOS as shown in Fig. 3. One has the linear dispersion relation around the zero-energy. Other has the flat dispersion relation at the mid-gap energy ($E/\Delta_0 \sim 0.5$). We find that the zero-energy bound states consist of only up-spin quasiparticles and the mid-gap bound states consist of down-spin quasiparticles at a vortex core. This result means that the vortex for the gap function Δ_2 is spin-polarized. Far from a vortex, the intensity of the down-spin components rises as shown in Fig. 4. Thus, the partial LDOS has the oscillating function along the radial axis. We can explain the origin of this spin-polarized vortex with the use of the following analytical calculation.

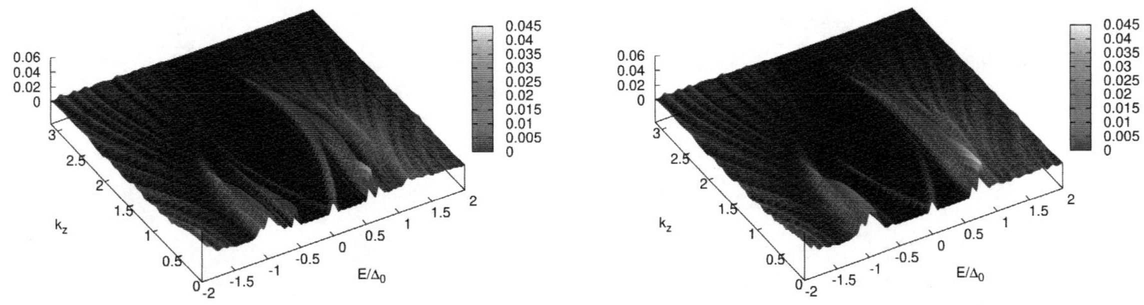

Figure 4. The energy dependence of the k_z-resolved partial LDOS with a vortex in the superconductor with the gap function Δ_2 far from a vortex core $(i_x, i_y) = (L_x/2 + 7, L_y/2)$. (Left panel) the partial LDOS for the orbital 1 and up-spin component. (Right panel) the partial LDOS for the orbital 1 and down-spin component.

5. Analytical Results

In the case of the gap function $\hat{\Delta}_2 = \gamma^2\gamma^0$, we can obtain two zero-energy solutions[27] expressed as

$$u_\uparrow(\boldsymbol{r}) = \frac{e^{-\int_0^r |f(r')|dr'}}{\sqrt{\lambda_\uparrow}} \begin{pmatrix} \sqrt{\mu + M_0} J_0(\bar{r}) \\ 0 \\ 0 \\ ie^{i\theta}\sqrt{\mu - M_0} J_1(\bar{r}) \end{pmatrix}, \quad u_{c\uparrow}(\boldsymbol{r}) = \frac{e^{-\int_0^r |f(r')|dr'}}{\sqrt{\lambda_\uparrow}} \begin{pmatrix} e^{-i\theta}\sqrt{\mu - M_0} J_1(\bar{r}) \\ 0 \\ 0 \\ -i\sqrt{\mu + M_0} J_0(\bar{r}) \end{pmatrix}, \tag{14}$$

$$u_\downarrow(\boldsymbol{r}) = \frac{e^{-\int_0^r |f(r')|dr'}}{\sqrt{\lambda_\downarrow}} \begin{pmatrix} 0 \\ e^{i\theta}\sqrt{\mu + M_0} J_1(\bar{r}) \\ -i\sqrt{\mu - M_0} J_0(\bar{r}) \\ 0 \end{pmatrix}, \quad u_{c\downarrow}(\boldsymbol{r}) = \frac{e^{-\int_0^r |f(r')|dr'}}{\sqrt{\lambda_\downarrow}} \begin{pmatrix} 0 \\ -\sqrt{\mu - M_0} J_0(\bar{r}) \\ -ie^{-i\theta}\sqrt{\mu + M_0} J_1(\bar{r}) \\ 0 \end{pmatrix}, \tag{15}$$

with

$$\lambda_\uparrow = 4\pi \int_0^\infty drr \left[(\mu + M_0) J_0(\bar{r})^2 + (\mu - M_0) J_1(\bar{r})^2 \right] e^{-2\int_0^r |f(r')|dr'}, \tag{16}$$

$$\lambda_\downarrow = 4\pi \int_0^\infty drr \left[(\mu - M_0) J_0(\bar{r})^2 + (\mu + M_0) J_1(\bar{r})^2 \right] e^{-2\int_0^r |f(r')|dr'}. \tag{17}$$

Assuming the solution $\psi(\boldsymbol{r})^T = c_\uparrow(u_\uparrow(\boldsymbol{r}), u_{c\uparrow}(\boldsymbol{r})) + c_\downarrow(u_\downarrow(\boldsymbol{r}), u_{c\downarrow}(\boldsymbol{r}))$, the eigenvalue equations are written as

$$\begin{pmatrix} 0 & ivk_z \\ -ivk_z & 0 \end{pmatrix} \begin{pmatrix} c_\uparrow \\ c_\downarrow \end{pmatrix} = E \begin{pmatrix} c_\uparrow \\ c_\downarrow \end{pmatrix}, \tag{18}$$

with

$$v \equiv \frac{2\pi}{\sqrt{\lambda_\uparrow \lambda_\downarrow}} \int_0^\infty drre^{-2\int_0^r |f(r')|dr'} \sqrt{\mu^2 - M_0^2} \left(J_0(\bar{r})^2 - J_1(\bar{r})^2 \right). \tag{19}$$

Therefore, the dispersion relations are given by

$$E = \pm v k_z, \tag{20}$$

and the coefficients $(c_\uparrow, c_\downarrow) = (1, \pm i)/\sqrt{2}$ do not depend on k_z. We define spin-resolved LDOS expressed as $n_\uparrow(E = \pm v k_z, r) \equiv |\psi_1|^2 + |\psi_3|^2$ and $n_\downarrow(E = \pm v k_z, r) \equiv |\psi_2|^2 + |\psi_4|^2$. The spin-resolved LDOS $n_\uparrow$ and $n_\downarrow$ are written as

$$n_\uparrow(E = \pm v k_z, r) \equiv \left(\frac{\mu + M_0}{2\lambda_\uparrow} + \frac{\mu - M_0}{2\lambda_\downarrow} \right) J_0(\bar{r})^2 e^{-2\int_0^r |f(r')| dr'}, \tag{21}$$

$$n_\downarrow(E = \pm v k_z, r) \equiv \left(\frac{\mu - M_0}{2\lambda_\uparrow} + \frac{\mu + M_0}{2\lambda_\downarrow} \right) J_1(\bar{r})^2 e^{-2\int_0^r |f(r')| dr'}. \tag{22}$$

These equations show that $n_\uparrow$ is localized around a vortex core, since the Bessel function J_1 becomes zero at the origin ($J_1(0) = 0$). The numerical results can be reproduced by the above equations.

Let us discuss the reason of the spin-polarization. Equations (14) and (15) mean that the angular momentum is a good quantum number, since the Bessel functions are the eigenstates of the angular momentum. The rotational symmetry does not break in a superconducting phase, since the gap function Δ_2 is a "pseudo-scalar" gap function. It should be noted that $i\partial_x\gamma_1 + i\partial_y\gamma_2$ includes a raising and lowering operators $\partial_x \pm i\partial_y$ in the normal-state Hamiltonian (13). Thus, the up-spin and down-spin components have the different angular momentum. If the zero-energy states exist, a certain components of the wave function is proportional to the Bessel function $J_0(r)$, since the zero-angular momentum state is the lowest energy state in the eigenstates of the angular momentum. The component with the zero-angular momentum depends on the direction of the vortex. On the other hand, for the gap function Δ_{4a}, the angular momentum is not a good quantum number, since the polar-vector breaks the rotational symmetry around z-axis. Thus, the wave-functions of zero-energy bound states consist of the linear combination of the eigenstates of the angular momentum, and then the vortex is not spin-polarized for the gap function Δ_{4a}. We conclude that a vortex of the gap function Δ_2 is a spin-polarized vortex in the wide parameter range, as shown in the numerical and analytical calculations.

6. Conclusion

In conclusion, we proposed how to identify the pairing state for $Cu_xBi_2Se_3$. We numerically and analytically calculated the LDOS with a single vortex for the gap function Δ_2 which has the isotropic odd-parity spin-triplet Cooper pairs. We adopted the tight-binding model for the numerical calculation and the massive Dirac model for the analytical calculation. In the numerical calculation, we found two bound states in the k_z-resolved LDOS as shown in Fig. 3. One has the linear dispersion relation around the zero-energy with the up-spin dominant quasiparticles. The other has the flat dispersion relation at the mid-gap energy ($E/\Delta_0 \sim 0.5$) with the down-spin dominant quasiparticles. In the analytical calculation, we obtained two degenerated solutions at the zero-energy around a vortex core. The spin-resolved LDOS shows that there is a spin-polarized vortex for the gap function Δ_2. We showed that the solutions around a vortex for this gap function are the eigenstates of the angular momentum but that for the gap function Δ_{4a} can not be the eigenstates of the angular momentum because of the order parameter represented by the polar-vector. According to the numerical and analytical calculations, we concluded that a vortex of the gap function Δ_2 is a spin-polarized vortex in the wide parameter range.

References

[1] Kane C L and Male E J 2005 *Phys. Rev. Lett.* **95** 146802

[2] Hasan M Z and Kane C L 2010 *Rev. Mod. Phys.* **82** 3045, and references therein
[3] Bernevig B A, Hughes T L and Zhang S C 2006 *Science* **314** 1757.
[4] Fu L, Kane C L and Mele E J 2007 *Phys. Rev. Lett.* **98** 106803
[5] Moore J E and Balents L 2007 *Phys. Rev.* B **75** 121306(R)
[6] König M, Wiedmann S, Brüne C, Roth A, Buhmann H, Molenkamp L W, Qi X L and Zhang S C, 2007 *Science* **318** 766
[7] Fu L and Kane C L 2007 *Phys. Rev.* B **76** 045302
[8] Nishide A, Taskin A A, Takeichi Y, Okuda T, Kakizaki A, Hirahara T, Nakatsuji K, Komori F, Ando Y and Matsuda I 2010 *Phys. Rev.* B **81** 041309(R)
[9] Sato T, Segawa K, Guo H, Sugawara K, Souma S, Takahashi T and Ando Y 2010 *Phys. Rev. Lett.* **105** 136802
[10] Kuroda K, Ye M, Kimura A, Eremeev S V, Krasovskii E E, Chulkov E V, Ueda Y, Miyamoto K, Okuda T, Shimada K, Namatame H and Taniguchi M 2010 *Phys. Rev. Lett.* **105** 146801
[11] Chen Y L, Liu Z K, Analytis J G, Chu J H, Zhang H J, Yan B H, Mo S K, Moore R G, Lu D H, Fisher I R, Zhang S C, Hussain Z and Shen Z X 2010 *Phys. Rev. Lett.* **105** 266401 (2010)
[12] Hor Y S, Williams A J, Checkelsky J G, Roushan P, Seo J, Xu Q, Zandbergen H W, Yazdani A, Ong N P and Cava R J 2010 *Phys. Rev. Lett.* **104** 057001
[13] Wray L A, Xu S Y, Xia Y, Hor Y S, Qian D, Fedorov A V, Lin H, Bansil A, Cava R J and Hasan M Z 2010 *Nature Phys.* **6** 855
[14] Kriener M, Segawa K, Ren Z, Sasaki S and Ando Y 2011 *Phys. Rev. Lett.* **106** 127004
[15] Fu L and Berg E 2010 *Phys. Rev. Lett.* **105** 097001
[16] Sasaki S, Kriener M, Segawa K, Yada K, Tanaka Y, Sato M and Ando Y 2011 *Phys. Rev. Lett.* **107** 217001
[17] Kirzhner T, Lahoud E, Chaska K B, Salman Z, Kanigel A 2012 *Phys. Rev.* B **86** 064517
[18] Tanaka T and Kashiwaya S 1995 *Phys. Rev. Lett.* **74** 3451
[19] Nagai Y and Hayashi N 2009 *Phys. Rev.* B **79** 224508
[20] Sigrist M, Furusaki A, Honerkamp C, Matsumoto M, Ng K K and Okuno Y 2000 *J. Phys. Soc. Jpn.* **89** Suppl. B 127
[21] Nagai Y, Ota Y and Machida M 2012 *J. Phys. Soc. Jpn.* **81** 024710
[22] Nagai Y, Nakai N and Machida M, 2012 *Phys. Rev.* B **85** 092505
[23] Covaci L, Peeters F M, and Berciu M 2010 *Phys. Rev. Lett.* **105**167006
[24] Weisse A, Wellein G, Alvermann A and Fehske H 2006 *Rev. Mod. Phys.* **78** 275
[25] Nagai Y, Nakamura H, Machida M 2012 *Phys. Rev.* B **86** 094507
[26] Nishida Y 2010 *Phys. Rev.* D **81** 074004
[27] Nagai Y, Nakamura H and Machida M, in preparation

Dubna-Nano 2012 IOP Publishing
Journal of Physics: Conference Series **393** (2012) 012018 doi:10.1088/1742-6596/393/1/012018

Chiral Majorana fermion edge states in a heterostructure of superconductor and semiconductor with spin-orbit coupling

Long-Hua Wu[1,2], Qi-Feng Liang[1], Zhi Wang[1] and Xiao Hu[1,2]

[1] International Center for Materials Nanoarchitectonics (WPI-MANA), National Institute for Materials Science, Tsukuba 305-0044, Japan

[2] Graduate School of Pure and Applied Sciences, University of Tsukuba, Tsukuba 305-8571, Japan

E-mail: Hu.Xiao@nims.go.jp

Abstract. We study low-energy excitations in a device, consisting of spin-orbit coupling semiconductor, *s*-wave superconductor and ferromagnetic insulator. A vortex in the device generates zero-energy Majorana Fermions at both the edge of a finite sample and the vortex core. In this work, we identify the chirality of edge excitation states, and investigate how it is affected by Zeeman field and spin-orbit coupling. We prove the commutation relation of total angular momentum with Bogoliubov-de Gennes (BdG) Hamiltonian. With the help of the relation, we get analytically the angular part of the solution to the BdG equation for a circular sample. We perform numerical simulations to evaluate the radial part of wave-function, and derive distribution of Majorana fermion state and energy dispersion for low-energy excitations.

1. Introduction

Majorana Fermions (MFs) are particles with an intriguing property that they are their own anti-particles [1]. It is predicted theoretically that MFs obey non-Abelian statistics [2], which have potential applications in constructing topological quantum computers [3]. With such a promising future, MFs are searched among numerous systems. However, there is no firm evidence that MFs can be elementary particles up until now. It's more likely to have them as quasi-particle excitations in condensed matter systems. Many potential MF hosts have been proposed, including spinless *p*-wave superconductors [4], Pfaffian $\nu = 5/2$ fractional quantum Hall state [5] and *p*-wave superfluid of cold atoms [6]. Recently, there is a surge of discussions on realizing MFs in heterostructures, such as the surface state of a topological insulator with proximity-induced *s*-wave superconductivity [7], spin-orbit coupled semiconductor sandwiched by *s*-wave superconductor and ferromagnetic insulator (S/SM/FI) [8]. In these devices, the *p*-wave pairing is superseded by the interplay between proximity-induced *s*-wave superconductivity and spin-orbit coupling. One advantage that they offer is a relatively easier experimental accessibility.

In a finite S/SM/FI device, it has been shown that MFs appear as zero-energy excitations in the presence of vortex, one at the sample edge while the other bounded by vortex core [10]. Although there are numerous theoretical studies on the existence of MFs, not much attention has been paid to detailed properties of the MF at the edge, which is capable of being manipulated efficiently and follows non-Abelian statistics for braiding process [11]. In this sense, more complete understanding of the MFs is indispensable before one can make use of them. In the present work, we study the chirality and the spin direction of the MF at the edge and their dependence on Zeeman field and spin-orbit coupling. The total angular momentum for quasi-particles is proved to be a conserved quantity. Based on this, we get analytically the angular part of the solution to the Bogoliubov-de Gennes (BdG) equation. We solve the radial part numerically for several low-energy excitation states in a finite circular sample. With the full numerical wave-functions, we are able to study the distribution of MF state and the energy dispersion for low-energy excitations.

2. Bogoliubov-de Gennes equation and solution

We start from the Hamiltonian of a semiconductor with sizable Rashba-type spin-orbit coupling in proximity to a ferromagnetic insulator

$$H_0 = \int d\vec{r} c^\dagger(\vec{r}) \left[\frac{\vec{p}^2}{2m^*} - \mu + \alpha_R(\vec{\sigma} \times \vec{p}) \cdot \hat{z} + V_z \hat{\sigma}_z \right] c(\vec{r}), \quad (1)$$

where m^*, μ, α_R and V_z are effective electron mass, chemical potential, Rashba spin-orbit coupling strength and Zeeman field respectively, $c = (\hat{c}_\uparrow \; \hat{c}_\downarrow)^T$ is electron annihilation operator, $\vec{\sigma} = (\hat{\sigma}_x \; \hat{\sigma}_y \; \hat{\sigma}_z)$ are Pauli matrices.

The proximity-induced s-wave superconductivity is described by

$$H_{SC} = \int d\vec{r} \left[\Delta(\vec{r}) \hat{c}_\uparrow^\dagger \hat{c}_\downarrow^\dagger + h.c. \right], \quad (2)$$

where $\Delta(\vec{r})$ is the effective s-wave pairing potential. Therefore, the total Hamiltonian is $H = H_0 + H_{SC}$. The system behaves effectively as a 2D spinless $p_x + ip_y$ superconductor for $\sqrt{\Delta^2 + \mu^2} < V_z$ [8, 9], which gives rise to MFs.

With respect to H, we can write down the BdG equation

$$\begin{pmatrix} H_0 & \Delta \\ \Delta^\dagger & -\hat{\sigma}_y H_0^* \hat{\sigma}_y \end{pmatrix} \Psi(\vec{r}) = E \Psi(\vec{r}), \quad (3)$$

with the Nambu spinor $\Psi(\vec{r}) = [u_\uparrow(\vec{r}), u_\downarrow(\vec{r}), v_\downarrow(\vec{r}), -v_\uparrow(\vec{r})]^T$. Provided Ψ is an eigenstate of the BdG equation with eigenenergy E, we always have another eigenstate $\Xi\Psi$ with eigenenergy $-E$, where Ξ is the particle-hole transformation operator.

In the presence of n vortices, the induced s-wave pairing potential is $\Delta = \tilde{\Delta}_0(\vec{r}) e^{in\theta}$. The matrix form of the total Hamiltonian H in polar coordinates is given by

$$H = \left(-\frac{\hbar^2}{2m^*} \nabla^2 - \mu \right) \tau_z + V_z \sigma_z + i\frac{\alpha}{2} (\sigma_+ p_- - \sigma_- p_+) \tau_z + \tilde{\Delta}_0 \left[\cos(n\theta) \tau_x + \sin(n\theta) \tau_y \right], \quad (4)$$

where τ_x, τ_y, τ_z are 4×4 unitary matrices in the particle space with their explicit representations given by Eq.(A.3) in Appendix, and $\sigma_\pm$ and $p_\pm$ are defined as $\sigma_x \pm i\sigma_y$ and $p_x \pm ip_y$ respectively.

Total angular momentum J_z defined by [10]

$$J_z = L_z + \frac{\hbar}{2}(\sigma_z - n\tau_z), \quad (5)$$

where L_z, σ_z are orbit and spin angular momentum operators respectively, can be proven to commute with the total Hamiltonian H (see Appendix for detailed proof). Due to the commutation relation, the eigenfunction of H, denoted by $\Psi(r,\theta)$ in polar coordinates, can be written in terms of eigenfunction of J_z with additional radial part. Since

$$J_z\psi(\theta) = \left[L_z + \frac{\hbar}{2}(\sigma_z - n\tau_z)\right]\psi(\theta) = m_J\hbar\psi(\theta) \tag{6}$$

for eigenvalue $m_J\hbar$ and corresponding eigenstate $\psi(\theta)$ of operator J_z, we have

$$L_z\psi(\theta) = -i\hbar\frac{\partial}{\partial\theta}\psi(\theta) = \left(m_J\hbar - \frac{\hbar\sigma_z}{2} + \frac{n\hbar\tau_z}{2}\right)\psi(\theta). \tag{7}$$

We can then compose the eigenstate of the BdG Hamiltonian

$$\Psi(r,\theta) = \begin{pmatrix} u_{\uparrow,m_J}(r)e^{i(m_J+\frac{n-1}{2})\theta} \\ u_{\downarrow,m_J}(r)e^{i(m_J+\frac{n+1}{2})\theta} \\ v_{\downarrow,m_J}(r)e^{i(m_J-\frac{n+1}{2})\theta} \\ -v_{\uparrow,m_J}(r)e^{i(m_J-\frac{n-1}{2})\theta} \end{pmatrix}, \tag{8}$$

where $(u_{\uparrow,m_J}(r), u_{\downarrow,m_J}(r), v_{\downarrow,m_J}(r), -v_{\uparrow,m_J}(r))^T$ is the radial part of $\Psi(r,\theta)$. Analyzing the radial part of $\Psi(r,\theta)$, one can derive the condition for the existence of MFs [10]. However, physical quantities of edge states in finite systems are not accessible analytically. Therefore, we rely on numerical solutions to the BdG equation to study the distribution of MF state and the energy dispersion of low-energy excitations.

3. Low-energy excitations at the sample edge

We now solve the BdG equation numerically by employing tight binding model [11]

$$\begin{aligned} \tilde{H} = & -t_0\sum_{\mathbf{i},\mathbf{j},\sigma}\hat{c}^\dagger_{\mathbf{i}\sigma}\hat{c}_{\mathbf{j}\sigma} - \mu\sum_{\mathbf{i},\sigma}\hat{c}^\dagger_{\mathbf{i}\sigma}\hat{c}_{\mathbf{i}\sigma} + \sum_{\mathbf{i}}V_z\left(\hat{c}^\dagger_{\mathbf{i}\uparrow}\hat{c}_{\mathbf{i}\uparrow} - \hat{c}^\dagger_{\mathbf{i}\downarrow}\hat{c}_{\mathbf{i}\downarrow}\right) \\ & + it_\alpha\sum_{\mathbf{i},\delta}\left(\hat{c}^\dagger_{\mathbf{i}+\delta_x}\sigma_y\hat{c}_{\mathbf{i}} - \hat{c}^\dagger_{\mathbf{i}+\delta_y}\sigma_x\hat{c}_{\mathbf{i}} + h.c.\right) + \sum_{\mathbf{i}}\left(\Delta\hat{c}^\dagger_{\mathbf{i}\uparrow}\hat{c}^\dagger_{\mathbf{i}\downarrow} + h.c.\right), \end{aligned} \tag{9}$$

where $t_0 = \hbar^2/2m^*a^2$ and $t_\alpha = \alpha_R\hbar/2a$ are hopping rates with respect to spin-reservation and spin-flipping between nearest neighbors; a is the grid spacing; $\mathbf{i},\mathbf{j}$ are two dimensional mesh indices. The radius of the circular sample we take is $75a$, which leads the dimension of the Hamiltonian matrix to the order of 10^5. The discretized Hamiltonian $\tilde{H}$ is diagonalizable by applying the numerical technique provided in [12].

For a circular sample (see Fig.1(a)), where one vortex is induced at the center with pairing potential $\Delta(\vec{r}) = \Delta_0[1-\exp(-(r/s)^2)]e^{i\theta}$ ($s = 4a$), we find a zero-energy MF state at the sample edge, as shown in Fig.1(b). The chirality of the MF is clockwise for $V_z > 0$, and the in-plane spin component of the MF points away from the center of circle with additionally $\alpha > 0$, as shown in Fig. 2. The chirality and the in-plane spin component depend on the Zeeman field and the spin-orbit coupling in the way summarized in Table 1.

By diagonalizing the Hamiltonian H, we are able to calculate the energy and the total angular momentum $\langle J_z\rangle$ of quasi-particles at the edge of a finite circular sample. As shown in Fig. 3, the excitation energy and the total angular momentum are in linear relationship with slope $-E_0$ for $V_z > 0$, where E_0 is the excitation gap between the first excitation state with finite positive energy and the zero-energy MF state, and the sign of the slope can be changed by turning the Zeeman field to the opposite direction.

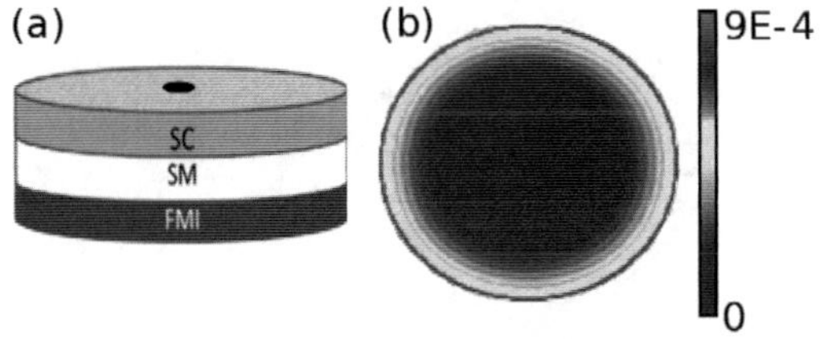

Figure 1. (a) Heterostructure of s-wave superconductor with a vortex, spin-orbit coupled semiconductor and ferromagnetic insulator. (b) Distribution of the MF at the sample edge. Results are obtained in a sample with radius $75a$ for $\Delta_0 = 0.5t_0, V_z = 0.8t_0, \mu = 0$ and $\alpha = 0.9$, which satisfy the relation $V_z > \sqrt{\mu^2 + \Delta^2}$.

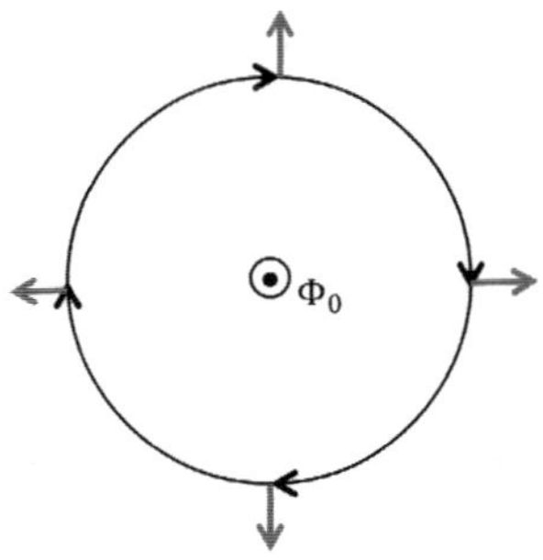

Figure 2. Chirality (black arrow) and in-plane spin component (blue arrow) of the MF at the edge. The parameters used are the same as those given in Fig. 1.

Table 1. Chirality and in-plane spin component of the MF as function of Zeeman field and spin-orbit coupling

V_z	α	Chirality	In-plane spin component
+	+	clockwise	away from center
+	-	clockwise	towards center
-	+	anticlockwise	towards center
-	-	anticlockwise	away from center

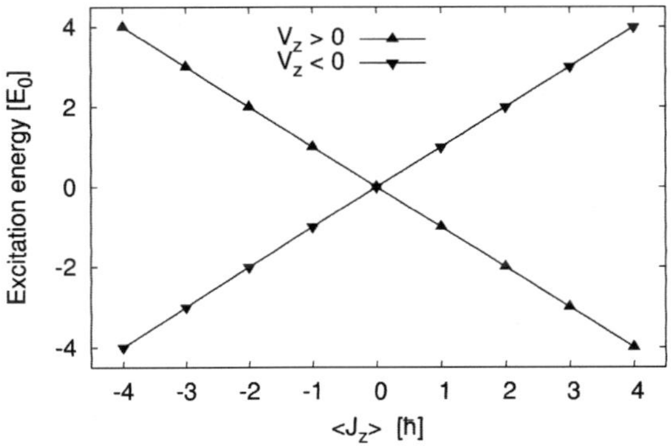

Figure 3. Energy dispersion relation of the edge states for both $V_z > 0$ and $V_z < 0$. $E_0 = 1.55 \times 10^{-2} t_0$ for parameters given in Fig. 1.

4. Conclusion

We have studied the edge states in a finite sample of circular shape. The commutation relation between the total angular momentum and the full Hamiltonian has been proven. Based on the relation, the angular part of the solution to the Bogoliubov-de Gennes equation has been derived analytically. For the radial part, we solve the Bogoliubov-de Gennes equation numerically. With the full solutions, we have derived the energy dispersion of low-energy excitations at the edge. We have revealed the dependence of the chirality and the in-plane spin component of the MF on the Zeeman field and the spin-orbit coupling.

Acknowledgments
This work was supported by WPI initiative on Materials Nanoarchitectonics, MEXT of Japan, and Grants-In-Aid for Scientific Research (No.22550377), JSPS.

Appendix

In this Appendix, we prove the commutation relation between the total angular momentum J_z and the Bogoliubov-de Gennes (BdG) Hamiltonian H. Let's define $\sigma_+ = \sigma_-^\dagger = \sigma_x + i\sigma_y$, $p_+ = p_x + ip_y = e^{i\theta}(-i\partial_r + \frac{1}{r}\partial_\theta)$ and $p_- = p_x - ip_y = e^{-i\theta}(-i\partial_r - \frac{1}{r}\partial_\theta)$. The matrix σ_x, σ_y and σ_z are given by

$$\sigma_x = \begin{pmatrix} 0 & 1 & 0 & 0 \\ 1 & 0 & 0 & 0 \\ 0 & 0 & 0 & 1 \\ 0 & 0 & 1 & 0 \end{pmatrix}, \quad \sigma_y = \begin{pmatrix} 0 & -i & 0 & 0 \\ i & 0 & 0 & 0 \\ 0 & 0 & 0 & -i \\ 0 & 0 & i & 0 \end{pmatrix}, \quad \sigma_z = \begin{pmatrix} 1 & 0 & 0 & 0 \\ 0 & -1 & 0 & 0 \\ 0 & 0 & 1 & 0 \\ 0 & 0 & 0 & -1 \end{pmatrix}. \tag{A.1}$$

After substituting $\vec{p}$ in Eq.(3) by $p_\pm$, we can form the full BdG Hamiltonian in polar coordinates

$$H_{BdG} = \int d\vec{r} \left(-\frac{\hbar^2}{2m^*}\nabla^2 - \mu\right)(c_\uparrow^\dagger c_\uparrow + c_\downarrow^\dagger c_\downarrow - c_\uparrow c_\uparrow^\dagger - c_\downarrow c_\downarrow^\dagger) + V_z(c_\uparrow^\dagger c_\uparrow - c_\downarrow^\dagger c_\downarrow - c_\uparrow c_\uparrow^\dagger + c_\downarrow c_\downarrow^\dagger)$$
$$+ i\alpha\left(p_- c_\uparrow^\dagger c_\downarrow - p_+ c_\downarrow^\dagger c_\uparrow + p_- c_\downarrow c_\uparrow^\dagger - p_+ c_\uparrow c_\downarrow^\dagger\right) + \tilde{\Delta}_0\left(e^{in\theta} c_\uparrow^\dagger c_\downarrow^\dagger + e^{-in\theta} c_\downarrow c_\uparrow - e^{in\theta} c_\downarrow^\dagger c_\uparrow^\dagger - e^{-in\theta} c_\uparrow c_\downarrow\right). \tag{A.2}$$

By defining unitary matrix τ_x, τ_y and τ_z in the particle space as

$$\tau_x = \begin{pmatrix} 0 & 0 & 1 & 0 \\ 0 & 0 & 0 & 1 \\ 1 & 0 & 0 & 0 \\ 0 & 1 & 0 & 0 \end{pmatrix}, \quad \tau_y = \begin{pmatrix} 0 & 0 & i & 0 \\ 0 & 0 & 0 & i \\ -i & 0 & 0 & 0 \\ 0 & -i & 0 & 0 \end{pmatrix}, \quad \tau_z = \begin{pmatrix} 1 & 0 & 0 & 0 \\ 0 & 1 & 0 & 0 \\ 0 & 0 & -1 & 0 \\ 0 & 0 & 0 & -1 \end{pmatrix}, \tag{A.3}$$

we have the matrix form of full Hamiltonian in Nambu space

$$H_{BdG} = \left(-\frac{\hbar^2}{2m^*}\nabla^2 - \mu\right)\tau_z + V_z\sigma_z + i\frac{\alpha}{2}(\sigma_+ p_- - \sigma_- p_+)\tau_z + \tilde{\Delta}_0\{\cos(n\theta)\tau_x + \sin(n\theta)\tau_y\}. \tag{A.4}$$

Let $\tau_+ = \frac{\tau_x + i\tau_y}{2}$ and $\tau_- = \frac{\tau_x - i\tau_y}{2}$, the following basic commutation relations are established.

$$[\sigma_a, \sigma_b] = 2i\varepsilon_{abc}\sigma_c; \quad [\tau_a, \tau_b] = -2i\varepsilon_{abc}\tau_c; \quad [\sigma_z, \sigma_+] = 2\sigma_+; \quad [\tau_z, \tau_+] = -2\tau_+; \tag{A.5}$$

$$[\sigma_z, \sigma_-] = -2\sigma_-; \quad [\tau_z, \tau_-] = 2\tau_-; \quad [\sigma_z, \tau_+] = 0; \quad [\tau_z, \sigma_+] = 0; \tag{A.6}$$

$$[L_z, p_-] = [xp_y - yp_x, p_x - ip_y] = [x, p_x]p_y + i[y, p_y]p_x = i\hbar p_y - \hbar p_x = -\hbar p_-; \tag{A.7}$$

$$[L_z, p_+] = [xp_y - yp_x, p_x + ip_y] = [x, p_x]p_y - i[y, p_y]p_x = i\hbar p_y + \hbar p_x = \hbar p_+. \tag{A.8}$$

The first two terms in Eq.(A.4) remain invariant under rotation. Therefore, we have

$$\left[J_z, \left(\frac{\hbar^2}{2m^*}\nabla^2 - \mu\right)\tau_z + V_z\sigma_z\right] = 0. \tag{A.9}$$

Considering the spin-orbit coupling part in Eq.(A.4), we have

$$[J_z, \sigma_+ p_-] = [L_z, \sigma_+ p_-] + \frac{\hbar}{2}[\sigma_z - n\tau_z, \sigma_+ p_-] \tag{A.10}$$

$$= [L_z, \sigma_+]p_- + \sigma_+[L_z, p_-] + \frac{\hbar}{2}\{[\sigma_z, \sigma_+]p_- + \sigma_+[\sigma_z, p_-]\} + \frac{\hbar}{2}[-n\tau_z, \sigma_+ p_-] \tag{A.11}$$

$$= -\hbar\sigma_+ p_- + \hbar\sigma_+ p_- = 0. \tag{A.12}$$

Similarly, $[J_z, \sigma_- p_+] = 0$. Thus,

$$\left[J_z, i\frac{\alpha}{2}\left(\sigma_+ p_- - \sigma_- p_+\right)\tau_z\right] = 0. \tag{A.13}$$

As for the superconducting part $(e^{-in\theta}\tau_+ + e^{in\theta}\tau_-)$ in Eq.(A.4), we have

$$\begin{aligned}
[J_z, e^{-in\theta}\tau_+] &= [L_z, e^{-in\theta}\tau_+] + \frac{\hbar}{2}[\sigma_z - n\tau_z, e^{-in\theta}\tau_+] && \text{(A.14)}\\
&= [L_z, e^{-in\theta}]\tau_+ + e^{-in\theta}[L_z, \tau_+] + \frac{\hbar e^{-in\theta}}{2}[\sigma_z, \tau_+] - \frac{n\hbar}{2}e^{-in\theta}\cdot(-2\tau_+) && \text{(A.15)}\\
&= [-i\hbar\partial_\theta, e^{-in\theta}]\tau_+ + 0 + 0 + n\hbar e^{-in\theta}\tau_+ && \text{(A.16)}\\
&= -n\hbar e^{-in\theta}\tau_+ + n\hbar e^{-in\theta}\tau_+ = 0. && \text{(A.17)}
\end{aligned}$$

Similarly, $[J_z, e^{in\theta}\tau_-] = 0$. Therefore, We get

$$\left[J_z, \tilde{\Delta}_0\left\{\cos(n\theta)\tau_x + \sin(n\theta)\tau_y\right\}\right] = 0. \tag{A.18}$$

Thus, the BdG Hamiltonian H commutes with the total angular momentum J_z

$$[J_z, H] = 0. \tag{A.19}$$

References

[1] Wilczek F 2009 *Nat. Phys.* **5** 614
[2] Ivanov D A 2001 *Phys. Rev. Lett.* **86** 268
[3] Nayak C, Simon S H, Stern A, Freedman M and Das Sarma S 2008 *Rev. Mod. Phys.* **80** 1083
[4] Read N and Green D 2000 *Phys. Rev.* B **61** 10267
[5] Das Sarma S, Freedman M and Nayak C 2005 *Phys. Rev. Lett.* **94** 166802
[6] Tewari S, Das Sarma S, Nayak C, Zhang C W and Zoller P 2007 *Phys. Rev. Lett.* **98** 010506
[7] Fu L and Kane C L 2008 *Phys. Rev. Lett.* **100** 096407
[8] Sau J D, Lutchyn R M, Tewari S and Das Sarma S 2010 *Phys. Rev. Lett.* **104** 040502
[9] Lutchyn R M, Sau J D and Das Sarma S 2010 *Phys. Rev. Lett.* **105** 077001
[10] Sau J D, Tewari S, Lutchyn R M, Stanescu T and Das Sarma S 2010 *Phys. Rev.* B **82** 214509
[11] Liang Q F, Wang Z and Hu X 2011 *arXiv:*1112.6109
[12] Bollhöfer M and Notay Y 2007 *Comp. Phys. Com.* **177** 951

Enhancement of tunnel conductivity by Cooper pair fluctuations in electron-hole bilayer

D K Efimkin[1], **Yu E Lozovik**[1,2]

[1] Institute of Spectroscopy RAS , 142190, Troitsk, Moscow Region, Russian Federation.
[2] Moscow institute of Physics and Technology, 141700, Dolgoprudniy, Moscow Region, Russian Federation.

E-mail: `lozovik@isan.troitsk.ru`

Abstract. Influence of Cooper pair fluctuations that are precursor of pairing of electrons and holes located on opposite surfaces of topological insulator film on tunnel conductivity between the surfaces is investigated. Due to restrictions caused by momentum and energy conservation dependence of tunnel conductivity on external bias voltage has peak that becomes more prominent with decreasing of disorder and temperature. We have shown that Cooper pair fluctuations considerably enhance tunneling and height of the peak diverges in vicinity of critical temperature with critical index $\nu = 2$. Width of the peak tends to zero in proximity of critical temperature. Pairing of electrons and holes can be suppressed by disorder and in vicinity of quantum critical point height of the peak also diverges as function of Cooper pair damping with critical index $\mu = 2$.

1. Introduction

Recently pairing of spatially separated electrons and holes located on opposite surfaces of thin film of topological insulator was predicted [1, 2, 3]. Pairing can lead to a number of interesting effects including superfluidity [4], anomalous drag effect [5] and anomalous Andreev reflection [6]. Condensate formed by electron-hole Cooper pairs is topological one [1] and can host Majorana fermions on topological defects. But the most prominent manifestation of pairing is internal Josephson effect.

Internal Josephson effect leads to colossal enhancement of tunnel conductance between electron and hole layers. It was observed [7] in a system of composite electrons and composite holes that was realized in semiconductor GaAs structures in strong magnetic field. Its quantitative description is complicated one due to disorder, strong phase fluctuations that take place in two-dimensional superfluids and topological defects ([8] and references therein).

Here we propose new physical effect that can be observed in normal state in vicinity of critical temperature. In the normal state pairing areas can appear as thermodynamical fluctuations and can transfer coherent Josephson current. We have shown that pairing fluctuations lead to considerable enhancement of tunneling in normal state in vicinity of critical temperature. In spite of the fact that described effect is universal and can be realized in different realizations of electron-hole bilayers, for example in graphene bilayer [9], we considered it for definiteness in thin film of topological insulator.

The rest part of the article is organized in following way. In section 2 we briefly discuss electronic structure of topological insulator film, microscopical approach for description of

Journal of Physics: Conference Series **393** (2012) 012019 doi:10.1088/1742-6596/393/1/012019

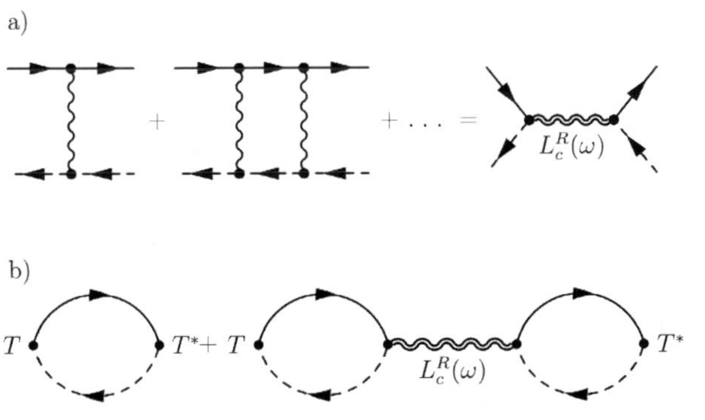

Figure 1. a) Introduction of Cooper propagator $L_c^R(\omega)$ for electron-hole bilayer b) Diagrams for tunnel conductivity including contribution of Cooper pair fluctuations

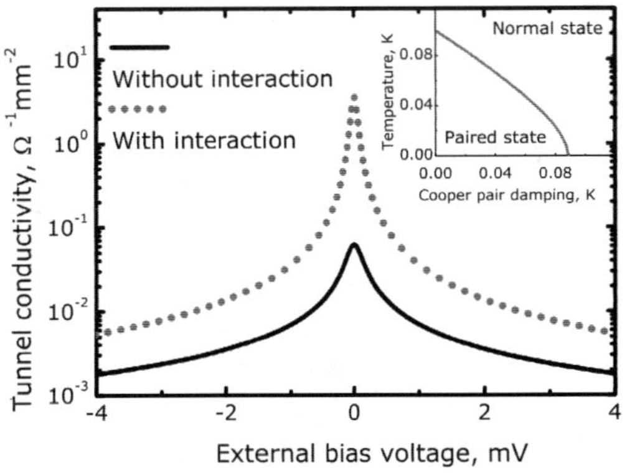

Figure 2. Dependence of tunnel conductivity on external bias voltage for $T = 0.2$K and $\gamma = 0.2$K. Inset: Phase diagram of the system in presence of disorder.

Cooper pair fluctuations and calculation of tunnel conductivity. In section 3 we discuss results of calculation and conclude. Section 4 is devoted to brief summary.

2. Calculation of tunneling conductivity

The considered system consists of topological insulator thin film. Surface states of its opposite surfaces are populated by electrons and holes controlled by external gates. In weak coupling regime [3] Hamiltonian of the system is given by

$$H = \sum_k (\xi_{kT} a^+_{\mathbf{k}T} a_{\mathbf{k}T} + \xi_{\mathbf{k}T} a^+_{\mathbf{k}B} a_{kB}) + \frac{\lambda}{\nu_F} \sum_{\mathbf{kk'q}} a^+_{\mathbf{k+q}T} a^+_{\mathbf{k'-q}B} a_{\mathbf{k'}B} a_{\mathbf{k}T} \tag{1}$$

where λ is dimensionless coupling constant in Bardeen-Cooper-Schrieffer (BCS) approximation; a_{kT} and a_{kB} are annihilation operators for electrons on top and bottom surfaces of topological insulator film; $\xi_{kT} = v_F k - \mu$ and $\xi_{kB} = -v_F k + \mu$ are their dispersion law (electrons from bottom layer have negative velocity indicating that spectrum belong hole type); v_F is Fermi velocity and ν_F is density of states on Fermi level. Pairing is very sensitive to mismatch of concentrations of electrons and holes so here we consider only balanced case. Intralayer Coulomb interaction leads only to renormalization of quasiparticle spectrum and can be omitted. Definition of dimensionless coupling constant λ in BCS approximation including interaction screening and specifics of topological insulator surface states can be found in [3]. Also theory of electron-hole pairing in more complicated realistic model that includes screening, disorder and hybridization caused by tunneling is developed in [3].

In microscopic theory of fluctuations in superconductors [10] Cooper pair propagator is usually introduced. It corresponds to vertex function that is sum of infinite series of scattering diagrams between electrons in Cooper channel. On Fig.1a introduction of propagator for Cooper pairs $\mathrm{L}_c^R(\omega)$ formed by electrons and holes in film of TI is represented. Its analytical value is given by

$$L_c^R(\omega) = \frac{\lambda}{1 - \lambda/\nu_F \Pi_c^R(\omega)}, \tag{2}$$

Journal of Physics: Conference Series **393** (2012) 012019 doi:10.1088/1742-6596/393/1/012019

where $\Pi_c^R(\omega)$ correspond to bubble diagram on Fig. 1 b in and is given by

$$\Pi_c^R(\omega) = \nu_F(\frac{1}{\lambda} - [\ln\frac{T}{T_0} + \Psi(\frac{1}{2} + \frac{\gamma}{2\pi T} - i\frac{\omega}{4\pi T}) - \Psi(\frac{1}{2})]). \quad (3)$$

Here T_0 is critical temperature of pairing in absence of disorder, γ is Cooper pair damping that is average value of damping of electron and hole and $\Psi(x)$ is logarithmic derivative of Gamma function. At critical temperature denominator of (2) become zero indicating Cooper instability of normal state of the system against electron-hole pairing.

For calculation of tunnel conductivity we used linear response approach [11]. We showed that if momentum of electron is conserved in tunneling process then in absence of interactions between electrons and holes tunneling conductivity is given by

$$\sigma_T(V) = \frac{e^2}{h}\frac{4\pi|t|^2}{eV}\text{Im}[\Pi_c^R(eV)], \quad (4)$$

where V is external bias voltage, t is amplitude of tunneling process. This term corresponds to first term on diagram on Fig 1b. Contribution of Cooper pair fluctuation correspond second term on diagram on Fig 1b. These contribution can be interpreted as fluctuation internal Josephson effect. The value of tunneling conductivity in presence of Cooper pair fluctuations is given by

$$\sigma_T(V) = \frac{e^2}{h}\frac{4\pi|t|^2}{eV}\text{Im}[\frac{\Pi_c^R(eV)}{1 - \lambda/\nu_F\Pi_c^R(eV)}]. \quad (5)$$

In vicinity of critical temperature denominator of expression tends to zero hence Cooper pair fluctuations leads to critical behavior of tunnel conductivity.

3. Conclusions

For numerical calculation we used the following set of parameters [3] $\mu = 0.02$eV, $T_0 = 0.1$K, $\lambda = 0.13$, $t = 0.01$K. In BCS regime disorder can easily suppress pairing since it acts differently on components of Cooper pair if its damping γ exceed critical value $\gamma_0 = 0.89T_0$. The phase diagram of the system in presence of disorder is presented on inset of Fig.2.

The dependence of tunnel conductivity on external bias is plotted for $T = 0.2$K and $\gamma = 0.2$K on Fig.2. For other values of these parameters the dependence is qualitatively the same. Peak is caused by conservation laws for momentum and energy that should be satisfied for each tunneling process. Coulomb interaction considerably enhances tunneling between layers.

Dependence of tunnel conductivity peak height and peak width as function of temperature is presented on Fig. 3 and Fig. 4 for two values of Cooper pair damping. At $\gamma_1 = 0.06$K disorder only reduce critical temperature and at $\gamma_2 = 0.1$K pairing is suppressed. In vicinity of critical temperature peak height diverges and its width tends to zero. In critical region corresponding analytical dependence is given by

$$\sigma_T(V = 0) \sim (T/T_{0d} - 1)^{-\nu},$$

where $\nu = 2$ is critical index and T_{0d} is critical temperature in presence of disorder. Dependence of height on temperature is universal — it depends only on proximity to critical temperature T_{0d} and does not depend on disorder explicitly. We found that even if pairing of electrons and holes is suppressed by disorder the tunneling is strongly enhanced in region $\gamma < \gamma_0/\lambda$. Width of peak is decreasing with decreasing of temperature and its minimal value depends on Cooper pair damping. In vicinity of quantum phase transition tunneling conductivity also diverges and is given by

$$\sigma_T(V = 0) \sim (\gamma/\gamma_0 - 1)^{-\mu},$$

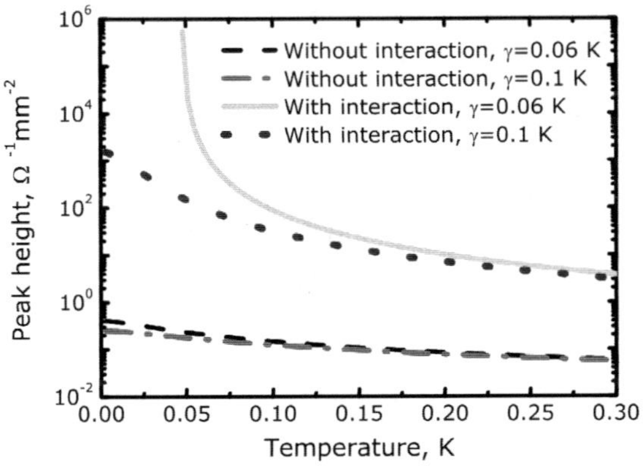

Figure 3. Dependence of height of tunnel peak on temperature.

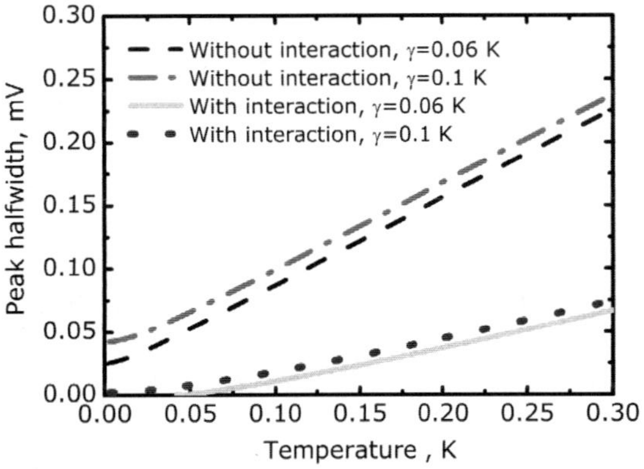

Figure 4. Dependence of half-width of tunnel peak on temperature.

where $\mu = 2$ is corresponding critical index.

Interlayer Coulomb interaction considerably influence tunneling in electron-hole bilayer even if Cooper pairing is suppressed by disorder. The situation is completely different from one in electron-electron bilayer [12] where interlayer interaction in ladder approximation does not influence height of the peak and insignificantly decreases its width.

4. Summary

Influence of fluctuation of Cooper pairs formed by electron and hole in normal state in thin film of topological insulator on tunneling between its surfaces is considered. We find critical behavior of tunneling conductivity in vicinity of classical phase transition with critical index $\nu = 2$. In vicinity of quantum phase transition dependence of tunnel conductivity on damping of Cooper pairs is also critical with critical index $\mu = 2$.

Acknowledgments

The work was supported by Grant of the President of Russian Federation MK-5288.2011.2 and by scholarship from Dynasty Foundation.

[1] Seradjeh B, Moore J E and Franz M 2009 *Phys. Rev. Lett.*. **103**, 066402.
[2] Tilahun D, Lee B, Hankiewicz E M and MacDonald A H 2011, *Rev. Lett.* **107**, 246401.
[3] Efimkin D K, Lozovik Yu E and Sokolik A A, in print.
[4] Lozovik Yu E and Yudson V I 1975, *JETP Lett.* **22**, 274.
[5] Vignale G and MacDonald A H 1996*Phys. Rev. Lett.*. **76**, 2786.
[6] Pesin D and MacDonald A H 2011 *Phys. Rev.* B. **84**, 075308.
[7] Spielman I B, Eisenstein J P, Pfieffer L N and West K W 2000 *Phys. Rev. Lett.*. **84**, 5808.
[8] Lozovik Yu E and Poushnov A V 1997 *Phys. Lett.* A. **228**, 399. Fogler M and Wilczek F 2001 *Phys. Rev. Lett.*. **86**, 183. Stern A, Girvin S M, MacDonald A H and Ma N 2001 *Phys. Rev. Lett.*. **86**, 1829.
[9] Lozovik Yu E and Sokolik A A 2007 *JETP Lett.*. **87**, 55. Efimkin D K, Kulbachinskii V A and Lozovik Yu E 2011 *JETP Lett.*. **113**, 880. Efimkin D K and Lozovik Yu E 2011 *JETP*. **113**, 880.
[10] Larkin A I and Varlamov A 2005 *Theory of fluctuations in superconductors* (Oxford:Clarendon Press).
[11] Mahan G D 1981 *Many-particle physics* (New York: Plenum Press)
[12] Zheng L and MacDonald A H 1993 *Phys.Rev* B. **47**, 10620.

Resonance features of coupled Josephson junctions: radiation and shunting

Yu. M. Shukrinov[a,b,c], **P. Seidel**[d], **E. Il'ichev**[e], **W. Nawrocki**[f], **M. Grajcar**[g] **P. A. Plecenik**[g], **I. R. Rahmonov**[a,h], **K. Kulikov**[a,c]

[a] BLTP, Joint Institute for Nuclear Research, Dubna, Moscow Region, 141980, Russia
[b] Max-Planck-Institute for the Physics of Complex Systems, 01187 Dresden, Germany
[c] Department of Theoretical Physics, International University of Dubna, Dubna, 141980, Russia
[d] Institut für Festkörperphysik, Jena, D-07743 Jena, Germany
[e] Institute of Photonic Technology, P.O. Box 100239, D-07702 Jena, Germany
[f] Poznan University of Technology, Poznan, Poland
[g] Department of Experimental Physics, Comenius University, Bratislava, Slovakia
[h] Umarov Physical Technical Institute, TAS, Dushanbe, 734063 Tajikistan

E-mail: shukrinv@theor.jinr.ru

Abstract. We study the phase dynamics and the resonance features of coupled Josephson junctions in layered superconductors and their manifestations in the current- voltage characteristics and temporal dependence of the electric charge in the superconducting layers. Results on the effect of the external radiation and shunting of the stack of Josephson junctions by LC-elements are presented. We discuss the ideas concerning the experimental observation of these resonances.

1. Introduction

Strongly anisotropic high-T_c superconductor (HTSC) forms a natural stack of intrinsic Josephson junctions (JJ) and shows the intrinsic Josephson effect [1]. A model for describing the physical properties of intrinsic JJs in HTSCs, including nonlinear effects and various nonequlibrium phenomena, is offered by a system of coupled JJs [2, 3]. It should be emphasized that the system of coupled JJs is a promising object in HTSC electronics that has been extensively studied in recent years [4, 5, 6, 7]. In particular, broad possibilities for various applications are offered by the recently discovered coherent electromagnetic radiation of noticeable power in a terahertz frequency range, which is generated by the intrinsic JJ system [8]. An interesting aspect of this phenomenon is that the emission is related to a certain region on the current-voltage characteristic (CVC), namely, that it corresponds to parametric resonance (PR) in the JJ system [9, 10]. Numerical modeling the CVC predicts important properties of the system of coupled JJs and describes its phase dynamics in the branching (hysteresis) region of interest [11].

2. Model and method

A system of $N + 1$ superconducting layers in an anisotropic HTSC, which is characterized by the order parameter $\Delta_l(t) = |\Delta| \exp(i\theta_l(t))$ with the time dependent phase $\theta_l(t)$, comprises N

Josephson junctions [1]. Figure 1(a) shows a schematic diagram of this layered system.

Superconducting layers, which are numbered by l running from 0 to N and characterized by the time-dependent order parameters with moduli Δ_l and phases θ_l, form a system of JJs with phase differences $\varphi_l = \theta_l - \theta_{l-1}$; d_s and d are the thicknesses of the superconducting and dielectric layers, respectively.

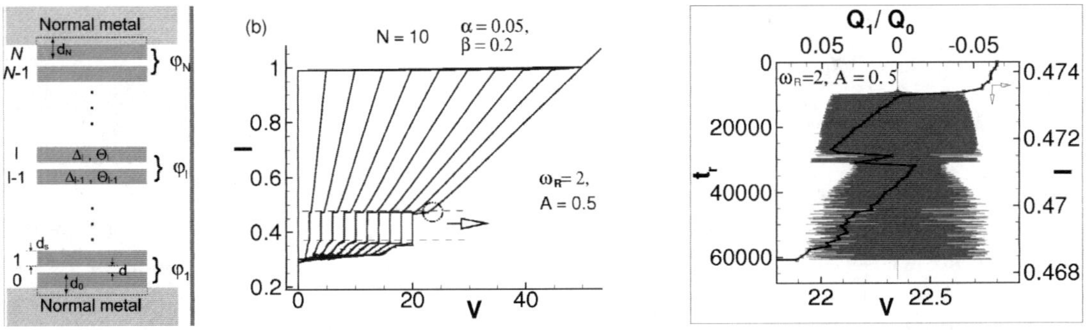

Figure 1. (a) Schematic diagram of a system of intrinsic JJs in HTSCs (see text); (b) Total CVC of the stack with 10 JJs under radiation at $\omega_R = 2$ and amplitude $A = 0.5$. The part of the outermost branch of CVC marked by circle shown enlarged in (c) together with charge-time dependence, demonstrating the "bump" structure in CVC and charging of S-layers above Shapiro step.

The thickness of superconducting layers (about 3 Å) in an HTSC is comparable with the Debye length r_D of electric charge screening. Therefore, there is no complete screening of the charge in the separate layers, and the electric field induced in each JJ penetrates into the adjacent junctions. Thus, the electric neutrality of superconducting layers is dynamically broken and, in the case of the alternating current (ac) Josephson effect, a capacitive coupling appears between the adjacent junctions [12]. The absence of complete screening of charge in the superconducting layer leads to the formation of a generalized scalar potential Φ_l of the layer, which is defined in terms of the scalar potential ϕ_l and the derivative of phase θ_l of the superconducting order parameter as follows: $\Phi_l(t) = \phi_l - V_0 \frac{d\theta_l}{dt}$, where $V_0 = \hbar\omega_p/(2e)$, $\omega_p = \sqrt{2eI_c/\hbar C_j}$ is the plasma frequency, I_c is the critical current, and C_j is the capacitance of the JJ. The generalized scalar potential is related to the charge density Q_l on the superconducting layer $Q_l = -\frac{1}{4\pi r_D^2}\Phi_l$ [12, 13]. The existence of a relationship between the electric charge Q_l of the lth layer and the generalized scalar potential Φ_l of this layer reflects a nonequilibrium nature of the ac Josephson effect in layered HTSCs [13].

When an external current passes through the stack of coupled Josephson junctions, the system appears under nonequilibrium conditions [13]. The Josephson relation is generalized in this case. Also the diffusion contribution to the quasiparticle current arises due to the generalized scalar potential difference

$$I^l_{dif} = \frac{\Phi_l - \Phi_{l-1}}{I_c R_j} = -\frac{(Q_l - Q_{l-1})}{4\pi r_D^2 I_c R_j} = -\frac{(Q_l - Q_{l-1})}{2e^2 N(0) I_c R_j} = \beta\dot{\varphi}_l - \beta V_l = -\alpha\beta(V_{l+1} + V_{l-1} - 2V_l) \quad (1)$$

where R_j is the resistance of the Josephson junction in the resistive state and $N(0)$ denotes the density of states on the Fermi level. The structure of the CVC in the capacitively coupled Josephson junction model with diffusion current (CCJJ+DC model) is equidistant [14] in agreement with the experiments [15]. In Ref. [16] was demonstrated that the diffusion current plays an important role in the formation of the CVC, particularly, determining the width of the hysteresis region.

To investigate the phase dynamics of an intrinsic JJ we use the one-dimensional CCJJ+DC model with the gauge-invariant phase differences $\varphi_l(t)$ between S-layers l and l+1 in the presence of electromagnetic irradiation described by the system of equations:

$$\begin{cases} \dfrac{\partial \varphi_l}{\partial t} = V_l - \alpha(V_{l+1} + V_{l-1} - 2V_l) \\ \dfrac{\partial V_l}{\partial t} = I - \sin\varphi_l - \beta\dfrac{\partial \varphi_l}{\partial t} + A\sin\omega_R t + I_{noise} \end{cases} \quad (2)$$

where t is dimensionless time, normalized to the inverse plasma frequency ω_p^{-1}, $\beta = \dfrac{1}{R_j}\sqrt{\dfrac{\hbar}{2eI_cC_j}} = \dfrac{1}{\sqrt{\beta_c}}$, β_c is McCumber parameter, α gives the coupling between junctions [12], A is the amplitude of the radiation. To find the CVC of the stack of intrinsic JJs we solve the system of nonlinear second-order differential equations (1) using the fourth order Runge-Kutta method. In our simulations we measure the voltage in units of V_0, the frequency in units of ω_p, the bias current I and the amplitude of radiation A in units of I_c. We note that different kind of couplings between junctions, like inductive coupling in the presence of magnetic field [17], capacitive [12, 18], charge-imbalance [19] and phonon [20, 21] couplings determine a variety of CVCs observed in HTSC. The influence of these couplings on the parametric resonance in the system is still an open problem.

The important information concerning the resonance features of intrinsic JJ in HTSC can be obtained by detailed investigation of the charge dynamics of superconducting layers. To study the time dependence of the electric charge in the S-layers, we use the Maxwell equation $\text{div}(\varepsilon\varepsilon_0 E) = Q$, where ε and ε_0 are relative dielectric and electric constants, respectively. The charge density Q_l in the S-layer l is proportional to the difference between the voltages V_l and V_{l+1} in the neighbor insulating layers $Q_l = Q_0\alpha(V_{l+1} - V_l)$, where $Q_0 = \varepsilon\varepsilon_0 V_0/r_D^2$. The details of the simulation procedure are presented in Refs. [10, 22].

3. Results and discussions

An explanation of coherent radiation from the stack of intrinsic JJ in BSCCO at zero magnetic field [8] is one of the important unsolved problem today. Association of the emission with conventional Fiske steps [23, 24, 25] is problematic because their amplitude is zero at zero magnetic field [26]. Different mechanisms of emission are discussed in literature [23, 26, 27]. A question if the parametric resonance play any role in the radiation physics was not investigated yet.

In this paper we discuss the resonance features of coupled Josephson junctions related to the effects of the external radiation and shunting the system of junctions by LC-elements. First, we demonstrate the effect of microwave's amplitude variation on the CVC of this system and time dependence (temporal oscillations) of the electric charge in the superconducting layers.

In Fig. 1(b) we show the results of simulation of the total CVC for the stack with 10 JJ under external radiation with frequency $\omega_R = 2$ and amplitude $A = 0.5$. The simulation was done at coupling parameter $\alpha = 1$ and the dissipation parameter $\beta = 0.2$. The CVC demonstrates the first Shapiro step on each branch. Results are in qualitative agreement with the experimental ones presented in Ref. [28]. Circle marks a part of the outermost branch demonstrating an interesting behavior of the above steps. Its enlarged part shown in Fig. 1(c).

To clarify this feature, we study a charging of superconducting layers in this interval of bias current. The charge-time dependence together with CVC are shown in Fig. 1(c), too. We see that an irradiation changes the character of the charge-time dependence essentially and brings about a "bump" structure on the outermost branch of CVC, as shown in Fig. 1(c). So, we observe here an additional parametric resonance. We call this resonance as radiation related

PR (rrPR) [29]. The results of simulations at smaller amplitude show some "excess" current in CVC in this region [30, 31]. The "bump" structure and "excess current" structure were recently observed experimentally [4, 15, 32, 33]. We stress the importance of these features and necessity of the additional experimental and theoretical studies of "excess current" and "bump" structures. The question if the charging of the S-layers appears at other types of resonances in coupled JJs [24, 25, 26] has not yet been investigated.

Another important problem on the way of using the intrinsic Josephson junctions in HTSC as terahertz electromagnetic waves sources [34] is a synchronization of all junctions in a stack to increase a power of radiation. Intensive attempts to solve this problem are based on using LC-shunting which leads to such synchronization [34, 35, 36].

Let us consider the system, presented in Fig. 2(a). In normalized units the system of equations, describing this electric scheme, can be written in the form [37]

$$\begin{cases} \frac{\partial \varphi_l}{\partial t} = V_l - \alpha(V_{l+1} + V_{l-1} - 2V_l) \\ \frac{\partial V_l}{\partial t} = I - \sin\varphi_l - \beta\frac{\partial \varphi_l}{\partial t} - C\frac{\partial u_c}{\partial t} \\ \frac{\partial^2 u_c}{\partial t^2} = \frac{1}{LC}\left(\sum_{l=1}^{N} V_l - u_c\right) \end{cases} \tag{3}$$

Here u_c is the voltage at the capacitance, which is normalized to V_0. The bias current I is normalized to the critical current I_c of JJ, time - to the inverse plasma frequency ω_p; shunt capacitance C_{sh}- to the capacitance of the JJ C_j and denote as C, and shunt inductance L_{sh} - to $(C_j\omega_p^2)^{-1}$ and denote as L.

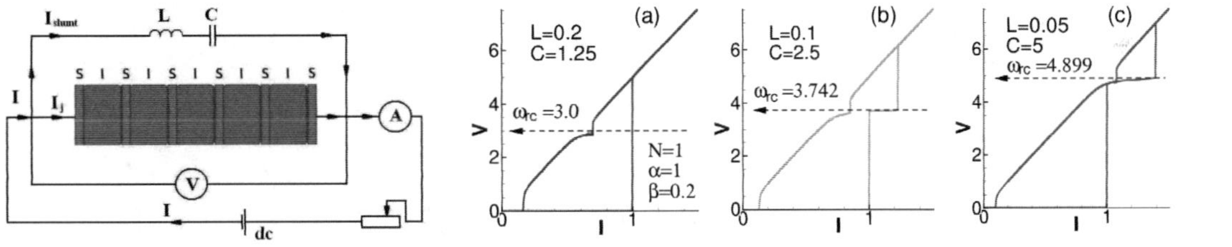

Figure 2. (a) Schema of the JJs stack with LC shunting elements; (b) CVC of the single JJ at different values of shunting parameters L and C.

Now we study the phase dynamics of this system based on the model (3) and discuss the influence of LC shunting on CVC and parametric resonance in the system. We note that the stack of JJs together with LC-elements form a resonance circuit with its eigenfrequency

$$\omega_{rc} = \sqrt{\frac{1 + NC}{LC}} \tag{4}$$

which depends on the number of junctions in the stack N. To simplify analysis we consider first a case of single junction instead of stack of JJ shown in Fig. 1(a). We see that the LC shunting leads to a step-like structure in the oneloop CVC, when the value of Josephson frequency approaches to the eigenfrequency of the LC resonance circuit [38, 39]. The location of the step depends on parameters of this LC circuit. For hysteretic JJ these steps are rc-branches, and they might be obtained completely by changing sweeping current direction after jump on it in the current decreasing procedure [37].

In Fig. 2 we present CVC of a single JJ shunted by the LC-circuit at different values of L and C. The simulation was done in the framework of the RCSJ model, $\beta = 0.2$. It means that the calculations were done by Eqs. (3), but at $\alpha = 1$ and $N = 1$. When Josephson frequency ω_J coincides with the resonance circuit frequency ω_{rc}, the CVC demonstrates a step in CVC at corresponding voltage value. We see such step at $\omega_{rc} = 3.0$ in Fig. 2(a). An increase in ω_{rc} brings this step up on the voltage scale. We see that in Fig. 2(b). Here we observe a jump to this branch at $I = I_c$ as well. So, this rc-branch is separated for two parts in one loop CVC simulation procedure. Then, in some frequency interval we observe a jump at $I = I_c$ to the total rc-branch, as shown at $\omega_{rc} = 4.899$ in Fig. 2c.

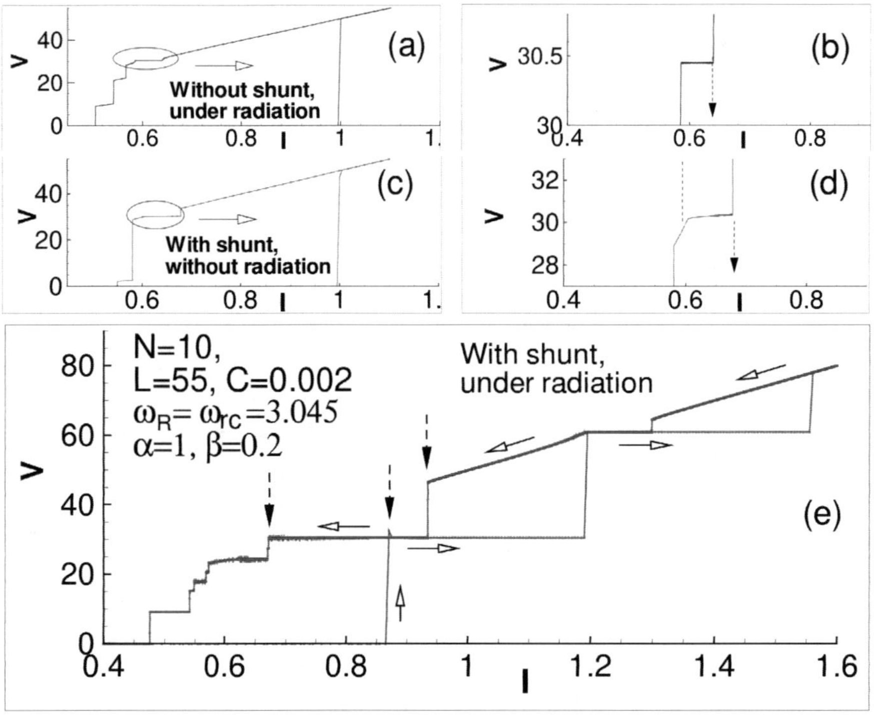

Figure 3. Demonstration of the effect of radiation on shunted coupled JJ at double resonance conditions: (a, b) CVC under radiation, but without shunting; (c, d) CVC with shunting by $L = 55$ and $C = 0.002$, but without radiation; (e) CVC at both effects simultaneously

Let us now consider the influence of the external electromagnetic radiation on the CVC of the investigated system. In the pioneering work of Werthamer and Shapiro [40] was shown that a JJ in a cavity will show additional self-induced resonant steps which lead to subharmonics and some additional structures also in combination with those steps induced by an external microwave frequency. In some cases besides subharmonics also chaotic dynamics was observed [41, 42, 43]. The situation becomes much more complex if an external microwave is taking into account and can result in situations where the chaotic dynamics can be controlled and synchronization of JJ can be improved [44, 45, 46]. Thus we restrict ourself to the question: what would happen with the CVC at double resonance conditions: $\omega_J = \omega_R = \omega_{rc}$. To make it clear we show first CVC of coupled JJs under radiation, but without shunting. In Fig. 3a we present such CVC of ten coupled JJ without shunting under external radiation with frequency $\omega_R = 3.045$ and amplitude $A = 0.5$. As we see, this CVC has a Shapiro step at $V = 30.45$, indicating the external frequency $\omega_R = V/10 = 3.045$. Fig. 3b enlarges this step.

Then we show in Fig. 3c the CVC of ten coupled JJ with shunting by $L = 55$ and $C = 0.002$, but without of external radiation. At this parameters the eigenfrequency of the resonance circuit according to the formula $\omega_{rc} = \sqrt{\frac{1+NC}{LC}}$ is equal to 3.045, so CVC has a corresponding rc-branch. Fig. 3d enlarges this rc-branch. Finally we demonstrate both effects simultaneously. In Fig. 3e we see a crucial changes in CVC when shunting and radiation are both taken into account. Resonance of Josephson and circuit oscillations triggers the appearance of the Shapiro step harmonics and subharmonics. At this resonance the width of Shapiro steps and its harmonics are sharply increased because of the changes in the CVC induced by the rc-branch structure. An interesting question concerns the influence of parametric resonance on this feature which will be investigated separately [47].

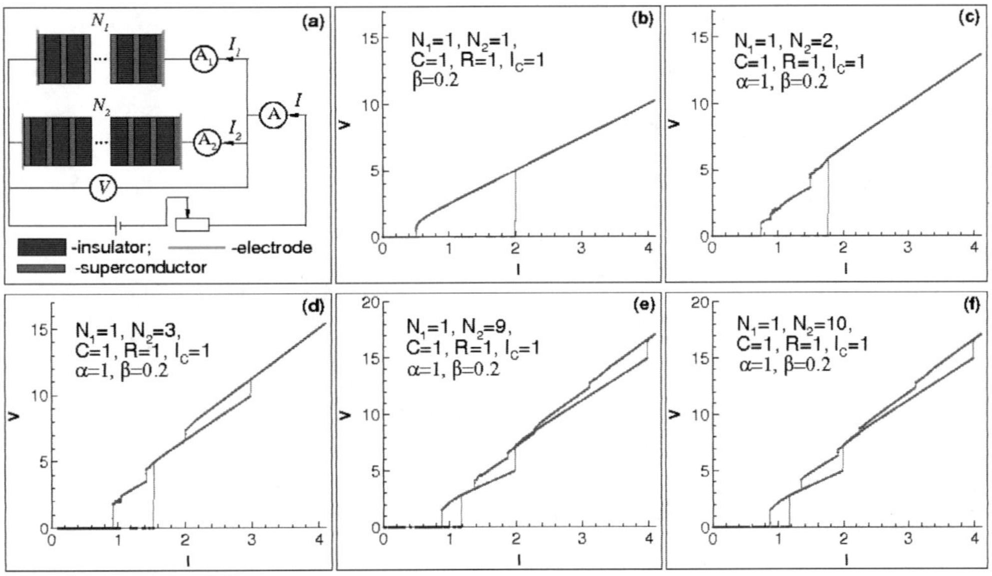

Figure 4. (a) Schema of parallel connections of two stacks with N_1 and N_2 coupled JJ; CVC for the different cases: (b) $N_1 = 1$, $N_2 = 1$; (c) $N_1 = 1$, $N_2 = 2$; (d) $N_1 = 1$, $N_2 = 3$; (e) $N_1 = 1$, $N_2 = 9$; (f) $N_1 = 1$, $N_2 = 10$.

For some applications it is interesting to consider the circuits which consist of two stacks of coupled JJ in parallel connection shown in Fig. 4(a). In the simplest case N1=N2=1 this is the well-known superconducting quantum interference device (DC-SQUID). For DC-SQUIDs the appearance of resonance features corresponding to the inductance L and capacitance C of the circuit like separate rc-branches were reported [48, 49, 50]. There is an additional strong influence of the external magnetic field leading to so-called "beating branches". Two junction SQUIDs can be extended to symmetric as well as asymmetric multi-junction loops, see [35, 51, 52] and references therein. Such circuits have a very complex dynamics. Multi-junction SQUIDs based on intrinsic Josephson junctions were investigated e.g. by [53, 54, 55]. In Ref.[55] Shapiro steps in a DC-SQUID with multiple identical junctions in each arm were investigated.

Here we shortly discuss the cases $N_1 = 1$ with different N_2 (N_2=1, 2, 3, 9, 10). System of equations which describes the phase dynamics in this circuit can be written in the form

$$\begin{cases} \dot{\psi} = U \\ \dot{\varphi}_l = V_l - \alpha(V_{l+1} + V_{l-1} - 2V_l) \\ \dot{V}_l = \sum_{k=1}^{N_2} B_{lk}^{-1}\Big(I - I_c \sin\psi - \frac{\beta}{R}U - \sin\varphi_l - \beta\dot{\varphi}_l\Big) \\ U = \sum_{l=1}^{N_2} V_l \end{cases} \tag{5}$$

where ψ and U is the phase difference and voltage in the single JJ, relatively. We use the same notations for JJ in the stack and notations with index 1 for single JJ: C_1-capacitance, I_{c1}-critical current, R_1-resistance for single JJ, and C_{st}-capacitance, $I_{c,st}$-critical current, R_{st}-resistance for JJ in the stack.

We introduced dimensionless parameters $C = \frac{C_1}{C_j}$, $I_c = \frac{I_{c1}}{I_c}$ and $R = \frac{R_1}{R_j}$. Matrix B is determined by

$$\hat{B} = \begin{pmatrix} C+1 & C & \dots & & C \\ C & C+1 & C & \dots & C \\ \vdots & \vdots & \vdots & \vdots & \vdots \\ C & \dots & C & C+1 & C \\ C & & \dots & C & C+1 \end{pmatrix};$$

We solve the system of nonlinear second-order differential equations (5) using the fourth order Runge-Kutta method. Results for the following cases (b) $N_2 = 1$; (c) $N_2 = 2$; (d) $N_2 = 3$; (e) $N_2 = 9$; (f) $N_2 = 10$. We see strengthening of branching with increase in N_2, see Fig. 4(c) to (f). Detailed analysis of these CVC will be published elsewhere.

4. Summary

We studied the resonance features of coupled Josephson junctions and demonstrated a strong effect of the external radiation on the CVC in the parametric resonance region. A crucial changes in CVC found at the double resonance condition when radiation frequency coincides with Josephson and resonance circuit frequencies.

5. Acknowledgments

The work was supported by Heisenberg-Landau and Bogoliubov-Infeld Programs, Collaboration between JINR, Dubna and Slovakia, by the Slovak Research and Development Agency under the contract No. APVV-0494-11, Russian Fund for Basic Research (RFBR) under grant 12-02-90901-mob-sng-st., and JINR under grant 12-302-03. E.I. acknowledges a support by the German Ministry for Education and Science under Grant No. 13N9307 (project THz-Videocam).

6. References

[1] Kleiner R, Steinmeyer F, Kunkel G and Müller P 1992 *Phys. Rev. Lett.* **68** 2394
[2] Krasnov V M 2011, *Phys. Rev. B* **83** 174517
[3] Kurter C, Zhuravel A P, Ustinov A V and Anlage S M 2011 *Phys. Rev. B* **84** 104515
[4] Benseman T M, Koshelev A E, Gray K E, Kwok W -K, Welp U, Kadowaki K, Tachiki M and Yamamoto T 2011 *Phys. Rev. B* **84** 064523
[5] Koshelev A E 2010 *Phys. Rev. B* **82** 174512
[6] Pfeiffer J, Abdumalikov Jr A A, Schuster M and Ustinov A V 2008 *Phys. Rev. B* **77** 024511
[7] Yurgens A A 2000 *Supercond. Sci. Technol.* **13** R85

[8] Ozyuzer L *et al.* 2007 *Science* **318** 1291
[9] Shukrinov Yu M and Mahfouzi F 2007 *Phys. Rev. Lett.* **98** 157001
[10] Shukrinov Yu M, Mahfouzi F and Suzuki M 2008 *Phys. Rev. B* **78** 134521
[11] Shukrinov Yu M and Rahmonov I R 2010 *JETP Lett.* **92** 327; *Pis'ma v ZhETF* **92** 364
[12] Koyama T and Tachiki M 1996 *Phys. Rev. B* **54** 16183
[13] Ryndyk D A 1998 *Phys. Rev. Lett.* **80** 3376
[14] Shukrinov Yu M, Mahfouzi F and Seidel P 2006 *Physica C* **449** 62
[15] Irie A, Shukrinov Yu M and Oya G 2008 *Appl. Phys. Lett.* **93** 152510
[16] Shukrinov Yu M and Rahmonov I R 2012 *JETP* **115** 289; *ZhETF* **142** 323
[17] Bulaevskii L N, Dominguez D, Maley M, Bishop A and Ivlev B 1996 *Phys. Rev. B* **53** 14601
[18] Machida M, Koyama T and Tachiki M 1999 *Phys. Rev. Lett.* **83** 4816
[19] Ryndyk D A, Keller J and Helm C 2002 *J. Phys.: Condens. Matter* **14** 815
[20] Helm Ch, Preis Ch, Forsthofer F, Keller J, Schlenga K, Kleiner R and Müller P 1997 *Phys. Rev. Lett.* **79** 737; Preis Ch, Helm Ch, Schmalzl K, Keller J, Kleiner R and Müller P 2001 *Physica C* **362** 51
[21] Shukrinov Yu M, Nasrulaev Kh, Sargolzaei M, Oya G and Irie A 2002 *Supercond. Sci. Technol.* **15** 178
[22] Shukrinov Yu M, Mahfouzi F, Pedersen N F 2007 *Phys. Rev. B* **75** 104508
[23] Tsujimoto M *et al.* 2012 *Phys. Rev. Lett.* **108** 107006
[24] Lin S -Z, Hu X and Bulaevskii L 2011 *Phys. Rev. B* **84** 104501
[25] Koshelev A E 2010 *Phys. Rev. B* **82** 174512
[26] Krasnov V M 2011 *Phys. Rev. B* **83** 174517
[27] Grib A and Seidel P 2009 *Physica Status Solidi RRL* **3** 302
[28] Wang H B, Wu P H and Yamashita T 2001 *Phys. Rev. Lett.* **87** 107002
[29] Shukrinov Yu M, Rahmonov I R and Gaafar M A 2012 *Cond-Mat ArXiv* 0485172
[30] Shukrinov Yu M, Rahmonov I R and El Demery M 2010 *Journal of Physics: Conference Series* **248** 012043
[31] Shukrinov Yu M, Rahmonov I R and El Demery M 2010 *The 7th International Symposium on Intrinsic Josephson Effects and Plasma Oscillations in High-Tc Superconductors (PLASMA 2010)* Hirosaki Japan April 25-28
[32] Irie A, Shukrinov Yu M and Oya G 2008 *Journal of Physics: Conference Series* **129** 012029
[33] Turkoglu F, Koseoglu H, Demirhan Y, Ozyuzer L, Preu S, Malzer S, Simsek Y, Muller P, Yamamoto T and Kadowaki K *unpublished*
[34] Likharev K K 1985 *Introduction to the dynamics of Josephson junctions* Moscow, Izdatel'stvo Nauka 320 pp (In Russian)
[35] Darula M, Beuven S, Siegel M, Darulova A and Seidel P 1995 *Appl. Phys. Lett.* **67** 1618
[36] Grib A N, Seidel P, Scherbel J 2002 *Phys. Rev. B* **65** 094508
[37] Shukrinov Yu M, Rahmonov I R and Kulikov K 2012 *unpublished*
[38] Tachiki M, Ivanovic K, Kadowaki K and Koyama T, 2011 *Phys. Rev. B* **83** 014508
[39] Zhou T, Mao J, Cui H, Zhao X, Fang L and Yan S 2009 *Physica C* **469** 785
[40] Werthamer N R and Shapiro S 1967 *Phys. Rev.* **164** 523
[41] Dalsgaard Jensen H, Larsen A and Mygind J 1990 *Physica B* **165-166** 1661
[42] Dempsey D G, Levinsen M T and Ulrich B T 1975 *IEEE Trans. Mag. MAG-11* 811
[43] Calander N, Claeson T and Rudner S 1982 *Physica Scripta* **25** 837
[44] Bauer M, Krüger U and Martienssen W 1989 *Europhys. Lett.* **9** 191
[45] Dana S K, Sengupta D C and Edoh K D 2001 *IEEE Trans Circuits Systems* **48** 990
[46] Dana S K, Roy P K, Sethia G C, Sen A and Sengupta D C 2006 *IEE Proc.-CircuitsDevices Syst.* **153** 453
[47] Shukrinov Yu M, Seidel P, Ilichev E, Rahmonov I R and Kulikov K *unpublished*
[48] Ben-Jacob E and Imry Y 1981 *J. Appl. Phys.* **52** 6806
[49] Voss R F, Laibowitz R B, Broers A N, Raider S I, Knoedler C M and Viggiano J M 1981 *IEEE Trans. Magnetics* **17** 395
[50] Schmidt W -D, Seidel P and Heinemann S 1985 *Phys. Stat. Sol. (a)* **91** K155
[51] Darula M, Seidel P, Busse F and Benacka S 1993 *J. Appl. Phys.* **74** 2674
[52] Grib A N, Seidel P and Darula M 1998 *J. Low Temp. Phys.* **112** 323
[53] Krasnov V M 2002 *Physica C* **368** 246
[54] Irie A and Oya G 2005 *IEEE Trans Appl Supercond* **15** 813
[55] De Luca R and Romeo F 2005 *J Appl Phys* **98** 073904

Dubna-Nano 2012 IOP Publishing
Journal of Physics: Conference Series **393** (2012) 012021 doi:10.1088/1742-6596/393/1/012021

Shapiro and parametric resonances in coupled Josephson junctions

Ma A Gaafar[a,c], **Yu M Shukrinov**[a,b] **and A Foda**[d]

a BLTP, JINR, 141980 Dubna, Moscow Region, Russia
b Max-Planck-Institute for the Physics of Complex Systems, 01187 Dresden, Germany
c Faculty of Physics, Philipps-Universität Marburg, 35032 Marburg, Germany
d Nile University, Smart Village, Cairo, Egypt

E-mail: mgaafar@theor.jinr.ru

Abstract. The effect of microwave irradiation on the phase dynamics of intrinsic Josephson junctions in high temperature superconductors is investigated. We compare the current-voltage characteristics for a stack of coupled Josephson junctions under external irradiation calculated in the framework of CCJJ and CCJJ+DC models.

1. Introduction

Since the discovery of the intrinsic Josephson effect in high temperature superconductors (HTSC) such as $Bi_2Sr_2CaCu_2O_y$ (BSCCO), numerous studies of the basis nature, nonlinear dynamics and device applications of the superconductors have been extensively developed. The high-frequency response of Intrinsic Josephson junctions (IJJs) have been constantly updated. One of the most spectacular indications of the Josephson effect is locking of Josephson oscillations to the frequency of external microwave (MW) irradiation. As the Josephson frequency is proportional to voltage, the locking leads to appearance of steps at quantized voltages, called Shapiro steps [1, 2]. The steps appear in the current-voltage characteristic (CVC) at discrete voltage values $V_n = nh\omega/2e$ (ω, the frequency of the applied MW signal; h, the Planck constant; e, the elementary charge; n, integer number).

It was shown by Koyama and Tachiki [3] that the system of equations for capacitively coupled Josephson junctions has a solution corresponding to the longitudinal plasma wave (LPW) propagating along the c-axis. So, the Josephson oscillations can excite the LPW by their periodical actions. The frequency of Josephson oscillations ω_J is determined by the voltage value in the junction, and at $\omega_J = 2\omega_{LPW}$, where ω_{LPW} is LPW frequency, the parametric resonance (PR) is realized. External radiation essentially changes the physical picture of the coupled JJs. As known, the one-dimensional models with coupling between junctions capture the main features of real IJJs, like hysteresis and branching of the CVC, and help to understand their physics. So it is of interest to investigate this system under MW irradiation using these models.

In this article we investigate the phase dynamics of the coupled Josephson junctions under MW irradiation in the framework of both CCJJ and CCJJ+DC models.

2. Model and Method of Calculations

Because of the thickness of the S-layers in HTSC is extremely small and comparable to the Debye length r_D of the electric charge screening, therefore, there is no complete screening in a separate S-layer. It leads to the generalized scalar potential Φ_l of S-layer. The Φ_l is expressed through the electric scalar potential ϕ_l and the derivative of order parameter's phase $\dot{\theta}_l$ of S-layer l by $\Phi_l(t) = \phi_l - V_0\dot{\theta}_l$.

Without screening of the charge in S-layer the Josephson equation is generalized $\dot{\varphi}_l = V_l + \alpha(2V_l - V_{l-1} - V_{l+1})$ (where α is the coupling between junctions) and together with the expression for total current through the stack $I = \dot{V}_l + \sin\varphi_l + \beta V_l$ describe the phase dynamics of the coupled system of JJs in the framework of CCJJ model [3]. In CCJJ-model, the system of equations for the gauge-invariant phase differences $\varphi_l(\tau)$ between S-layers l and $l+1$ in the presence of MW irradiation is described by:

$$\frac{d^2}{d\tau^2}\varphi_l = (I - \sin\varphi_l - \beta\frac{d\varphi_l}{d\tau}) + \alpha(\sin\varphi_{l+1} + \sin\varphi_{l-1} - 2\sin\varphi_l) - A\sin\omega\tau \tag{1}$$

where I is the external bias current, A is radiation amplitude, ω is MW radiation frequency and β is the dissipation parameter (β related to the McCumber parameter β_c by $\beta_c = 1/\beta^2$). The CVC in CCJJ model are characterized by branching at $I = I_c$ (where I_c is the Josephson critical current) and strong branching in the hysteresis region: the number of branches in general case is equal to 2^N, where N is the number of junctions in the stack.

In CCJJ+DC model [4] the diffusion current $J_D^l = \Phi_{l+1} - \Phi_l$ is added and the expression for current through the stack of junctions in this model is $I = \dot{V}_l + \sin\varphi_l + \beta\dot{\varphi}_l$. Second order differential equation for φ_l has additional terms in compare with CCJJ model which are proportional to the product of coupling and dissipation parameters

$$\frac{d^2}{d\tau^2}\varphi_l = (I - \sin\varphi_l - \beta\frac{d\varphi_l}{d\tau}) + \alpha(\sin\varphi_{l+1} + \sin\varphi_{l-1} - 2\sin\varphi_l)$$
$$+\alpha\beta(\frac{d\varphi_{l+1}}{d\tau} + \frac{d\varphi_{l-1}}{d\tau} - 2\frac{d\varphi_l}{d\tau}) - A\sin\omega\tau \tag{2}$$

To find the CVC of the stack we solve the system of dynamical equations for phase differences using the fourth order Runge-Kutta method. We use a dimensionless time $\tau = t\omega_p$, where ω_p is the plasma frequency $\omega_p = \sqrt{2eI_c/\hbar C}$ and C is the capacitance. In our simulations we measure the voltage in units of $V_0 = \hbar\omega_p/(2e)$, the frequency in units of ω_p, the current and the amplitude of radiation in units of I_c. To study the time dependence of the charge in the S-layers, we use the Maxwell equation $\varepsilon\varepsilon_0\nabla.\vec{E}$ =Q, where ε and ε_0 are relative dielectric and electric constants. The charge density Q_l in the S-layer l is proportional to the difference between the voltages V_l and V_{l+1} in the neighbor insulating layers $Q_l = Q_0\alpha(V_{l+1} - V_l)$, where $Q_0 = \varepsilon\varepsilon_0 V_0/r_D^2$.

We solve the system of dynamical equations for phase differences at fixed value of bias current I in some time interval $(0, T_m)$ with the time step $\delta\tau$. This interval is used for time averaging procedure. Then we change the bias current by δI, and repeat the same procedure for the current $I + \delta I$ in new time interval $(T_m, 2T_m)$. The values of the phase and its time derivative in the end of the first time interval are used as initial conditions for second time interval and so on. The total recorded time is calculated as $t_r = t\omega_p + T_m(I_0 - I)/\delta I$, where I_0 is an initial value of the bias current for time dependence recording. In our simulations we put $T_m = 1000$, $\delta\tau = 0.05$, $\delta I = 0.0001$ and we add to the bias current a small noise with amplitude in the interval $(+10^{-8}, -10^{-8})$. The details concerning the numerical procedure are given in Ref. [5].

3. Results and Discussion

In our present work, we describe the case $\omega > 2\omega_{LPW}$, when the SS is above PRR in CVC. It was shown in Ref. [6] that an increase of the amplitude of irradiation leads to the appearance of an additional PR above SS called as radiation related parametric resonance (rrPR).

In Fig. 1(a) we show the CVC of the stack with 10 coupled JJs under microwave irradiation with $\omega = 2$ and $A = 0.5$ simulated in the framework of both the CCJJ and CCJJ+DC models at $\alpha = 0.05$, $\beta = 0.2$ and periodic boundary conditions. At $\omega = 0$, PR is characterized by breakpoint current $I_{bp} \simeq 0.28$, and breakpoint voltage $V_{bp} \simeq 11.5$, corresponding to the Josephson frequency $\omega_J = 1.15$.

As we can see, the first SS is developed on the outermost branch of CVC at $V = \omega_J * N = 20$. Dashed line stresses this fact. Both models are coincide with the increase in bias current. The difference between CCJJ and CCJJ+DC models appears in the branching part of CVC in hysteretic region. The CCJJ model demonstrates more intensive branching and larger hysteresis region in compare with CCJJ+DC model. This feature could be explained by the role of diffusion current [7].

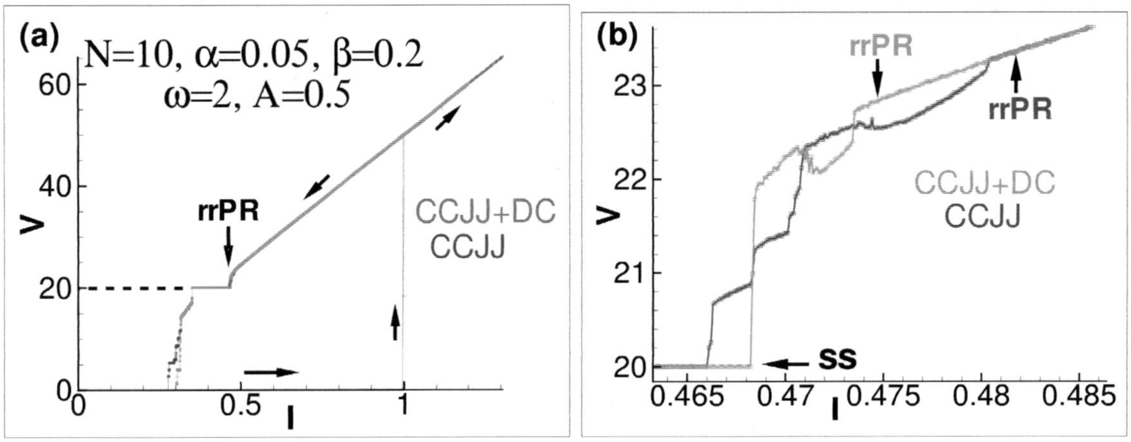

Figure 1. CVC of the stack with 10 coupled JJs under microwave irradiation with $\omega = 2$ and $A = 0.5$ simulated in the framework of the CCJJ and CCJJ+DC models at $\alpha = 0.05$, $\beta = 0.2$ and periodic boundary conditions; (b) The enlarged part of the rrPR above SS on the outermost branch of CVC.

As we mentioned previously, an increase of the amplitude of irradiation leads to the appearance of rrPR. Fig. 1(b) shows (for both models) an enlarged view of the rrPR region on the outermost branch of the CVC (the temporal oscillations of the charge in this region are shown in Fig. 2). We see that rrPR is started at higher currents for CCJJ than CCJJ+DC model. In addition, we observe an interesting structure before SS in CCJJ-model.

Fig. 2(a) demonstrates the temporal oscillations of the charge in the growing region of rrPR for the first S-layer of the stack with ten coupled JJs at $\alpha = 0.05$, $\beta = 0.2$, $\omega = 2$ and $A = 0.5$, combined with CVC of the outermost branch, calculated in the framework of CCJJ+DC model. The filled squares mark the bias current steps in CVC. We see in this figure the correlation between the charge-time dependence and CVC. The irradiation change the character of the charge-time dependence essentially and bring about a "bump" structure on the outermost branch of CVC, as shown in the figure. Analogies features in CVC were recently observed experimentally [8]. The inset shows the charge distribution among the layers in the growing region. The charge on the neighbor layers is equal in magnitude and opposite in sign. This is true for all adjacent layers and corresponds to the π-mode i. e. the wavelength of the created LPW is equal to $\lambda = 2d$.

The time dependence of the charge in the rrPR region calculated in the framework of CCJJ-model is presented in Fig. 2(b). We see that the charge on the S-layer appears at a higher value of current ($I_{bp} = 0.4809$) in comparison with the current value in case of CCJJ+DC-model ($I_{bp} = 0.4739$). The inset shows the charge distribution among the layers in the growing region, illustrates that the wavelength of LPW is the same as for CCJJ+DC-model ($\lambda = 2d$).

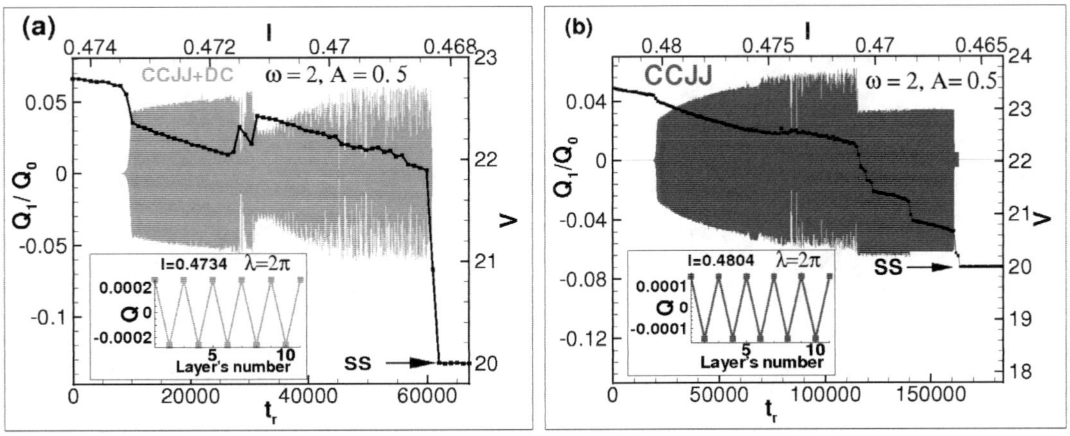

Figure 2. Charge time-dependence in the radiation related parametric resonance region of the stack with ten coupled JJs at $\alpha = 0.05$, $\beta = 0.2$, $\omega = 2$ and $A = 0.5$ calculated in the framework of (a) CCJJ+DC-model; (b) CCJJ-model. The insets show the charge distribution among the S-layers in the growing region.

4. Summary

We presented the CVC of the stack with 10 coupled Josephson junctions under microwave irradiation calculated in the framework of CCJJ and CCJJ+DC models. The CVC demonstrate different features in the radiation related parametric resonance region above SS. These results stress the essential role of the diffusion current in formation of the CVC of intrinsic Josephson junctions in high temperature superconductors. We showed that the radiation might bring to the "bump" structures in CVCs recently observed experimentally.

5. Acknowledgment

We would like to acknowledge the help received through the agreement between Egypt and Joint Institute for Nuclear Research, Dubna, Russia.

References

[1] Shapiro S 1963 *Phys. Rev. Lett.* **11** 80
[2] Tinkham M 1996 *Introduction to Superconductivity-2nd edition* New York: McGraw-Hill
[3] Koyama T and Tachiki M 1996 *Phys. Rev. B* **54** 16183
[4] Shukrinov Yu M, Mahfouzi F and Seidel P 2006 *Physica C* **449** 62
[5] Shukrinov Yu M, Mahfouzi F and Pedersen N 2007 *Phys. Rev. B* **75** 104508
[6] Shukrinov Yu M, Rahmonov I R and Gaafar M A 2012 *arXiv:1205.6923v1* [cond-mat.supr-con]
[7] Shukrinov Yu M and Rahmonov I R 2010 *JTEP letter* **92** 327
[8] Turkoglu F et al., *unpublished*

Dubna-Nano 2012 IOP Publishing
Journal of Physics: Conference Series **393** (2012) 012022 doi:10.1088/1742-6596/393/1/012022

Mathematical modeling of intrinsic Josephson junctions with capacitive and inductive couplings

I. R. Rahmonov[a,b], **Yu. M. Shukrinov**[a,b,c], **E. V. Zemlyanaya**[a], **I. Sarhadov**[a] **and O. Andreeva**[d]

[a] *Joint Institute for Nuclear Research, 141980 Dubna, Russia*
[b] *Umarov Physical Technical Institute, 734063 Dushanbe, Tajikistan*
[c] *Dubna International University for Nature, Society, and Man, 141980 Dubna, Russia*
[d] *OAO "Teplo Tyumeni", 625023 Tumen, Russia*

E-mail: rahmonov@theor.jinr.ru, ilhom-tj@inbox.ru

Abstract. We investigate the current voltage characteristics (CVC) of intrinsic Josephson junctions (IJJ) with two types of couplings between junctions: capacitive and inductive. The IJJ model is described by a system of coupled sine-Gordon equations which is solved numerically by the 4th order Runge-Kutta method. The method of numerical simulation and numerical results are presented. The magnetic field distribution is calculated as the function of coordinate and time at different values of the bias current. The influence of model parameters on the CVC is studied. The behavior of the IJJ in dependence on coupling parameters is discussed.

1. Introduction

Intrinsic Josephson effect arising in tunneling of Cooper pairs between superconducting layers CuO_2 inside of HTSC [1] gives a base to consider the HTSC as a system of coupled Josephson junctions (JJ). A stack of IJJ is one of the promising objects of superconducting electronics and studied intensively now [2, 3, 4, 5]. Recently discovered coherent electromagnetic radiation from the IJJ stack in the terahertz frequency range [6] opens a possibility for various applications. The coupled JJ system is described by the system of nonlinear and nonequilibrium equations [7, 8, 9, 10] which should be solved numerically. An important problem is related to the coupling between JJ [8, 9].

In this paper we study the CVCs and spatiotemporal features of long Josephson junctions with inductive and capacitive couplings within the framework of the model proposed by Machida and Sakai [9].

2. Model and method of simulation

We consider the IJJ stack of a length L where the width of the junction is much less than Josephson penetration depth λ_j. The superconducting and insulating layers have thicknesses d_s and d_i, respectively. Bias current I flows perpendicularly to the layers. The l-th Josephson junction is formed by l-th and $l-1$-th superconducting layers. It is described by the gauge-invariant phase difference $\varphi_l(t)$. The system of equations describing the phase dynamics of Josephson junctions with inductive and capacitive couplings has a form

$$\left\{\begin{aligned} \frac{\partial \varphi_l}{\partial \tau} &= \sum_{j=1}^{N} C_{lj} V_j \\ \frac{\partial V_l}{\partial \tau} &= \sum_{j=1}^{N} \left(\pounds_{lj}^{-1} \frac{\partial^2 \varphi_j}{\partial x^2} \right) - \beta V_l - \sin \varphi_l + I \end{aligned}\right. \tag{1}$$

where V_l is the dimensionless voltage of l-th JJ normalized to $V_0 = \hbar\omega_p/(2e)$, $\hbar$ – Planck's constant, e – charge of electron, $\omega_p = \sqrt{8\pi d_i e j_c/(\hbar\varepsilon)}$ – the plasma frequency, j_c – the critical current density, ε – the dielectric constant of insulating layer, $\beta = \sigma V_0/(d_i j_c)$ – the dissipation parameter, σ – the electrical conductivity, I is the bias current normalized to j_c.

The matrixes C and $\pounds$ describe the capacitive and inductive couplings, respectively and have the following form:

$$C = \begin{pmatrix} D^C & s^C & 0 & \dots & & & s^C \\ \vdots & \vdots & \vdots & \vdots & \vdots & \vdots & \vdots \\ \dots & 0 & s^C & D^C & s^C & 0 & \dots \\ \vdots & \vdots & \vdots & \vdots & \vdots & \vdots & \vdots \\ s^C & & & \dots & 0 & s^C & D^C \end{pmatrix} \quad \text{and} \quad \pounds = \begin{pmatrix} 1 & S & 0 & \dots & & S & \\ \vdots & \vdots & \vdots & \vdots & \vdots & \vdots & \vdots \\ \dots & 0 & S & 1 & S & 0 & \dots \\ \vdots & \vdots & \vdots & \vdots & \vdots & \vdots & \vdots \\ S & & & \dots & 0 & S & 1 \end{pmatrix}$$

where $D^C = 1 + (2\lambda_e/d_i)[\coth(d_s/\lambda_e)]$ and $s^C = -\lambda_e/[d_i \sinh(d_s/\lambda_e)]$ are effective electrical length and capacitive coupling parameter, respectively normalized to d_i, λ_e – Debye screening length, $S = s^L/D^L$ – normalized inductive coupling parameter, $D^L = d_i + 2\lambda_L \coth(d_s/\lambda_L)$ – effective magnetic thickness, $s^L = -\lambda_L/\sinh(d_s/\lambda_L)$ and λ_L – London penetration depths. Coordinate x is normalized to the Josephson penetration depth $\lambda_j = \sqrt{\hbar c^2/(8\pi e j_c D^L)}$ and time τ is normalized to ω_p^{-1}.

For numerical simulation of the system (1) we use two independent numerical methods. First method is based on the standard explicit finite-difference algorithm. Within the framework of another approach we employ the 2nd order finite-difference approximation of (1) only in the coordinate x and then, we use the 4th order Runge-Kutta algorithm for numerical solution of the resulting system of time-dependent ordinary differential equations. Both algorithms provide the same numerical results. Below, we present results obtained in the framework of the second approach.

Introducing the uniform grid in the interval $0 < x < L$ with the stepsize Δx we approximate the 2nd derivative of the phase difference with respect to the coordinate x in (1) as follows:

$$\frac{\partial^2 \varphi_l^1}{\partial x^2} = \frac{2(\varphi_l^2 - \varphi_l^1)}{\Delta x^2} - \frac{2H_{ext}}{\Delta x}; \quad \frac{\partial^2 \varphi_l^m}{\partial x^2} = \frac{\varphi_l^{m+1} - 2\varphi_l^m + \varphi_l^{m-1}}{\Delta x^2}; \quad \frac{\partial^2 \varphi_l^M}{\partial x^2} = \frac{2(\varphi_l^{M-1} - \varphi_l^M)}{\Delta x^2} + \frac{2H_{ext}}{\Delta x}$$

where H_{ext} is the external magnetic field and $\varphi_l^m = \varphi_l(x_m)$ (m is the x-node number, $1 \le m \le M$, $x_1 = 0$, $x_M = L$).

We fix quantity I and put $V_l = \varphi_l = 0$ as the initial conditions of our numerical simulation. At each step of time of the Runge-Kutta process we average the phase difference and voltage in coordinate as $\bar{V}(t) = 1/L \int_0^L V_l(x,t)dx$ using the Simpson quadrature formula for numerical integration. Then the voltage is averaged in time as $V = 1/(T_{max} - T_{min}) \int_{T_{min}}^{T_{max}} \bar{V}(t)dt$ using the rectangle quadrature method.

The presented procedure allows us to obtain the value of V for the fixed value I, i.e. the point of CVC. Increasing the current I by the stepsize ΔI, we repeat the simulation in accordance to the above algorithm. The current I increases up to I_{max}, then I is decreased back to zero.

Our simulations have been done with $\beta = 0.2$ and $H_{ext} = 0$. The system (1) is solved in the time interval $[0, 300]$ with the time stepsize $\Delta\tau = 0.025$ and coordinate stepsize $\Delta x = 4\Delta\tau$. The bias current is increased from the value $I = 0.01$ till $I_{max} = 1.1$ with the stepsize $\Delta I = 0.005$ and then decreased back to $I_{min} = 0$. In the interval $I = 0.2 \div 1$ the current step is taken $\Delta I = 0.0001$. A noise with amplitude $\delta I = \pm 10^{-8}$ is added to the bias current.

3. Results and discussions

Let us start with the consideration of the single JJ. Fig.1 (a) shows the one-loop CVC of the single JJ with length $L = 3$. We see branches related to the fluxon states in JJ shown by hole arrows.

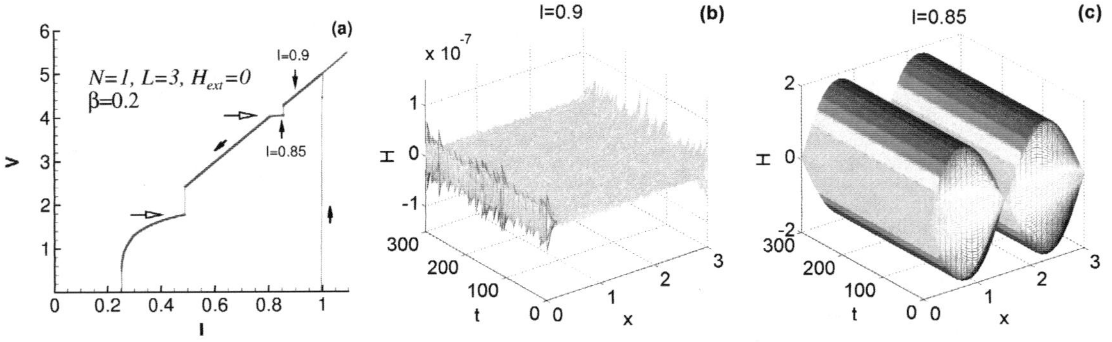

Figure 1. (a) CVC of the single JJ at $L = 3$, $\beta = 0.2$, $H_{ext} = 0$; (b) Spatiotemporal dependence of internal magnetic field $H(t, x)$ in JJ at $I = 0.9$; (c) The same at $I = 0.85$.

The fluxon branches are in the agreement with the similar results of the Refs [11, 12]. Position of branches is related to the number of fluxons in JJ. The spatiotemporal dynamics of the system is demonstrated in Fig.1 (b,c) where the distribution of the internal magnetic field $H(t, x)$ is presented for $I = 0.9$ and $I = 0.85$. At $I = 0.9$ the magnetic field strength is about 10^{-7}. The distribution of $H(t, x)$ at $I = 0.85$ is shown in Fig.1 (c) which demonstrates two fluxon states.

Let as now consider the system of JJ with the inductive coupling.

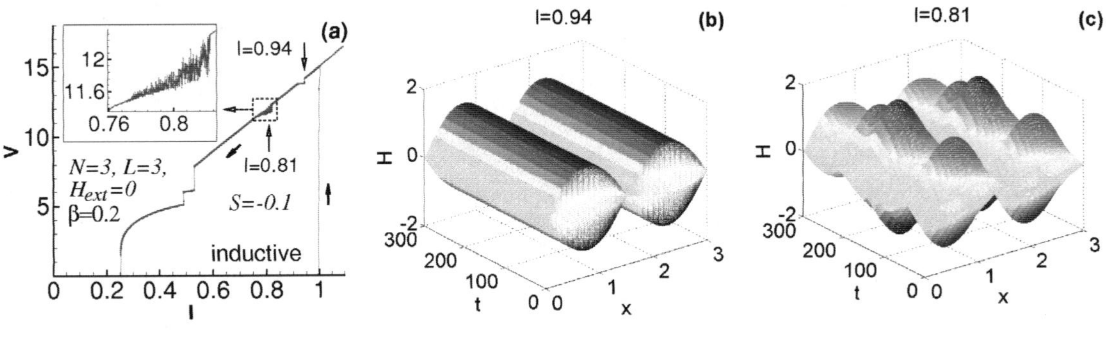

Figure 2. (a) CVC of the stack with $N = 3$ JJ with inductive coupling at $L = 3$, $\beta = 0.2$, $H_{ext} = 0$, $S = -0.1$; (b) Spatiotemporal dependence of magnetic field $H(t, x)$ in first JJ at $I = 0.9$; (c) The same at $I = 0.81$.

Fig.2 (a) demonstrates the CVC of the stack of 3 JJ with $L = 3$ taking into account the inductive coupling with coupling parameter $S = -0.1$. The value $I = 0.94$ corresponds to the additional branch in the CVC which is not observed in the case of the single junction. Fig.2 (b) shows the spatiotemporal dependence of magnetic field at $I = 0.94$. The two fluxon state is realized at this I value. In the interval of current $0.77 < I < 0.822$ the CVC demonstrates a chaotic behavior, while in the case of single junction chaotic feature is not observed. We assume that the chaotic behavior here is related to the appearing of unstable fluxons (see Fig.2 (c)).

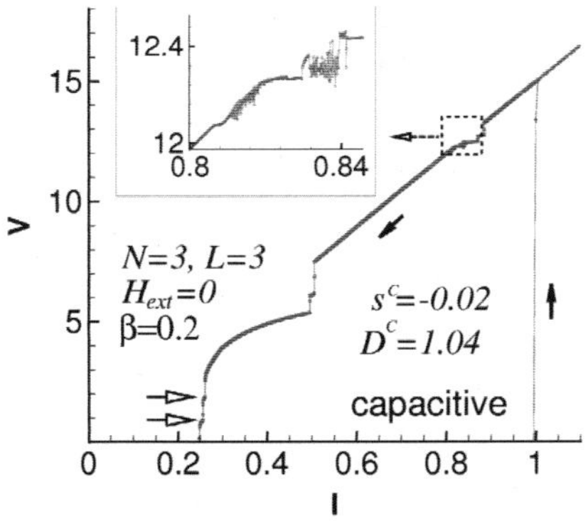

Figure 3. CVC of the stack with $N = 3$ JJ with capacitive coupling at $L = 3$, $\beta = 0.2$ $H_{ext} = 0$, $s^C = -0.02$, $D^C = 1.04$.

Fig.3 shows the CVC of the stack of 3 JJ with $L = 3$ taking into account the capacitive coupling between junctions with parameters $s^C = -0.02$, $D^C = 1.04$. CVC illustrates a chaotic behavior like in the inductive coupling case. We note that some branches in CVC are not related to the fluxons. In particular, branches in the CVC at $V = 1.87235$ and $V = 0.908239$, which shown with hollow arrows correspond to the states with one and two JJ in rotating state, respectively.

4. Conclusions

We developed the simulation method of CVC for the system of JJ with inductive and capacitive couplings and demonstrate the fluxon branch and chaotic behavior. We showed that chaotic behavior is related to the appearance of unstable fluxons.

5. Acknowledgment

This work is supported by the Russian Foundation for Basic Research (RFBR) under grant **12-02-90901** and JINR under grant **12-302-03**.

6. References

[1] Kleiner R, Steimmeyer F, Kunkel G and Muller P 1992 *Phys. Rev. Lett.* **68** 2394 .
[2] Yurgens A A 2000 *Supercond. Sci. Technol.* **13** R85
[3] Benseman T M, Koshelev A E, Gray K E, Kwok W -K, Welp U, Kadowaki K, Tachiki M and Yamamoto T 2011 *Phys. Rev. B* **84** 064523
[4] Koshelev A E 2010 *Phys. Rev. B* **82** 174512
[5] Pfeiffer J, Abdumalikov Jr A A, Schuster M and Ustinov A V 2008 *Phys. Rev. B* **77** 024511
[6] Ozyuzer L et al 2007 *Science* **318** 1291
[7] Koyama T and Tachiki M 1996 *Phys. Rev. B* **54** 16183
[8] Sakai S, Bodin P and Pedersen N F 1992 *Jorn Appl. Phys* **73** 2411
[9] Machida M and Sakai S 2004 *Phys. Rev. B* **70** 144520
[10] Ryndyk D A 1998 *Phys. Rev. Lett.* **80** 3376
[11] Kaplunenko V K, Goldobin E B, Khabipov M I, Larsen Britt H , Mygind J and Pedersen N F 1993 *J. Appl. Phys.* **74** 5854
[12] Goldobin E and Ustinov A V 1999 *Phys. Rev. B* **59** 11 532

Dubna-Nano 2012 IOP Publishing
Journal of Physics: Conference Series **393** (2012) 012023 doi:10.1088/1742-6596/393/1/012023

Interconnection between static regimes in the LJJs described by the double sine-Gordon equation

P.Kh.Atanasova[1], E.V.Zemlyanaya[2], Yu.M.Shukrinov[2]

[1] University of Plovdiv, FMI, Plovdiv 4003, Bulgaria
[2] Joint Institute for Nuclear Research, Dubna 14198, Russia.

E-mail: atanasova@uni-plovdiv.bg

Abstract. The second harmonic contribution to the current-phase relation changes the properties of the static magnetic flux distributions in the long Josephson junction (LJJ) and inspires new homogenious and fluxon static states. We study stability properties and bifurcations of these static regimes within the frame of a model described by the double sine-Gordon equation. The critical curves behavior and the interconnection between different types of magnetic flux distributions are analyzed.

1. Motivation and model

Long Josephson junctions (LJJs) are attractive objects of theoretical and experimental investigations because of a wide range of applications in nano- and quantum physics. Important step of their experimental study is the measurement of *static* properties including dependence of the critical current on the external magnetic field [1].

In this paper we investigate the static magnetic flux distributions in LJJ's model described by the boundary value problem for the double sine-Gordon equation (2SG) [2]:

$$-\varphi'' + a_1 \sin\varphi + a_2 \sin 2\varphi - \gamma = 0\,, \quad x \in (-l; l)\,, \quad \varphi'(\pm l) = h_e\,. \qquad (1)$$

Here the prime means a derivative with respect to the coordinate x, γ is the external current, l – the semilength of the junction, a_1 and a_2 – the normalized amplitudes of the first and second harmonics of the Josephson current [3, 4], h_e – the external magnetic field. All the magnitudes are assumed to be dimensionless [2].

As previously in [5, 6, 7, 8, 9], we analyze stability and bifurcations of static solutions $\varphi(x,p)$, $p = (l, a_1, a_2, h_e, \gamma)$ on the basis of numerical solution of the corresponding Sturm-Liouville problem [10]:

$$-\psi'' + q(x)\psi = \lambda\psi, \quad \psi'(\pm l) = 0, \quad q(x) = a_1 \cos\varphi + 2a_2 \cos 2\varphi, \quad \int_{-l}^{+1} [\psi(x)]^2\, dx = 1. \qquad (2)$$

The minimal eigenvalue $\lambda_0(p) > 0$ corresponds to the stable solution. In case $\lambda_0(p) < 0$ solution $\varphi(x,p)$ is unstable. The case $\lambda_0(p) = 0$ indicates the bifurcation with respect to one of the parameters p. Details of numerical approach are given in [5, 8].

In order to characterize the solutions of Eq.(1) we calculate two quantities:

- full magnetic flux of the distribution $\Delta\varphi = \varphi(l) - \varphi(-l)$;
- quantity $N(p) = \frac{1}{2l\pi}\int_{-l}^{l}\varphi(x)\,dx$ denoted as "number of fluxons" in [5].

The second harmonic contribution plays important role in the modeling of the SNINS and SFIFS junctions (where S is a superconductor, I is an insulator, N is a normal metal and F is a weak metallic ferromagnet) [3, 11]. The second harmonic factor in the LJJ current-phase relation deforms the shape and changes the properties of standard static magnetic flux distributions. Also, the nonzero a_2 inspires new homogenious and fluxon static states [1, 5, 9, 12].

In this contribution we study the effect of second harmonic factor a_2 on the interconnection between static magnetic field distributions. We show that nonzero a_2 gives a rise the stable mixed states of fluxons and leads to the "created-by-current" (CBC) states.

Here we present results for the case of negative factor a_2 only.

2. Results and discussion

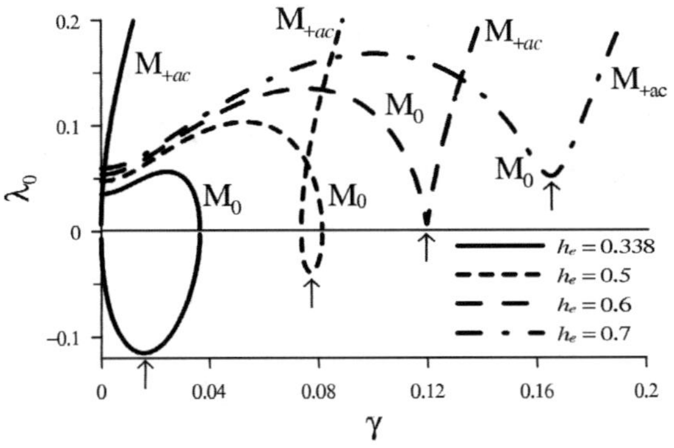

Figure 1. Dependence $\lambda_0(\gamma)$ for M_0 and M_{+ac} for $h_e \in [0.338; 0.7]$ and $a_1 = 1$, $a_2 = -0.7$, $2l = 10$. Vertical rows indicate the points of transformation of the branch M_0 to M_{+ac}.

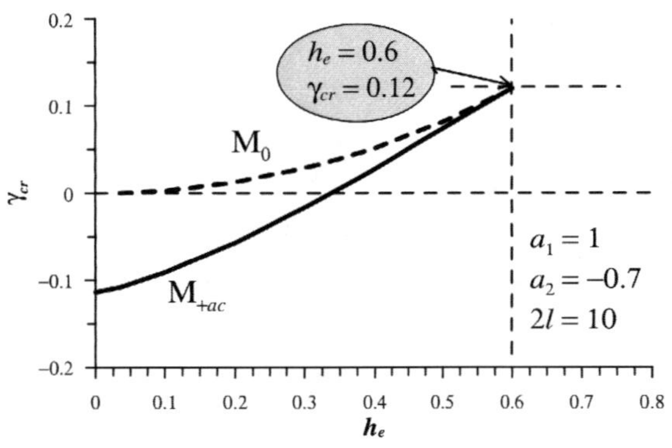

Figure 2. Fragment of $\gamma_{cr}(h_e)$ diagram: CBC-type critical curve for M_{+ac} distribution. Here $a_1 = 1$, $a_2 = -0.7$, $2l = 10$.

Together with the well-known distributions (standardly called M_π and M_0) the nonzero a_2 in Eq.(1) gives a rise another state (denoted as $M_{\pm ac}$ in [5]). Stability and bifurcations of $M_{\pm ac}$ have been investigated in [5, 9]. The states M_0, M_π and $M_{\pm ac}$ are constant for $h_e = 0$ and lose their uniformity as h_e is nonzero.

Interesting property of the a_2-inspired $M_{\pm ac}$ distributions is that for some values h_e they are unstable at $\gamma = 0$ and stabilize under nonzero current through the junction. This type of distribution was observed in LJJs with inhomogeneities [13] and denoted as "created-by-current" state (CBC). Let us demonstrate the CBC-properties of M_{+ac} state for the case $a_2 = -0.7$.

The M_{+ac} distribution is stable for $\gamma = 0$ as h_e is sufficiently small. However, as h_e is growing to $h_e > 0.338$ the solution M_{+ac} loses its stability at $\gamma = 0$ and stabilizes only for $\gamma = \gamma_{cr} > 0$. Corresponding dependence of minimal eigenvalue λ_0 on γ for several values h_e is given in Fig.1. It is seen that the curves $\lambda_0(\gamma)$ for M_0 and for M_{+ac} are interconnected. In the case $h_e = 0.5$, the stable portions of M_0- and M_{+ac}-curves overlap, i.e. stable M_0 and M_{+ac} states coexist in a narrow interval of h_e. For $h_e > 0.6$ the state M_0 smoothly transforms to M_{+ac}-distribution without a loss of stability, i.e. whole curve $M_0 \leftrightarrow M_{+ac}$ is stable. So, the CBC-type M_{+ac} state is observed at $0.338 < h_e < 0.6$.

Corresponding critical curves $\gamma_{cr}(h_e)$ of M_0 and M_{+ac} distributions are shown in Fig.2. Dashed curve shows the bifurcation points $\lambda_0 = 0$ of the branches M_0 on Fig.1 (these points are placed on the right of vertical rows). Solid curve in Fig.2 corresponds the points $\lambda_0 = 0$ on the left of vertical rows (see Fig.1) which belong the branches M_{+ac}. The solid and dashed critical curves in Fig.2 meet at the point $h_e = 0.6$, $\gamma_{cr} = 0.12$ where the state M_0 transforms to the M_{+ac} just at the bifurcation point $\lambda_0 = 0$, see the long-dashed curve on Fig.1.

Critical curves for M_{+ac} and φ^n $(n = 1, 2, 3, 4, 5)$ distributions in the case $a_1 = 1$, $a_2 = -0.7$, $2l = 10$ are shown in Fig.3. The curve demarcated M_{+ac} corresponds the critical points (h_e, γ_{cr}) where the solution M_{+ac} loses its stability as we continue the top branches $\lambda_0(h_e)$ (see Fig.2) in the growing h_e direction. We stress the strong overlapping of critical curves corresponding to the fluxon states φ^1, φ^2, φ^3, *etc.* as a result of the second harmonic contribution and the sufficiently large length of the junction. This means, at some values h_e we have two or more critical values of current where coexisting branches of fluxon distributions have a rise.

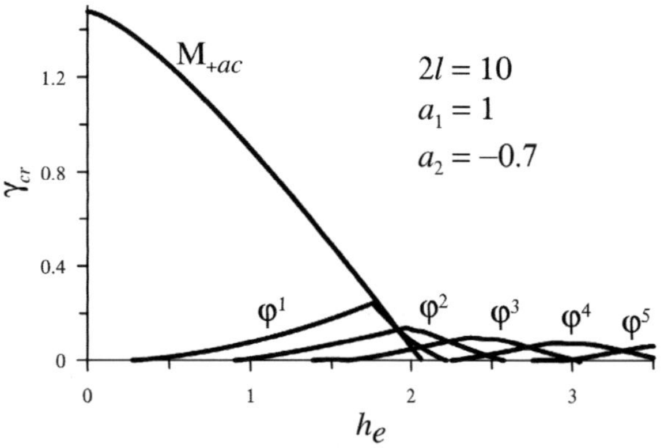

Figure 3. Critical curves for M_{+ac} and φ^n $(n = 1, 2, 3, 4, 5)$ distributions. Here $a_1 = 1$, $a_2 = -0.7$, $2l = 10$.

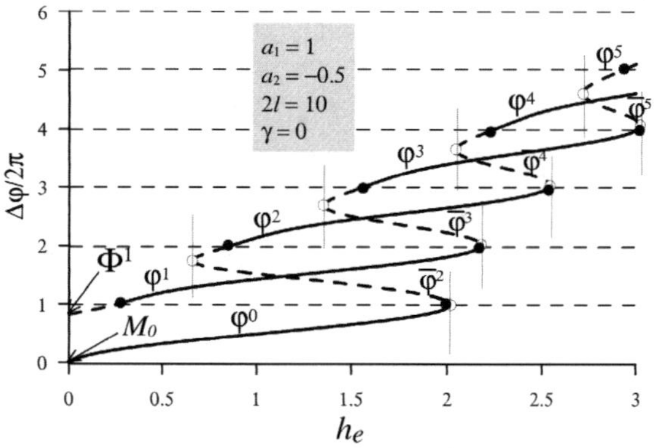

Figure 4. Full magnetic flux in dependence on h_e for M_0 and $\varphi^n, n \geq 1$; $a_2 = -0.5$, $a_1 = 1$, $2l = 10$, $\gamma = 0$.

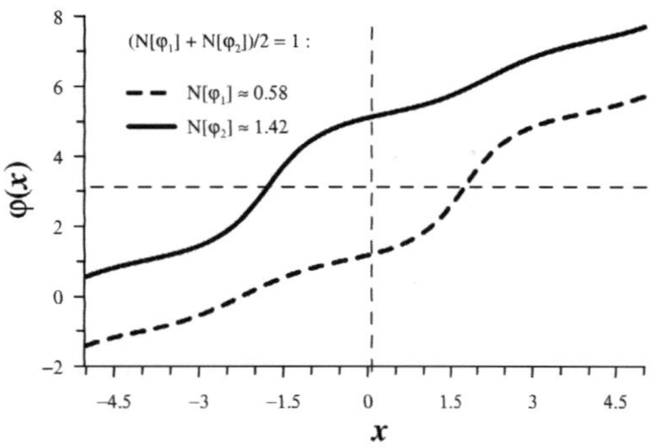

Figure 5. The magnetic flux distribution $\varphi(x)$ for the mixed stable bound state of fluxons with $h_e = 0.6$, $a_1 = 1$, $a_2 = -1$, $2l = 10$, $\gamma = 0$.

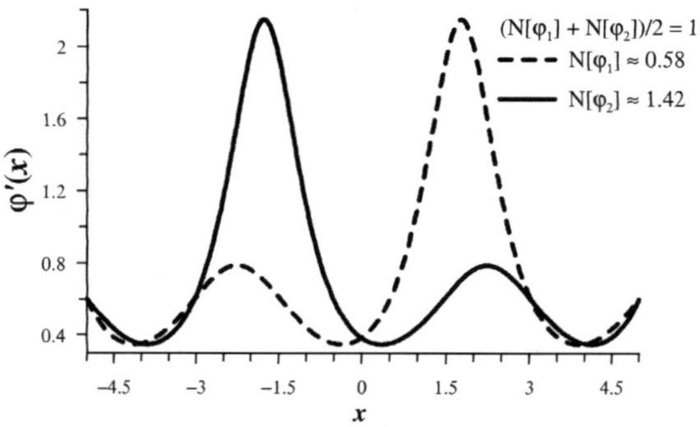

Figure 6. The internal magnetic field distribution $\varphi'(x)$ for the mixed stable bound state of fluxons with $h_e = 0.6$, $a_1 = 1$, $a_2 = -1$, $2l = 10$, $\gamma = 0$.

In our recent paper [8], we established, for $a_2 = 0$, the interconnection between the trivial solution $M_0|_{h_e=0}$ and the static fluxon distributions φ^n with even numbers n as well as the interconnection between the fluxon state $\Phi^1 = \varphi^1|_{h_e=0}$ and φ^n with odd numbers n. The same effect is observed in the case $a_2 \neq 0$. Interconnection between coexisting (stable and unstable) static distributions of Eq.(1) is demonstrated in Fig.4 where the full magnetic flux distribution $\Delta\varphi/2\pi$ is shown in dependence on h_e for $a_2 = -0.5$, $a_1 = 1$, $2l = 10$, $\gamma = 0$.

The $\Delta\varphi(h_e)$ branches in Fig.4 are obtained by numerical continuation of the solutions $M_0|_{h_e=0}$ and $\Phi^1 = \varphi^1|_{h_e=0}$ from the point $h_e = 0$ to the $h_e > 0$ direction. Stable and unstable branches are plotted, respectively, by solid and dashed curves. At the light circle points the curves $\Delta\varphi(h_e)$ turn back to another, upper, branches. The change of stability occurs at the dark circle points where the $\lambda_0(h_e)$ crosses zero. When the $\Delta\varphi(h_e)$ curve turns to the left ("$\supset$"-type turning) the quantity N is increased to $N+2$. So, the branch started from the $M_0|_{h_e=0}$ solution, joins the fluxons with the even N while the another branch (emanating from Φ^1 fluxon) connects fluxons with the odd quantity N.

The $\Delta\varphi(h_e)$ diagram in Fig.4 looks similar the corresponding diagram for $a_2 = 0$ in [8]. The further increasing of the a_2 contribution produces more complex interconnection between coexisting branches.

Beside of the mentioned above M_{ac} solution, one more a_2-inspired distribution is known, so called "small fluxon" [12, 9]. While the standard "large fluxon" Φ^1 has $N = 1$, the "small fluxon" is characterized by $N = 0$. Another a_2-inspired fluxon solution (denoted as Φ^{1*} in [7]) exists at $a_2 < -0.5$ and characterized by $N = 1$. Beside, our study shows that different types of fluxons can form the stable bound states. Mixed bound states φ_1 and φ_2 are shown in Figs.5,6 for $h_e = 0.6$, $a_1 = 1$, $a_2 = -1$, $2l = 10$, $\gamma = 0$. They are characterized by $(N[\varphi_1] + N[\varphi_2])/2 = 1$ with fractional quantities $N[\varphi_1]$ and $N[\varphi_2]$. Stability of the mixed states depends on a_2 and h_e. For the large a_2 these solutions stabilize at sufficiently small h_e.

3. Summary

We show that the second harmonic contribution inspires an increasing of a complexity of coexisting static distributions in the LJJs described by the 2SG equation. Nonzero a_2 gives a rise new (stable and unstable) static distributions and produces the CBC-type states. We expect, the numerically predicted effects can be confirmed in experimental observations.

We thank E.Goldobin for the stimulating discussions and important suggestions. The work was partially supported by the Program "JINR – Bulgaria". E.Z. is partially supported by RFFI under grant 12-01-00396. P.A. is partially supported by the project NI11-FMI-004.

References

[1] Goldobin E, Koelle D, Kleiner R, and Mints R G 2011 *Phys. Rev. Lett.* **107** 227001
[2] Likharev K K 1986 *Dynamics of Josephson junctions and circuits* Gordon and Breach, Philadelphia
[3] Golubov A A, Kypriyanov M Yu, Il'ichev E 2004 *Rev. Mod. Phys.* **76** 411–469
[4] Buzdin A, Koshelev A E 2003 *Phys. Rev. B.* **67** 220504(R)
[5] Atanasova P Kh, Boyadjiev T L, Shukrinov Yu M, Zemlyanaya E V 2011 *Math. Models and Computer Simulations* **3** No. 3 389-398
[6] Atanasova P Kh, Boyadjiev T L, Shukrinov Yu M, Zemlyanaya E V 2011 *Lecture Notes in Computer Sciences* **6046** 347-352
[7] Atanasova P Kh, Boyadjiev T L, Shukrinov Yu M, Zemlyanaya E V *e-print: cond-mat/1005.5691*; In: *Finite Difference Methods: Theory and Applications, Proc. 5th Intern. Conf.* (2010, Lozenetz, Bulgaria) Koleva M, Vulkov L (Eds), University of Rousse "Angel Kanchev" ISBN: 978-954-8467-44-5 (2011) 9-21
[8] Atanasova P Kh, Zemlyanaya E V, Shukrinov Yu M 2012 *Lecture Notes in Computer Sciences* **7125** 201-206
[9] Atanasova P Kh, Boyadjiev T L, Shukrinov Yu M, Zemlyanaya E V, Seidel P 2010 *J. Phys. Conf. Ser.* **248** 012044
[10] Galpern Yu S, Filippov A T 1984 *Sov. Phys. JETP.* **59** 894-903
[11] Ryazanov V V, Oboznov V A, Rusanov A Yu et al 2001 *Phys. Rev. Lett.* **36** 2427–2430
[12] Goldobin E, Koelle D, Kleiner R and Buzdin A 2007 *Phys. Rev. B.* **76** 224523
[13] Boyadjiev T L, Andreeva O Yu, Semerdjieva E G, Shukrinov Yu M 2008 *A Letters J. Exploring the Frontiers of Physics* **83** 47008

Plasmon nanooptics with individual single wall carbon nanotubes

I. V. Bondarev

Department of Physics, North Carolina Central University, 1801 Fayetteville Str., Durham, NC 27707, USA

E-mail: ibondarev@nccu.edu

M. F. Gelin and W. Domcke

Department of Theoretical Chemistry, Technical University of Munich, Lichtenbergstrasse 4, D-85747 Garching, Germany

Abstract. We discuss how low-energy collective interband plasmon excitations can be used to tune the optical properties of individual single wall carbon nanotubes. Two specific examples are considered. We show that interband plasmons can efficiently mediate enhanced electromagnetic absorption by pristine semiconducting carbon nanotubes and bipartite entanglement in hybrid metallic carbon nanotube systems. We develop a theory for (non-linear) optical monitoring and control of the above phenomena. Our findings pave the way for the development of a new generation of tunable optoelectronic device applications with individual carbon nanotubes.

1. Introduction

Carbon nanotubes (CNs) — graphene sheets rolled-up into cylinders of one to a few nm in diameter and up to one cm in length [1, 2] — have been successfully integrated into miniaturized electronic, electromechanical, and chemical devices [3], scanning probes [4, 5], and into nanocomposite materials [6, 7]. Over the past years, optical nanomaterials research has uncovered intriguing optical attributes of their physical properties, lending themselves to attractive device applications [8, 9, 10]. It is currently realized that pristine semiconducting CNs, which are direct-band semiconductors, can be used both to generate and to detect light [11, 12]. Recent studies of CN based optical nanoscale devices have shown that CNs are able to perform as single-photon sources for quantum computing, communication, or cryptography [13]. Hybrid CN systems containing extrinsic atomic type species (dopants) such as semiconductor quantum dots [14, 15], extrinsic atoms or ions [16, 17], are shown to be promising candidates for the development of the new generation of tunable nanooptoelectronic devices — both application oriented, e.g., photovoltaic devices of improved light-harvesting efficiency [18, 19], and for fundamental research including nanophotonics, nanoplasmonics, cavity quantum electrodynamics, and quantum information [13, 20, 21, 22]. Thus, nanotubes offer a great potential for building a unified optoelectronic technology based on the same material — new technology that, due to the outstanding mechanical and chemical stability of CNs, could be seamlessly integrated into state-of-the-art semiconductor nanotechnology.

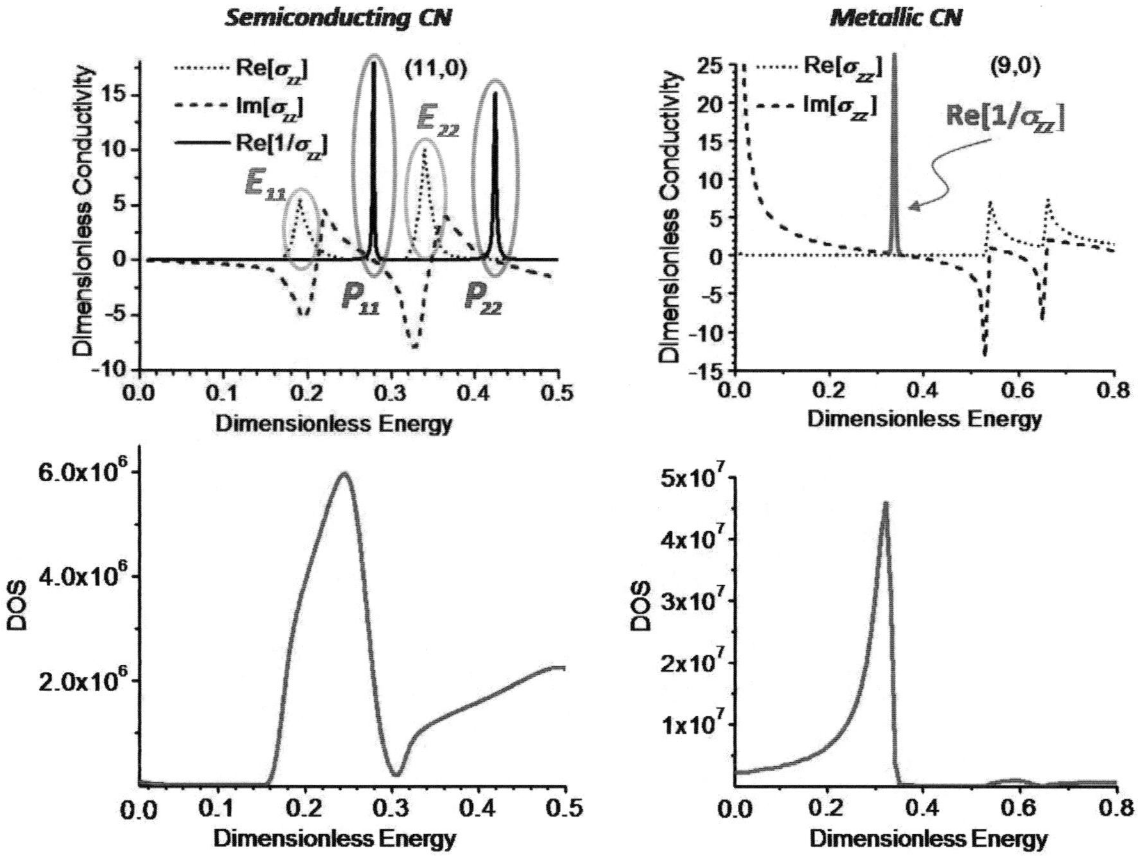

Figure 1. Surface dielectric response functions and local photonic density-of-states (DOS) functions for semiconducting (left panel) and metallic (right panel) CNs. *Top left*: Fragment of the energy dependence of the dimensionless (normalized by $e^2/2\pi\hbar$) axial surface conductivity σ_{zz} of the semiconducting (11,0) nanotube along the CN axis. Ovals mark exciton (E_{11}, E_{22}) and interband plasmon (P_{11}, P_{22}) excitations [peaks of $\mathrm{Re}\,\sigma_{zz}$ and $\mathrm{Re}(1/\sigma_{zz})$, respectively]. *Top right*: Same for the metallic (9,0) nanotube (plasmon resonance not to scale). *Bottom*: Local photonic DOS functions for the semiconducting (11,0) nanotube (left) and metallic (9,0) nanotube (right), when the TLS is placed on the nanotube symmetry axis. Dimensionless energy is defined as $[Energy]/2\gamma_0$, where $\gamma_0 = 2.7$ eV is the C-C overlap integral.

However, the true potential of CN based optoelectronic devices lies in the ability to tune their properties in a controllable way. Here, we discuss how low-energy collective interband plasmon excitations can be used to tune the optical properties of individual single wall CNs. We demonstrate that interband plasmons can efficiently mediate enhanced electromagnetic absorption by pristine semiconducting CNs and bipartite entanglement in hybrid metallic CN systems. We develop a theory for optical monitoring and control of the phenomena above.

2. Pristine semiconducting carbon nanotubes

In general, plasmons cannot be excited by light in optical absorption since they are longitudinal excitations, while photons are transverse. Small-diameter ($\sim$1 nm) semiconducting CNs, because of their quasi-one-dimensionality (1D), predominantly absorb the external

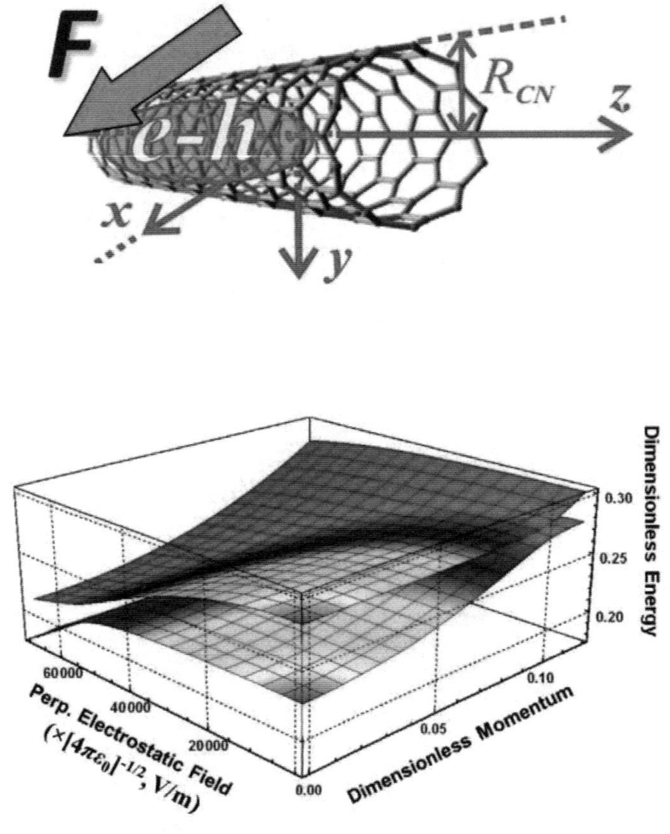

Figure 2. Exciton-plasmon dispersion relation (bottom) as a function of a perpendicular electrostatic field applied (schematic on the top) and longitudinal momentum for the lowest bright exciton [E_{11} in Fig. 1 (top left)] when coupled to the nearest interband plasmon [P_{11} in Fig. 1 (top left)] in the (11,0) nanotube. Dimensionless momentum is defined as $3b\,[Momentum]/2\pi\hbar$, where $b = 1.42$ Å is the C-C interatomic distance. See Fig. 1 caption for the dimensionless energy definition.

electromagnetic (EM) radiation polarized along the CN axis to excite excitons [23]. Figure 1 (top left) shows a typical dynamical dielectric response function for a semiconducting CN [24, 25]. One can see both of the collective excitations, both excitons and interband plasmons, to originate from the same electronic transitions and, therefore, to occur at similar energies (~ 1 eV), as opposed to bulk semiconductors, where they are separated by tens of eVs. These excitations do have different physical nature. Their coexistence in semiconducting CNs is a unique general feature of confined quasi-1D systems where the transverse electronic motion is quantized to form 1D bands and the longitudinal one is continuous.

Optically excited quasi-1D excitons have their transition dipole moments and translational quasi-momenta both directed along the CN axis (z axis of the problem — Fig. 2 (top); cylindrical coordinates are used). Therefore, they can couple to their neighboring longitudinal interband plasmon modes [24, 25]. The exciton-plasmon coupling can be controlled and tuned by means of an electrostatic field applied perpendicular to the CN axis [schematic in Fig. 2 (top)] via the quantum confined Stark effect (QCSE) [24, 26]. Figure 2 (bottom) shows a typical exciton-plasmon dispersion relation in this case [26]. We see the anti-crossing ~ 0.1 eV, both in the energy-momentum plane and in the energy-field plane, revealing the QCSE as an efficient tool to control the exciton-plasmon coupling strength in individual CNs.

The formation of coupled exciton-plasmon excitations can be viewed as an additional non-radiative channel (in addition to phonons [27] and defects [28]) for the exciton relaxation in CNs, where optically excited excitons decay into low-energy interband plasmons. In so doing, excitons generate the quanta of plasma oscillations on the CN surface, on the one hand, and this shortens their lifetime, on the other. Thus, by varying the exciton-plasmon coupling strength

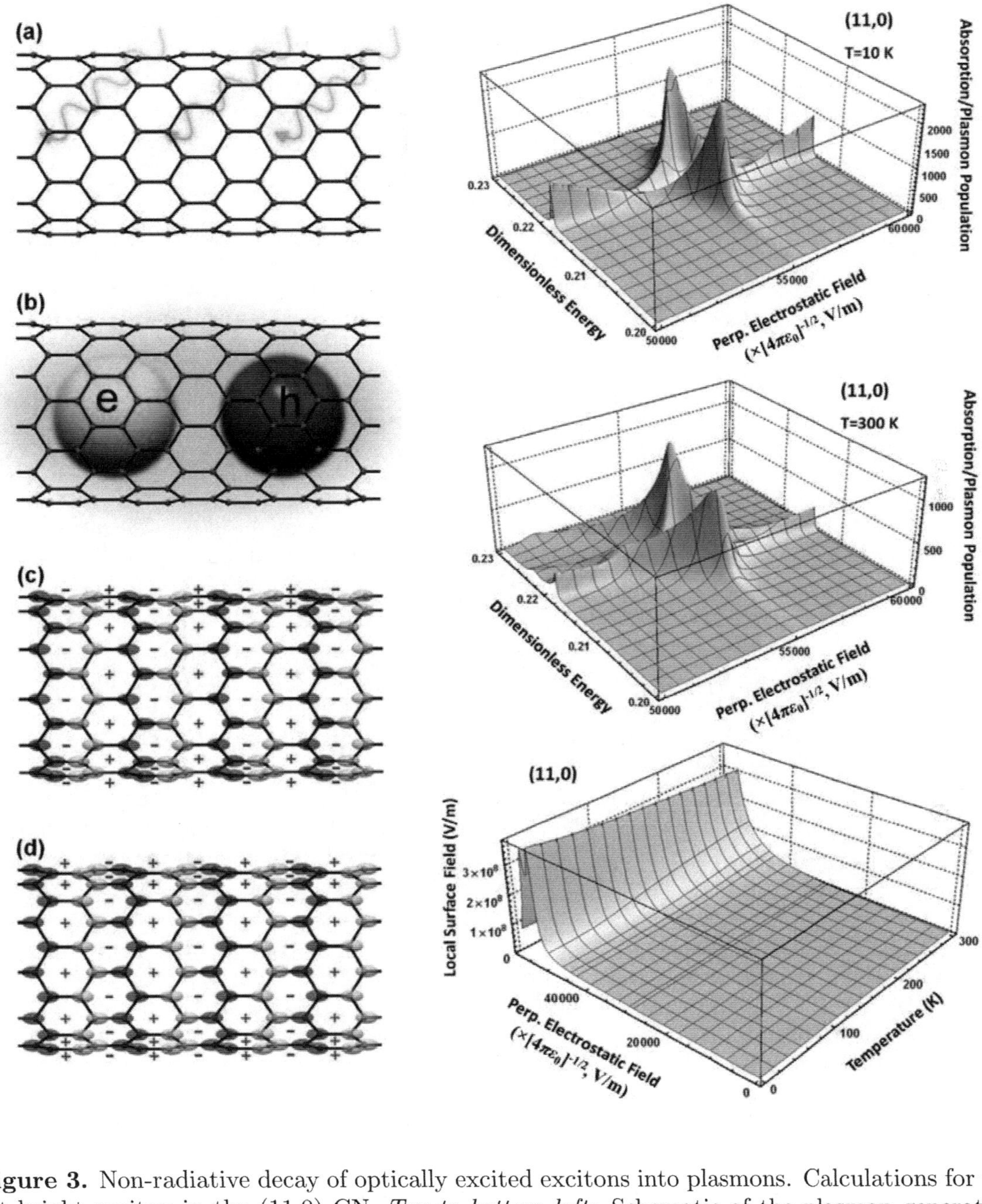

Figure 3. Non-radiative decay of optically excited excitons into plasmons. Calculations for the 1st bright exciton in the (11,0) CN. *Top to bottom left*: Schematic of the plasmon generation by the exciton. *(a),(b)* Exciton excitation by external EM radiation. *(c),(d)* Charge plasma oscillations (shown by + and − signs) produced by the non-radiative exciton decay are periodic opposite-phase displacements of electron shells relative to ion cores in the neighboring elementary cells (blue and yellow) of the CN. *Top to bottom right*: Plasmon population (also representing light absorption by excitons) and local surface field amplitude as functions of temperature and perpendicular electrostatic field applied. See Fig. 1 caption for the dimensionless energy.

using the QCSE, one controls both the radiative emission from an individual CN and surface electric field fluctuations associated with plasmons generated by excitons on the CN surface. This latter phenomenon is similar to the SPASER effect (Surface Plasmon Amplification by Stimulated Emission of Radiation) reported earlier for hybrid metal-semiconductor-dielectric nanomaterials [29]. In our case, the effect takes place in *individual* semiconducting CNs.

Plasma oscillations generated by the non-radiative exciton decay into the nearest interband plasmon mode are *standing* charge density waves due to the periodic opposite-phase displacements of the electron shells with respect to the ion cores in the neighboring elementary cells on the CN surface [Fig. 3 (left)]. Such periodic displacements induce local coherent oscillating electric fields of zero mean, but non-zero mean-square magnitude, concentrated locally across the diameter throughout the length of the nanotube. The mean-square longitudinal local field magnitude can be calculated as the expectation value $E_z^2(\mathbf{n})=\langle[-\boldsymbol{\nabla}_{\mathbf{n}}\,\hat{\varphi}(\mathbf{n})]_z^2\rangle$ of the quantum electrodynamical (QED) longitudinal-electric-field operator at the lattice site $\mathbf{n}=\{R_{CN},\varphi_n,z_n\}$, to result in [26]

$$E_z^2(\mathbf{n}) = \frac{\hbar L}{2\pi^2 c^2 R_{CN}} \sum_{\mathbf{k}=k_\varphi,k_z} f_{ex}(\mathbf{k},T) \int_0^\infty d\omega\,\omega^3\,\mathrm{Re}\frac{1}{\sigma_{zz}(\mathbf{k},\omega)}\left[N(\mathbf{k},\omega)+\frac{1}{2}\right]. \qquad (1)$$

Here, $N(\mathbf{k},\omega) = (4\pi c/L)I(\mathbf{k},\omega)$ is the number of plasmons generated by excitons with the momentum $\mathbf{k} = \{k_\varphi, k_z\}$ of the 1st Brillouin zone on the surface of the tubule of length L [circumferential confinement of the system results in k_φ being quantized to represent electron (e) and hole (h) subbands, while k_z remains continuous to characterize the longitudinal motion of the exciton with the effective mass $M_{ex}(k_\varphi) = m_e(k_\varphi) + m_h(k_\varphi)$], $I(\mathbf{k},\omega) = |C(\mathbf{k},\omega,t\to\infty)|^2$ is the exciton emission intensity distribution written in terms of the probability amplitude $C(\mathbf{k},\omega,t)$ to generate plasmons, and $f_{ex}(\mathbf{k},T) = \exp[-\hbar^2 k_z^2/2M_{ex}(k_\varphi)k_BT]/Q_{ex}$ is the exciton momentum distribution function with the partition function $Q_{ex}(T) = \sum_{\mathbf{k}}\exp[-\hbar^2k_z^2/2M_{ex}(k_\varphi)k_BT]$.

Equation (1) tells us that in order to produce strong local mean-square fields and related stronger optical absorption by excitons, the exciton emission resonance must overlap with the neighboring plasmon resonance. In other words, in terms of Fig. 1 (top left), the resonance of $\mathrm{Re}\,\sigma_{zz}(\omega)$ at $\omega\sim E_{11}$ [yielding the peak of $N(\mathbf{k},\omega\sim E_{11}) = (4\pi c/L)I(\mathbf{k},\omega\sim E_{11})$ in Eq. (1)] must overlap with the resonance of $\mathrm{Re}[1/\sigma_{zz}(\omega)]$ at $\omega\sim P_{11}$. We see in Fig. 2 (bottom) that this can be achieved by carefully tuning the exciton excitation energy and the nearest interband plasmon energy by means of the QCSE. Increasing the perpendicular electrostatic field brings E_{11} and P_{11} closer together, pushing the coupled exciton-plasmon system into the strong coupling regime where all of the optically excited excitons decay non-radiatively to generate interband (same-band) plasmons. This yields the peak optical absorption [Fig. 3 (top and middle right)] due to the efficient exciton-plasmon generation associated with the strong local mean-square fields induced by plasmons throughout the CN surface [Fig. 3 (bottom right)]. As the temperature grows, higher momenta excitons start contributing to the process, lowering the field necessary to achieve the strong coupling regime [blue contrast in Fig. 2 (bottom)].

3. Hybrid metallic carbon nanotube systems

In hybrid CN systems consisting of a metallic nanotube and extrinsic atomic type dopants, such as semiconductor quantum dots [14, 15], extrinsic atoms or ions [16, 17], the long-range EM interaction between the CN and the dopants [modeled by electronic two-level systems (TLS)] is theoretically demonstrated to enter the strong coupling regime as the TLS–CN distance decreases [30]. In the strong coupling regime, the two TLSs, that are properly positioned relative to each other and are close enough to the CN surface, can be significantly entangled even if they are a few tens of Angstroms apart [31]. This bipartite entanglement is due to the strong coupling of each individual TLS to the same surface plasmon resonance of the metallic

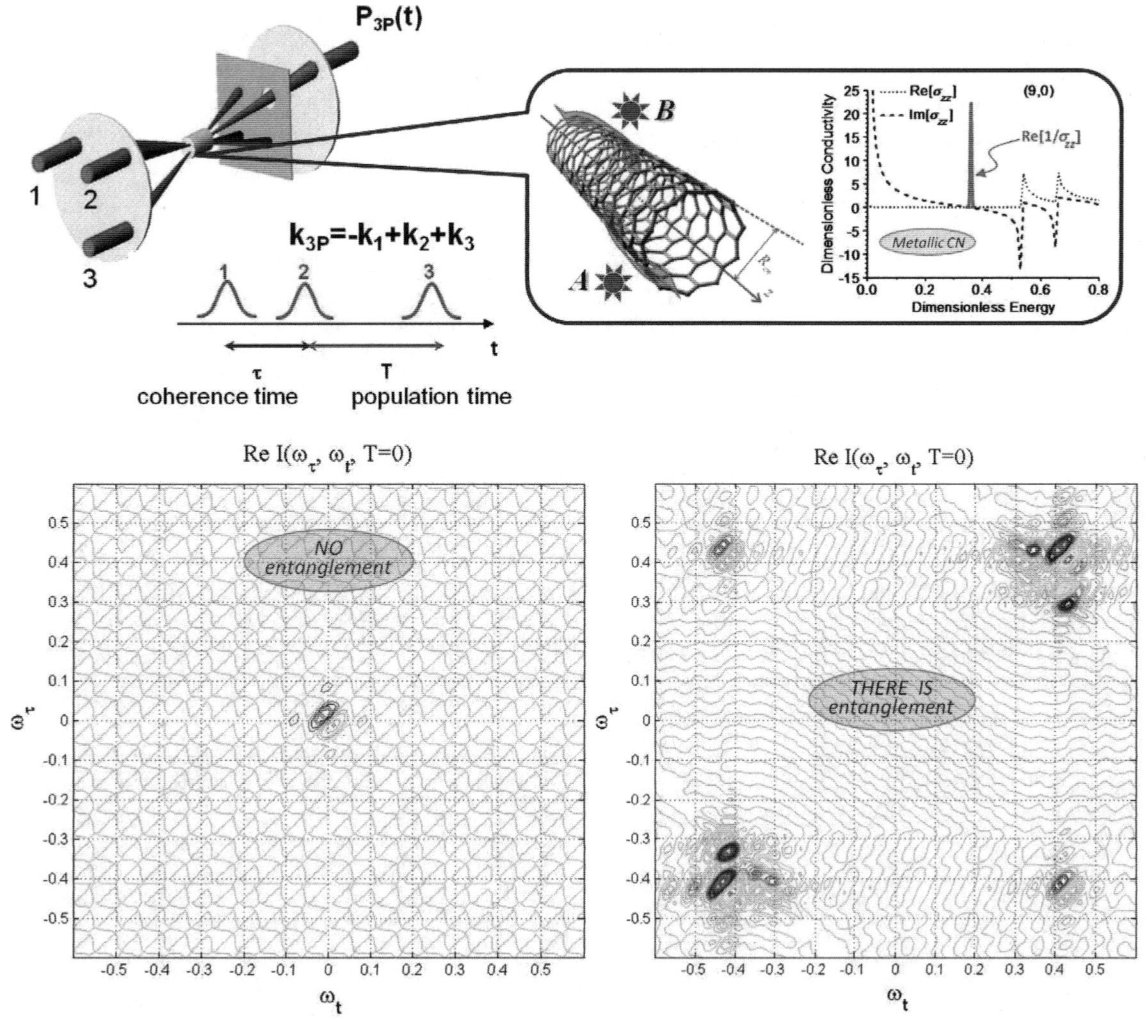

Figure 4. Calculations of the experimental 2D photon-echo signal for hybrid (atomically doped) metallic CN systems. *Top*: Schematic of the optical 2D photon-echo spectroscopy experiment for the detection of the bipartite entanglement between the two spatially separated dipole emitters A and B, which may be a pair of atoms, ions, or quantum dots, coupled to the same surface plasmon resonance mode of a metallic CN (sketch shown in the inset). *Bottom*: 2D optical photon-echo signals calculated in the cases where the bipartite entanglement is absent from (left graph), or present in (right graph) the system. The presence of the cross-peaks on the right panel indicates that the distant dipole emitters are coupled to the same CN surface plasmon resonance and "talk to each other" via the virtual plasmon exchange, being entangled.

CN [see an example in Fig. 1 (top right)]. The coupling is mathematically represented by the strong, sharply peaked, isolated resonance of the local density of photonic states (DOS) typical for metallic, and not for semiconducting, nanotubes (cf. left and right panels in Fig. 1).

Entanglement is a measure of strong quantum correlations (coherences) between quantum objects, which make the wave function of the correlated pair unseparable. To detect the entanglement by optical means, a spectroscopic technique is needed that would allow one to directly monitor coherences. 2D optical photon-echo spectroscopy has such a potential [32, 33]. To probe our system by means of the 2D optical spectroscopy, we have to consider its interaction with three laser pulses all polarized along the CN axis. It is this laser polarization that is needed to excite the two-level atomic dipole transition in order to couple it to the (same-energy) longitudinally polarized surface EM mode (the plasmon) of the CN. The three laser pulses will excite the TLSs and not the CN, since in metallic nanotubes there are no excitons in the optical spectral range (cf. top left and right panels in Fig. 1). Therefore, for the reasons discussed in Sec. 2 above, no optical laser excitation is possible of metallic CNs by the pulses in question.

Once one of the TLSs is laser-excited, it starts exchanging virtual surface plasmon excitations with its companion a distance away. This will make the spatially separated TLSs entangled (bipartite entanglement) through the coupling of each individual companion to a common coherent interaction energy exchange channel (the CN interband plasmon mode). To make this channel really coherent is not an easy, but doable, task. Indeed, if the TLS transition energy does not originally match the CN photonic DOS resonance [Fig. 1 (bottom right)], then it could still be possible to tune them up by using the QCSE. This narrows the metallic CN first excited band gap [34], shifting the photonic DOS resonance (the interaction energy exchange channel) to the red. At the same time, this leaves the TLS transition energies virtually unaffected (at least to the first order in the perpendicular electrostatic field applied), as they are selected to be excited by the laser field polarized along the CN axis and obey the selection rules that are different from those the perpendicular electrostatic field interaction follows. Thus, if the experiment is properly set up, the 2D photon-echo signal allows one to control and monitor the net response and the degree of the entanglement of the TLS dopants in the hybrid CN system.

In 2D photon-echo spectroscopy [32, 33], one is interested in the third-order non-linear polarization due to three external laser pulses, $P_{3P}(\tau,T,t)$, which obeys the phase-matching condition $\mathbf{k}_{3P}=-\mathbf{k}_1+\mathbf{k}_2+\mathbf{k}_3$ [see the schematic in Fig. 4 (top)]. The response signal of the system is then given by the double Fourier transform of the third order polarization

$$I(\omega_\tau,\omega_t,T)=i\int_{-\infty}^{\infty}d\tau dt\,\exp\left(-i\tau\omega_\tau+it\omega_t\right)P_{3P}(\tau,T,t). \tag{2}$$

Figure 4 (bottom) shows the 2D spectra for the (9,0) metallic nanotube with the pair of TLSs coupled to its low-energy plasmon resonance (sketch on the top of Fig. 4). The data are obtained using Eq. (2) and the methods of P_{3P} calculations developed in Ref. [35]. The left graph and the right graph show the spectra for the weak coupling and strong coupling regimes, respectively, with identical TLS–CN coupling constants for both TLSs and no relaxation. (The theory and more details on these calculations will be published elsewhere [36].) The presence of the cross-peaks along with the absence of the pronounced central peak in the 2D spectra are the signatures of the quantum entanglement in the hybrid CN system. The time evolution of the cross-peak intensities allows one to further monitor the entanglement dynamics in the system in the real time scale, thus demonstrating the superior diagnostic power of 2D photon-echo spectroscopy as applied to the phenomena discussed.

4. Conclusion

We consider two examples of how low-energy collective interband plasmon excitations in individual CNs can be used to tune their optical properties. For pristine semiconducting nanotubes, we show that the nonradiative exciton-to-plasmon energy transfer, whereby the external EM radiation absorbed to excite excitons transfers into the energy of surface plasmons, can efficiently mediate and greatly enhance the EM absorption. This enhancement is caused by the buildup of the macroscopic population numbers of coherent localized surface plasmons producing high-intensity local oscillating fields throughout the CN surface. The entire process can be controlled by means of the QCSE. Strong local coherent fields produced in this way can be used in various new technological applications of individual CNs, such as near-field sensing, optical switching, electromagnetic energy conversion, and materials nanoscale modification.

For hybrid metallic CN systems, we show that 2D photon-echo spectroscopy is a very sensitive tool for detecting the quantum entanglement of a pair of spatially separated, properly positioned dopant TLSs due to their coupling to the same surface plasmon resonance of a metallic CN. This bipartite entanglement can be identified through the presence of cross-peaks in the 2D photon-echo spectra.

Acknowledgments

I.V.B. is supported by the US DOE (DE-SC0007117) and US ARO (W911NF-11-1-0189). M.F.G. and W.D. acknowledge support from the DFG-Cluster of Excellence "Munich-Centre for Advanced Photonics (MAP)" (www.munich-photonics.de). I.V.B is partially supported by MAP.

References

[1] Dresselhaus M, Dresselhaus G and Avouris Ph (eds) 2001 *Carbon Nanotubes: Synthesis, Structure, Properties, and Applications* (Berlin: Springer-Verlag)
[2] Zheng L X, *et al* 2004 *Nature Materials* **3**, 673
[3] Baughman R H, Zakhidov A A and de Heer W A 2002 *Science* **297**, 787
[4] Popescu A, Woods L M and Bondarev I V 2008 *Nanotechnology* **19** 435702
[5] Popescu A and Woods L M 2009 Appl. Phys. Lett. **95** 203507
[6] Han J-H, *et al* 2010 Nature Materials **9** 833
[7] Dang X, Yi H, Ham M-H, Qi J, Yun D S, Ladewski R, Strano M S, Hammond P T and Belcher A M 2011 *Nature Nanotechnology* **6** 377
[8] Hertel T 2010 *Nature Photonics* **4** 77
[9] Avouris Ph, Freitag M and Perebeinos V 2008 *Nature Photonics* **2** 341
[10] Avouris Ph and Chen J 2006 *Materials Today* **9** 46 (2006)
[11] Mueller T, Kinoshita M, Steiner M, Perebeinos V, Bol A A, Farmer D B Avouris Ph 2010 *Nature Nanotechnology* **5** 27
[12] Xia F, Steiner M, Lin Y-M and Avouris Ph 2008 *Nature Nanotechnology* **6** 609
[13] Högele A, Galland Ch, Winger M and Imamoğlu A 2008, Phys. Rev. Lett. **100** 217401
[14] Olek M, Busgen T, Hilgendorff M and Giersig M 2006 *J. Phys. Chem.* B **110** 12901
[15] Haremza J M, Hahn M A and Krauss T D 2002 *Nano Lett.* **2** 1253
[16] Jeong G-H, Farajian A A, Hatakeyama R, Hirata T, Yaguchi T, Tohji K, Mizuseki H and Kawazoe Y 2003 *Phys. Rev.* B **68** 075410; *Thin Solid Films* **435** 307
[17] Khazaei M, Farajian A A, Jeong G-H, Mizuseki H, Hirata T, Hatakeyama R and Kawazoe Y 2004 *J. Phys. Chem.* B **108** 15529
[18] Robel I, Bunker B A and Kamat P V 2005 *Advanced Materials* **17** 2458
[19] Vietmeyer F, Seger B and Kamat P V 2007 *Advanced Materials* **19** 2935
[20] Steiner M, Freitag M, Perebeinos V, Naumov A, Small J, Bol A A and Avouris Ph 2009 *Nano Lett.* **9** 3477
[21] Bondarev I V 2010 *J. Comp. Theor. Nanosci.* **7** 1673
[22] Popescu A, Woods L M and Bondarev I V 2011 *Phys. Rev.* B **83** 081406(R)
[23] Ando T 2005 *J. Phys. Soc. Jpn.* **74** 777
[24] Bondarev I V, Woods L M and Tatur K 2009 *Phys. Rev.* B **80** 085407
[25] Bondarev I V, Tatur K and Woods L M 2009 *Opt. Commun.* **282** 661
[26] Bondarev I V 2012 *Phys. Rev.* B **85** 035448
[27] Perebeinos V, Tersoff J and Avouris Ph. 2005 *Phys. Rev. Lett.* **94** 027402
[28] Hagen A, Steiner M, Raschke M B, Lienau C, Hertel T, Qian H, Meixner A J and Hartschuh A 2005 *Phys. Rev. Lett.* **95** 197401
[29] Bergman D J and Stockman M I 2003 *Phys. Rev. Lett.* **90** 027402
[30] Bondarev I V and Vlahovic B 2006 *Phys. Rev.* B **74** 073401
[31] Bondarev I V and Vlahovic B 2007 *Phys. Rev.* B **75** 033402
[32] Cho M 2009 *Two-Dimensional Optical Spectroscopy* (Boca Raton: Taylor & Francis)
[33] Abramavicius D, Palmieri B, Voronine D V, Sanda F and Mukamel S 2009 *Chem. Rev.* **109** 2350
[34] O'Keeffe J, Wei C and Cho K 2002 *Appl. Phys. Lett.* **80** 676
[35] Gelin M F, Egorova D and Domcke W 2009 *Acc. Chem. Res.* **42** 1290
[36] Gelin M F, Bondarev I V and Meliksetyan A V 2012 *Chemical Physics* (at print)

Novel method of large area graphene fabrication

A. V. Frolov. Yu. I. Latyshev

IRE RAS, Mokhovaya 11-7, 101999 Moscow, Russia

fralek@mail.ru

Abstract. Studying of physical properties of graphene and graphene devices requires fabrication graphene samples of large area. Our fabrication method does not demand special expensive equipment and can be easily implemented. Single-layer graphene flakes with the size up to 200 μm and few-layer graphene flakes with the size up to 2000 μm were obtained by this technique. These flakes can be produced on various substrates including transparent ones. Atomically thin films of other layered materials can be also obtained using this method.

1. Introduction.

Graphene is a unique material, which has a whole spectrum of unusual physical properties. The discovery of graphene by A. Geim and K. Novoselov is a giant step in the development of micro- and nanoelectronics, optoelectronics, high-tensile materials and etc. It's possible to create in the future a large amount of devices, based on graphene. The main obstacle in the appearance of graphene devices is the following: the methods of fabrication of high-quality graphene flakes are complicated. This makes graphene scarce and expensive. A method described in this paper belongs to the group of graphene production methods by mechanical exfoliation of graphite (scotch-tape technique). One of these methods is the method invented by Manchester group of scientists headed by Andre Geim and Konstantin Novoselov in 2004 which is the prime method for graphene production. In comparison with the other methods of this group the method described in this article has the following advantages:

- *The size of graphene flakes.* The method of graphene fabrication by the mechanical exfoliation of graphite described in [1] allows to receive graphene flakes with the lateral size up to 200 μm. The method of reverse graphene exfoliation described in this article allows to increase the size of flakes. The lateral size of single-layer graphene flakes reaches up to 500 μm, the few-layer graphene up to 2000 μm.
- *The simplicity of graphene fabrication.* This method doesn't require special equipment, such as plasma reactor [2], MOCVD [3] or HPHT [4] apparata. Furthermore, the size of some flakes allows to make contacts even without electron beam lithography [1].
- *The broad spectrum of substrates.*The reversal exfoliation method allows to receive graphene flakes on various substrates, including the transparent ones. It allows to use graphene's unique optical properties. Graphene produce an unexpectedly high opacity for an atomic monolayer, with a startlingly simple value: it absorbs $\pi\alpha \approx 2.3\%$ of white light [5], where α is the fine-structure constant. Thus, it allows to estimate a number of graphene layers measuring absorption coefficient.

2. Fabrication technique.

The technique of graphene preparation, described below, has much in common with the original scotch-tape method invented by Konstantin Novoselov. However, there are quite significant differences between these approaches. Closest to our method is the method described in [6], where graphene is also fabricated using reverse exfoliation technique. The simple scheme of our method is shown at figure 1.

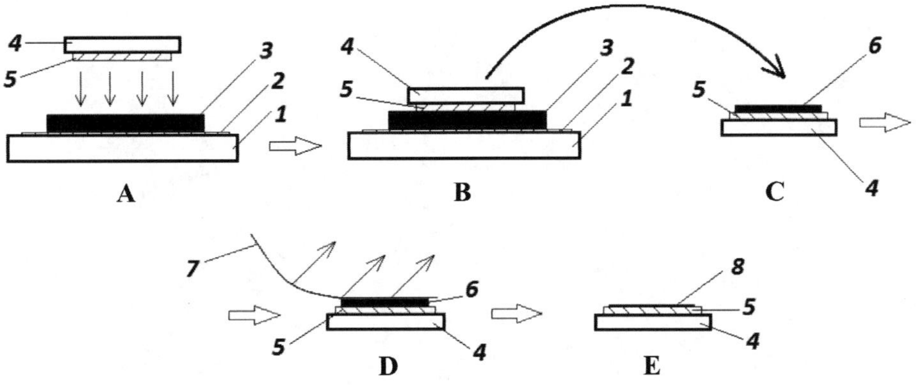

Figure1. The scheme of graphene production using reverse exfoliation technique.

The starting material for the production of graphene is highly oriented pyrolytic graphite (HOPG). At the first stage the original HOPG crystal 3 (Fig. 1) is attached to a substrate via an intermediate adhesive 2. As a substrate material is useful, for example, glass, silicon, polikor. As an adhesive 2 could use araldite or adhesive tape with two adhesive surfaces. Next, the surface of a layered crystal 3 must be clean and flat. To do so, the adhesive tape is attached to the surface of a HOPG crystal and then breaks away. If necessary, the cleanup procedure is repeated. Then on the clean surface HOPG crystal the basis substrate 4 is attached with glue 5 (fig.1B). After the glue has cured substrates are separated from each other (fig.1C). If the substrate is transparent, single-crystal fragments 6 (fig.1C), remaining on the substrate 4 may be investigated for the presence of light gray crystals of graphene with an optical microscope. The optical transparency of the corresponding single- or multiple-layer graphene is less than 20%. If the graphite crystals are still thick, the upper layers can be removed using an adhesive tape 7 (fig.1D). Removal of the upper layers occurs as long as at the substrate does not begin to appear single or multi-layer graphene flakes 8 (fig.1E).

3. The results obtained.

Single crystals of graphene, consisting of several layers, obtained as described above, are shown in figure 2. This photo was taken on an optical microscope, Carl Zeiss Axio Imager. Region 1 in Fig. 2 is a film of graphene, which consists of several layers. Area 2 is a section with no adhesive crystals of graphite. The value of light transmittance for a specific area was determined by the brightness of the image by using AxioVision. Change the light transmittance along the segment AB (Fig. 2) shown in Fig. 3. In the transition from area 2 in region 1 the transmission coefficient decreases by about 4%. From this it follows that the region 1 corresponds to approximately two layers of graphene.

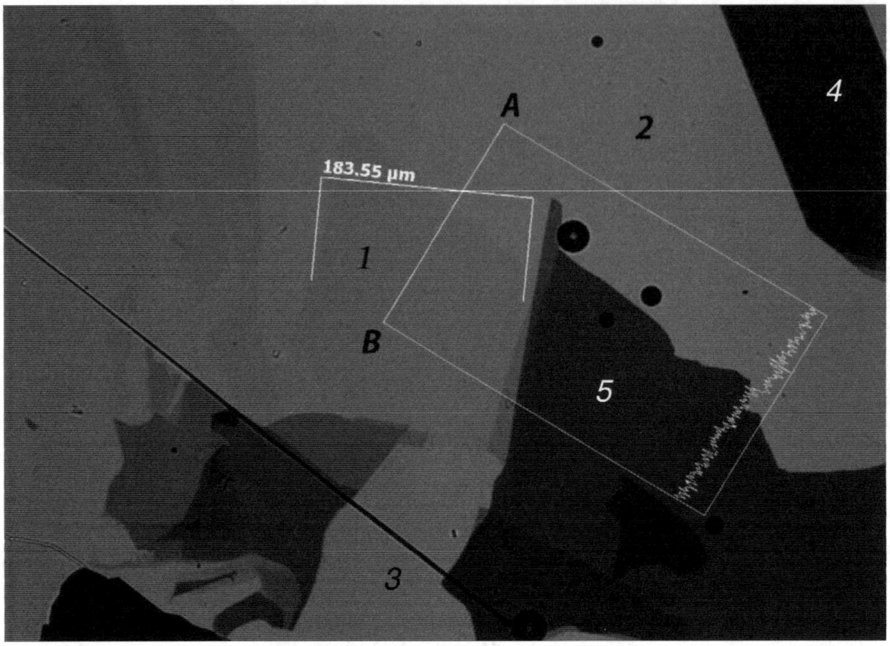

Figure 2. Microphotograph of the graphene film on the transparent substrate. Area 1-bilayer graphene, 2-epoxy glue, 3-monolayer graphene, 4-graphite, 5-few-layer (up to 30 layers) graphite.

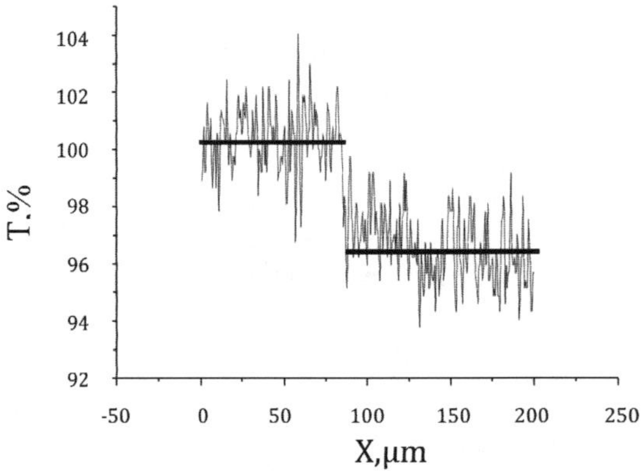

Figure 3. Transparency of the graphene along the line AB (Fig. 2).

The resulting graphene samples were investigated using Raman spectroscopy (RS). Fig. 4 shows the Raman spectrum for regions 3 and 4 (Fig. 2). For opaque graphite (area 4) is observed near the G-line is ~ 1582 cm^{-1} and 2D-line near ~ 2680 cm^{-1}. To the left of the 2D-line broadening is observed, characteristic of graphite. In addition, the amplitude of the G-line is much greater amplitude of the 2D-line. In area 3 of FIG. 2 2D-line is shifted by ≈ 50 cm^{-1} to lower frequencies and acquires a symmetric shape. Such a shift and shape of the 2D-lines are typical for single graphene. Also, the monolayer region 3 also indicates small amplitude of G-peak, which is comparable in magnitude to the amplitude of the 2D-line. These facts indicate that the area 3 of FIG. 2 with a lateral size of about 200 μm corresponds to a single-layer graphene. The additional lines appearing for graphene in the G-region, presumably correspond to the Raman scattering of glue by which the graphene is attached to the substrate. More detailed Raman spectroscopy of graphene is described in [7, 8].

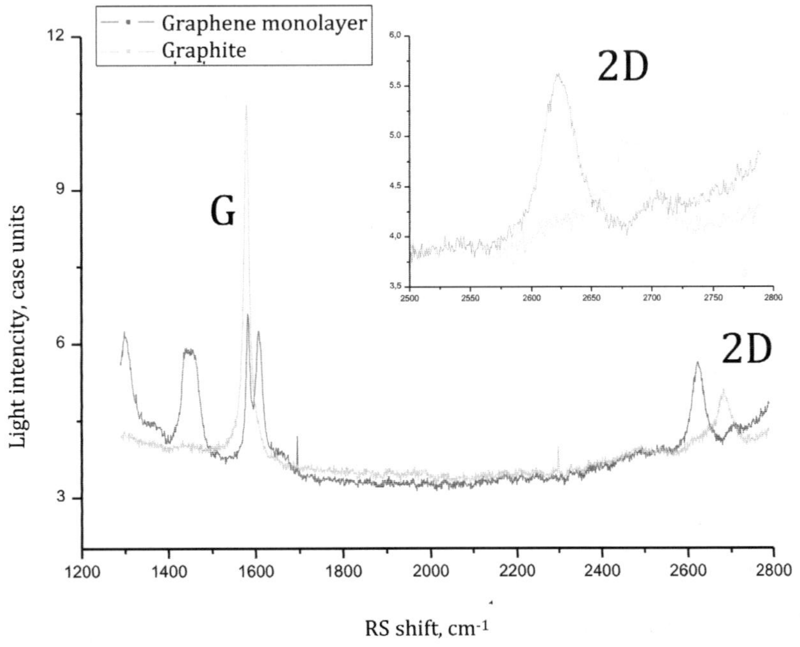

Figure 4. Raman spectrum for areas 3 and 4 (fig.2). Red plot corresponds to a graphene monolayer, green plot corresponds to a thin graphite.

4. Conclusion.

Using the method described in this paper, we can obtain monolayer and few-layer graphene samples. The large size of graphene flakes (up to 2000*2000 μm^2) allows to produce graphene structures without electron beam lithography. Also, due to the more powerful adhesive ability of epoxy glue we can separate more tightly-bonded materials up to monolayers. One of the major drawbacks of this method is the following: graphene is placed at the thin surface of epoxy glue, which physical properties hinders to change carriers concentration in graphene via the bottom gate. The problem could be solved, for example, by virtue of top gate deposition, for example, ionic liquid gate. At this moment research in that direction is carried out.

5. Acknowledgments.

The work has been supported by RFBR grants No 11-02-01379-a, No 11-02-90515-Ukr_f_a, 11-02-12167-ofi-m, state contracts No 16.740.11.0146 and No 16.513.11.3066.

6. References.

[1] Novoselov K S, Geim A K, Morozov S V 2004 *Science* **306** 666
[2] Yu. I. Latyshev,A. P. Orlov,V. V. Peskov,E. G. Shustin,A. A. Shchekin,V. A. Bykov 2012 *Doklady Akademii Nauk* **442** № 2 1–3
[3] Wang J. J.et. al. 2004 *Appl. Phys. Lett.* **85** 1265
[4] Parvizi F.,et. al. 2008 *Micro Nano Lett.* **3** 29
[5] Nair R. R.,et al. 2008 *Science* **320** 5881
[6] Huc V., et. al. 2008 *Nanotechnology* **19** 455601
[7] A.C.Ferrari, et al. 2006 *Phys. Rev. Lett.* **97** 187401
[8] Davy Graf et al. 2007 *Nano Lett.* **7**(2) 238-241

Journal of Physics: Conference Series **393** (2012) 012026 doi:10.1088/1742-6596/393/1/012026

Resonant Tunneling In Graphene-DNA Base-Graphene Junctions

V.L. Katkov, O.G. Isaeva and V.A. Osipov

Bogoliubov Laboratory of Theoretical Physics, Joint Institute for Nuclear Research, Dubna, Russia

E-mail: katkov@theor.jinr.ru

Abstract. The current-voltage characteristics in electrode-DNA-electrode system are calculated for three types of graphene electrodes: a monolayer, AA and AB bilayers. We found that AA bilayer electrodes provide higher currents in comparison with both graphene monolayer and AB bilayer thus making this material more favourable for fabrication of DNA base sequencing devices.

The electronic circuit consisting of two electrodes with molecule (quantum dot) accommodated between them (see figure 1) is the simplest object of molecular electronics which allows one to design such important devices as rectifiers [1], switches, transistors [2], and sensors [3]. Moreover, similar systems lie in the base of a new rapid and low-cost methods for whole-genome analysis where specific physical and electronic properties of DNA nucleotides are monitored. Recently, such DNA sequencing approach has been proposed by Zwolak and DiVentra [4]. According to this method gold electrodes are embedded within solid-state nanopore to measure the transverse current through single-strand DNA-molecule as it pulls through the pore by a voltage applied in parallel to DNA backbone axis. It has been suggested that the use of atomically thin graphene electrodes with planar geometry is able to essentially boost productivity of both DNA sequencing device [5] and an elementary device of molecular electronics [6] in comparison with the metallic ones. In both types of devices the productivity is defined by effectiveness of electron transport which in turn depends on electronic properties of junctions [7].

In this paper, we compare the transport properties of electrode-quantum dot-electrode device for three types of graphene electrodes: monolayer, AA bilayer, and AB bilayer. To this end, current-voltage characteristics of these devices are studied. A quantum dot in our system acts as an object with electronic properties similar to a certain DNA nucleotide. The current-voltage characteristics are calculated within the extended method of independent channels which was earlier used for description of field emission from AAA graphite, nanotubes, and few-layer (AB and ABC) graphene [8, 9, 10].

We assume that contacts have macroscopic sizes while a quantum dot is an atomic scale object. The total current is defined as the difference between the tunneling currents from the left and right electrodes in the form

$$I = I_L - I_R = \frac{2e}{h} \int_{-\infty}^{+\infty} [f_L(\varepsilon) - f_R(\varepsilon)] \overline{D}^{LR}(\varepsilon) d\varepsilon, \tag{1}$$

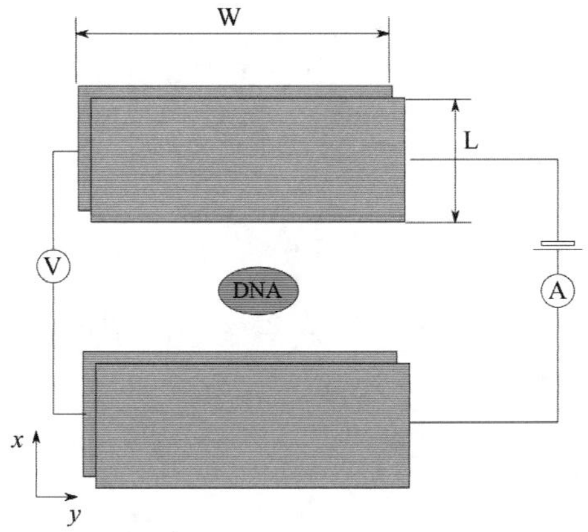

Figure 1. Schematical picture of the resonant tunneling device

where $I_{L,R} = \int d\varepsilon f_{L,R}(\varepsilon)\overline{D}^{LR}$ are currents from the left and right electrodes, respectively, $f_{L,R} = f(\varepsilon \mp V/2)$, V is a bias voltage, $f(\varepsilon)$ is the Fermi-Dirac distribution function, and $\overline{D}^{LR}$ is the transmission probability of an electron through a potential barrier.

The transmission probability is written as [11]

$$\overline{D}^{LR} = \frac{\overline{D}_L\overline{D}_R}{1 - 2\sqrt{\overline{R}_L\overline{R}_R}\cos\theta + \overline{R}_L\overline{R}_R} \qquad (2)$$

with θ being a phase shift acquired in one round-trip between the scatterers, and $\overline{R}_{L,R} = 1 - \overline{D}_{L,R}$. $\overline{D}_{L,R}$ can be expressed as a sum over modes (channels) at given energy $n_{L,R}(\varepsilon)$ normalized by the total number of channels $N_{L,R}$. Such normalization condition follows from the sum rule [11]. One gets

$$\overline{D}_{L,R}(\varepsilon) = P(\varepsilon)D_{L,R}(\varepsilon), \qquad (3)$$

where $D_{L,R}$ is the transmission probability for one channel and $P(\varepsilon) = n_{L,R}(\varepsilon)/N_{L,R}$. Here we assume that $D_{L,R}$ does not depend on a type of the channel. Indeed, within the framework of the WKB approximation $D_{L,R}$ depends only on a shape of the barrier and a distance between the vacuum energy level and given energy.

Let us consider a case of a single resonant level. This approximation can be used for low bias voltage (less than 0.5 V) because a distance between nearest levels in noninteracting DNA base is of an order of 1 eV (see [12]). Assuming that $\overline{R}_{L,R} \approx 1$, one can rewrite (2) via the Breit-Wigner formula

$$\overline{D}^{LR} = \frac{\Gamma_L\Gamma_R}{(\varepsilon - \varepsilon_r)^2 + \left(\frac{\Gamma_L+\Gamma_R}{2}\right)^2}, \qquad (4)$$

where $\Gamma_{L,R}/\hbar = \nu\overline{D}_{L,R}$ and ν is an attempt frequency [11]. Finally, the tunneling current through a single resonant level is written as

$$I = \frac{2e}{h}\int_{-\infty}^{\infty} \frac{\Gamma_L(\varepsilon - \varepsilon_r - V/2)\Gamma_R(\varepsilon - \varepsilon_r + V/2)}{(\varepsilon - \varepsilon_r)^2 + (\Gamma(\varepsilon)/2)^2}[f_L(\varepsilon) - f_R(\varepsilon)]d\varepsilon, \qquad (5)$$

where $\Gamma(\varepsilon) = \Gamma_L(\varepsilon - \varepsilon_r - V/2) + \Gamma_R(\varepsilon - \varepsilon_r + V/2)$. Notice that the same result can be obtained by using of either the sequential model or non-equilibrium Green function formalism [7, 13, 14].

Let the dot be located near the armchair edge. In this case, the localised edge states do not appear [15] and the channel density can be written as $P(\varepsilon) = 2a_0\Delta p_y(\varepsilon)/h$ with $2\Delta p_y(\varepsilon)$ being a "shadow" of the constant energy line on a plane normal to the tunneling direction in the momentum space [9, 16], and a_0 the interatomic distance. One has

$$\Gamma_{L,R} = \frac{a_0 \Delta p_y(\varepsilon)}{\pi} \nu D_{L,R}. \tag{6}$$

Our numerical calculations are based on (5) and (6). The difference between the resonant level and the Fermi energy of contacts is estimated as $\varepsilon_F - \varepsilon_r = \phi - A \equiv \Delta$, where A is an electron affinity of the DNA base. Thus, we have equated vacuum levels of DNA and the contact. In accordance with [12] $|\Delta| \approx 0.1$ eV for all DNA bases. We apply the WKB-type approximation for the transmission probability

$$D = \exp[-\lambda(d - d_0)], \qquad \lambda = -\frac{2\sqrt{2m\phi}}{\hbar} \tag{7}$$

with the work function $\phi = 4.7$ eV, $\lambda = 0.22$ nm^{-1}, and $d_0 = 0.135$ nm being a typical covalent bond length. One has $\hbar\nu \approx \hbar^2/(2md_0^2N)$ where N is a number of atoms in a DNA base having p_z electrons (N is taken to be equal to ten in average), so that $\nu \sim 10^{14}$ s^{-1}.

The "shadows" for graphene can be calculated as

$$\Delta p_y^A(\varepsilon) = 2\frac{\varepsilon}{v_F}, \tag{8}$$

$$\Delta p_y^{AB}(\varepsilon) = \frac{[\sqrt{|\varepsilon|(|\varepsilon| + \gamma_1)} + \theta(|\varepsilon| - \gamma_1)\sqrt{|\varepsilon|(|\varepsilon| - \gamma_1)}}{v_F}, \tag{9}$$

$$\Delta p_y^{AA}(\varepsilon) = \frac{|\varepsilon - \alpha_1| + |\varepsilon + \alpha_1|}{v_F}. \tag{10}$$

We assume that $\gamma_1 = \alpha_1 = 0.4$ eV, $d_L = d_R = 2d_0$, and $a_0 = 0.142$ nm. The results are shown in figures 2-5.

The current-voltage (I-V) characteristics for three types of electrodes are presented in figure 2. Figure 4 illustrates a smoothing role of the temperature factor. Notice that all curves are symmetric, nonlinear and have a step-like behaviour. As is seen from figure 3, at low voltage ($V < 2|\Delta|$) the current for AA bilayer grows more intensively in contrast to other electrodes. This finding can be explained by the energy dependence of the ratios Γ for different electrodes (see figure 5). Indeed, both Γ^A and Γ^{AB} grow with energy and have a similar behaviour while Γ^{AA} remains a constant up to the point $\gamma_1 = 0.4$ eV and has a bigger. This can be also directly seen from the approximate expression for conductivity in the small voltage limit

$$\sigma = \frac{2e^2}{h}\frac{1}{1 + \Delta^2/\Gamma^2}. \tag{11}$$

At $V > 2|\Delta|$ the curves for the monolayer and AB bilayer increase with the voltage while for AA bilayer the current tends to a saturation value (see figures 2 and 4). The behaviour of tunneling rates in figure 5 clarifies the saturation of the current at $V > 2|\Delta|$ for AA bilayer as well as its absence for graphene monolayer and AB bilayer.

In conclusion, we have calculated the current-voltage characteristics in electrode-quantum dot-electrode system for three types of electrodes: graphene monolayer, AA and AB bilayers. Our analysis shows that at the same bias voltage AA graphene bilayer electrodes provide a higher current in comparison with both graphene monolayer and AB bilayer. This finding can

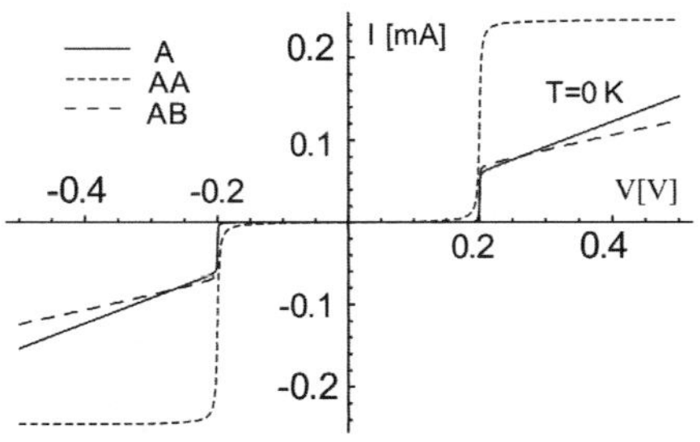

Figure 2. $I - V$ characteristics of tunneling device at zero temperature

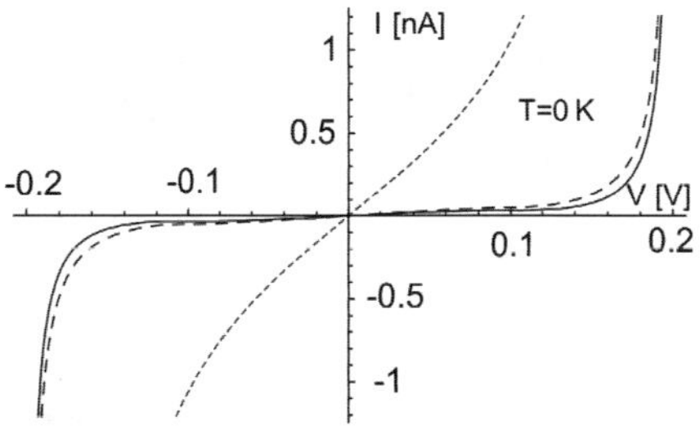

Figure 3. Scaled-up central part of Figure 2

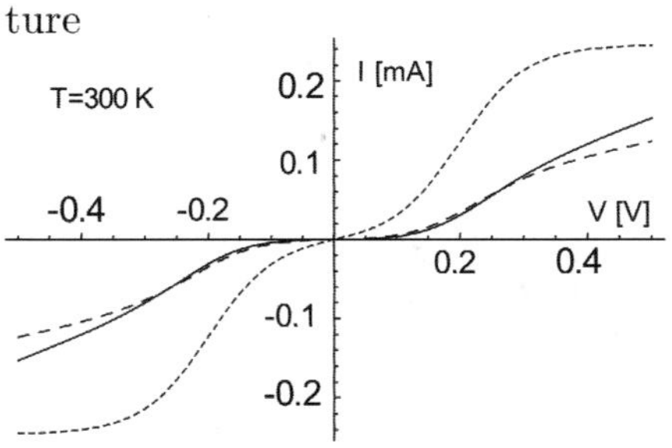

Figure 4. $I - V$ characteristics of tunneling device at room temperature

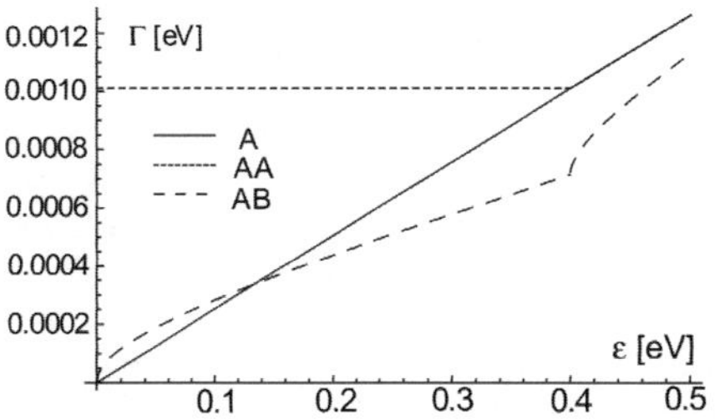

Figure 5. Energy dependence of tunneling rates. The Fermi energy is chosen to be zero

be explained by the fact that at the Fermi level the density of channels takes a nonzero value for AA bilayers while in AB bilayers and monolayers it tends to zero. Thus, the use of electrodes on the base of AA graphene bilayer is expected to be much more productive in such devices.

This work has been supported by the Russian Foundation for Basic Research under grant No. 12-02-01081.

References

[1] Aviram A and Ratner M A 1974 *Chem. Phys. Lett.* **29** 277
[2] Park H *et al* 2000 *Nature* **407** 57
[3] Guo X F *et al* 2006 *Science* **311** 356
[4] Zwolak M and Ventra D M 2004 *Nano Lett.* 5 **421**
[5] Prasongkit J *et al* 2011 *Nano Lett.* **11**, 1941
[6] Prins F et al. 2011 *Nano Lett.* **11** 4607
[7] Datta S 2005 *Quantum Transport: Atom to Transistor* (New York: Cambridge University Press)
[8] Katkov V L and Osipov V A 2008 *J. Phys.: Condens. Matter* **20** 035204
[9] Katkov V L and Osipov V A 2009 *JETP Lett.* **90** 304
[10] Katkov V L and Osipov V A 2010 *Physics of Particles and Nuclei* **41** 1027
[11] Datta S 1995 *Electronic transoirt in mesoscopic system* (Cambridge: Cambridge University Press)
[12] Hawke L G D, Kalosakas G and Simserides C 2010 *Eur. Phys. J. E* **32** 291
[13] Larkin A I and Matveev K A 1987 *Sov. Phys. JETP* **66** 580
[14] Gadzuk J W 1993 *Phys. Rev. B* **47** 12832
[15] Huaixiu Zheng *et al* 2007 *Phys.Rev. B* **75** 165414
[16] Harrison W A 1961 *Phys. Rev.* **123** 85

Journal of Physics: Conference Series **393** (2012) 012027 doi:10.1088/1742-6596/393/1/012027

Mechanical and emission properties of thinnest stable bamboo-like nanotubes

O.E.Glukhova, A.S.Kolesnikova

Saratov State University,410012, Astrakhanskaya, 83, Saratov, Russian Federation

E-mail: glukhovaoe@info.sgu.ru

Abstract. We investigate the stability of carbon bamboo-like nanotubes (BCNTs) with different diameters. It is shown that BCNTs with a diameter of 2.024 nm are the most stable bamboo-like nanotubes with the smallest diameter. It is shown that bamboo-like nanotubes with a certain distance between the bridges have superior emission properties than that the hollow nanotubes. Emission properties of the infinite bamboo-like nanotubes can be improved by adding the potassium atoms into the nanotube atomic structure. In this case the potassium atoms concentration should exceed 0.59%.

1. Introduction

Carbon bamboo-like nanotubes (BCNTs) find applications in various fields of science and technology due to their unique properties. In particular, bamboo-like nanostructures can be used as nanoindentors [1] to study the mechanical properties of the material, as well as nanoemitters in electronics [2].

Modeling of various properties of bamboo-like nanotubes has a large importance in the boundaries expansion of nanostructures application fields. Theoretical studies of electronic structure and topology of the atomic framework are carried out in order to identify the optimal configuration of bamboo-like nanotubes for a particular application scope.

It was found that the bamboo-like nanotubes have the improved mechanical properties compared to the hollow ones. It was experimentally established that the Young's modulus of bamboo-like nanotubes with a diameter of 10-20 nm and a length of 5-20mm is equal to 4.5 ± 0.8 GPa, and tensile strength of the tubes with indicated geometric characteristics equals to 150 ± 35MPa [3]. As a result of calculations performed in [4], it was found that the Young's modulus of nanotubes with a diameter of 10-45 nm and length of 1-20 mm is equal to 5.5 ± 0.8 GPa, while Young's modulus of the hollow nanotubes is equal to 2.7 ± 0.7 GPa.

Currently, researches were conducted in the direction of reducing the work function of carbon tubular structures. It was found that the alkali metal atoms in nanotubes structure help to increase the emission current from the surface of nanotubes and to shift their current-voltage characteristic towards lower voltages [5-8].

Researchers' attention has been focused recently on the nanotubes of complex shape in order to improve the emissivity of carbon nanostructures. It was established that the current density of bamboo-like nanotubes with a diameter of 50-100 nm in the absence of doping with alkali metal atoms reaches saturation (10-4 A/cm^2) at a field strength 4 V/mm [9].

The aim of this paper is a theoretical study of mechanical and emission properties of bamboo-like nanotubes. Investigation of mechanical properties was carried out for BCNT based on carbon nanotubes (CNT) (15,15) having a diameter of 2.024 nm and a length of 3.6 nm, and the CNT (10,10) having a diameter of 1.36 nm and a length of 3.728 nm. The study of the electronic properties was carried out for an infinite BCNT. We look for the stable configuration of BCNT using the tight-binding method [10]. Critical stress was found using the scheme presented in [11]. The modeling of the compression/expansion was carried out by empirical REBO method [12].

2. Modeling of the bamboo-like nanotubes and their geometrical parameters.

Bamboo-like structures were modeled by connecting fullerene fragments (bridge) to the inner surface of the hollow nanotube with the help of chemical bonds. The fragments of fullerenes C_{240} and C_{540} were used as bridges for the bamboo-like nanotubes constructed on the basis of the amchair nanotubes (10,10) and (15.15), respectively.

2.1. Study of the bamboo-like nanotubes stability

We look for the stable configuration of BCNT utilizing the optimization of atomic structure (i.e., minimization of the structure's total energy by the atoms coordinates). It was established that the length of chemical bond equals to 0.16 nm for bamboo-like nanotube with a diameter of 1.36 nm and the length of the chemical bond equals to 0.14 nm for the bamboo-like nanotube with a diameter of 2.024 nm. Thus, BCNT with the diameter of 1.36 are metastable.

The detailed study of the fracturing process and predicting of the areas of possible chemical bonds break was based on analysis of the local atomic stresses distribution. Calculation of local stresses was performed according to the scheme presented in [11]. The calculations revealed that bamboo-like nanotubes with a diameter of 2.024nm are stable nanotubes of the smallest diameter. The critical stresses, at which the destruction of bamboo-like nanotubes begins, are approximately 11 GPa. The greatest value of the stresses experienced by the atoms of BCNT with a diameter of 2.024nm was 8.77 GPa. Therefore, the further study of the emission and mechanical properties (tensile/compression) of bamboo-like nanostructures will be applied to the model of BCNT with a diameter of 2.024nm due to the stability of its atomic structure.

2.2. Modeling of tension and compression bamboo-like nanotubes

In this paper the search for the critical tensile strength for BCNT (15.15) and hollow nanotubes was carried out. The modeling of the compression/expansion is as follows:

1) The atoms at the tip and atoms on the opposite edge BCNT were fixed so that they will not return into their initial position after compression. Simulation of the microscope tip operation was achieved by fixing the atoms at the tip of the nanotube. In the modeling process of the interaction of structure surface with a microscope tip, the stretch of the indenter (nanotube) was observed.
2) The optimization of the atomic structure of BCNT by coordinates of all loose atoms was conducted by the empirical method REBO lead to the optimal structure of the system.
3) The shift of the sharp edge at 0,1 Å along the nanotube in 1000fsek was carried out. As result of the simulation, the Van der Waals interaction of the surface of the structure with a microscope tip was found.

Compression/expansion was carried out at a speed of 10 m / sec. It is shown that the destruction of BCNT occurs in the bridge area after 19 psec in tension, and the destruction the hollow nanotubes occurs after 14 ps. Areas of destruction of the atomic structure are shown in Figure 1. Tensile by 5% led to the BCNT destruction and tensile by 3% led to the hollow nanotubes destruction. Therefore one can conclude that the BCNTs are more mechanically resistant compared to the hollow nanotubes.

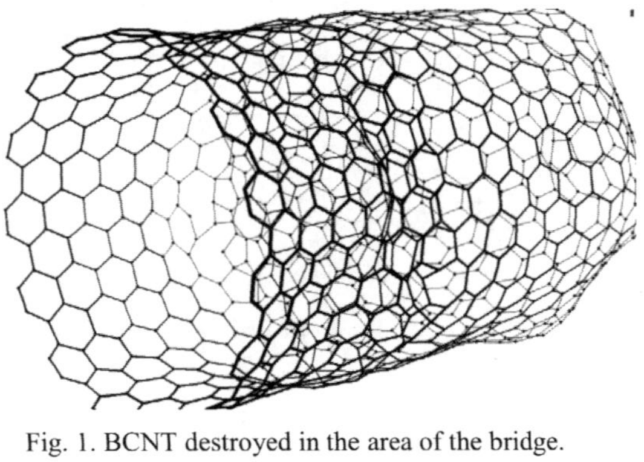

Fig. 1. BCNT destroyed in the area of the bridge.

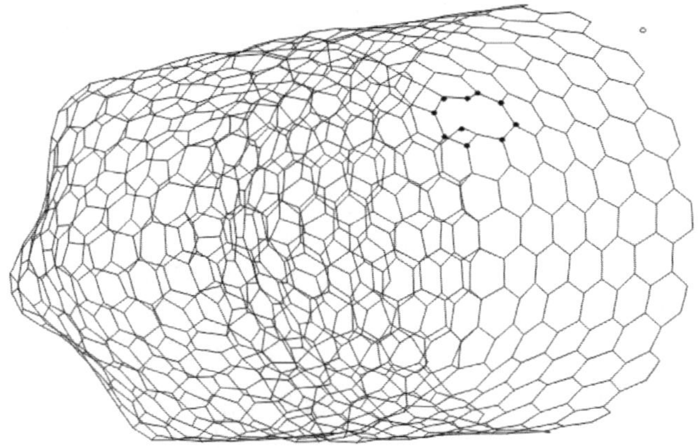

Fig. 2. BCNT destroyed in the area of the bridge.

The tip BCNT become concave in the compression process after 45psek. The chemical bond lengths reduced from 1.42Å to 1.38Å in the bridge and tip areas. The destruction of nanotubes occurred after 51 psec (Fig. 2).

2.3. Emission properties of bamboo-like nanotubes

The study of electronic properties of the infinite BCNT with increasing distance between the bridges requires a series of calculations. The optimization process, i.e. the finding the equilibrium structure by minimizing the total energy, was carried out for the bamboo-like nanotube with a length of 14.58 nm and a diameter of 2.024 nm with three bridges. This structure contains about 4000 atoms. The optimization process using the quantum-chemical method is problematic for such a structure due to the large number of atoms. The use of parallelized computing algorithm requires a large amount of computing resources and time and, therefore, is not rational. Optimization process was carried out by using the AMBER force field of molecular dynamics [13].

Infinite BCNTs were modeled with a translation vector [14]. Four types of cells, having a length of 2.071 nm 2.317 nm 2.564 nm and 2.811 nm, were chosen as the repeating BCNT cells. The distance between the bridges was being increased with elongation of the cell. The obtained ionization potential and the energy gap of bamboo-like nanotubes with different distances between the bridges are shown in Table 1. It was found that the ionization potential and emission properties of infinite bamboo-like nanotubes with the distance between bridges of the 2.811 nm exceed those of the infinite hollow nanotubes. Ionization potential of an infinite hollow nanotube is equal of 6.2 eV. In this regard, further we consider the change of the ionization potential of bamboo-like nanotubes with the distance between the bridges 2.811 nm and with an increase of potassium atoms concentration. The increase of the latter up to five in each unit cell with a length of 2.811 nm leads to a decrease in the value of the ionization potential by 0.2 eV. Consequently, we can assume that the emission properties can be controlled by potassium atoms adding or removal.

Table 1. Energy parameters of BCNT with different distances between bridges

Distance between bridges,nm	Ionization potential,eV	Energy gap,eV
2,071	6,436	0,593
2,317	6,349	0,508
2,564	6,348	0,519
2,811	6,065	0,019

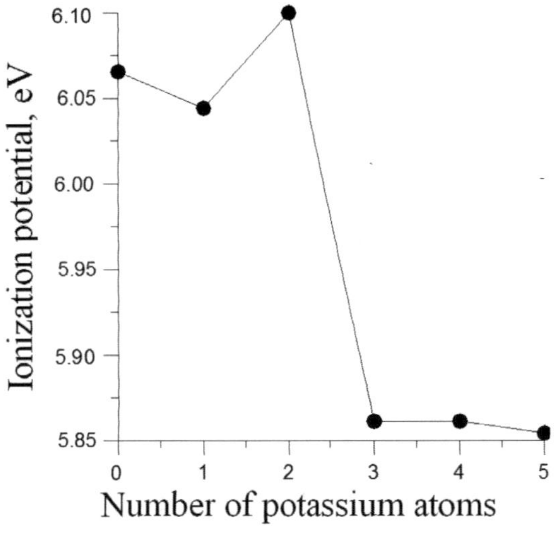

Fig.3. Change of the ionization potential depending on the number of potassium atoms in the repeating unit cell

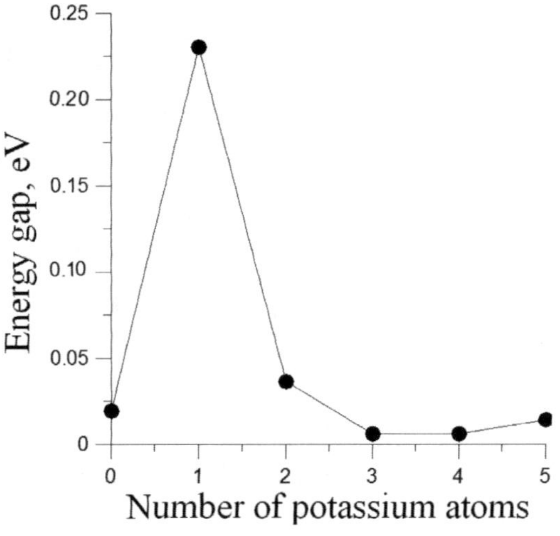

Fig.4. Change of the energy gap depending on the number of potassium atoms in the repeating unit cell

3. Conclusion

In the paper the properties of the bamboo-like nanotubes were investigated. First of all, it was established that 2.024 nm diameter of the bamboo-like nanotubes is the smallest one for the stable nanotubes (i.e. it is the smallest diameter of all the existing BCNTs).

Also, it was shown that the maximum stress appearing in the atomic framework of the BCNTs of 2.024nm, equals to 8.77 GPa. It is the stress at which no destruction for the bamboo-like nanotube is observed. On the other hand, under the dynamical stretching with the speed of 10 m/sec all the BCNTs are being destroyed after 19 psec. The stress at which the destruction is observed, equals to 11 GPa. It should be mentioned that under the similar conditions of the loading, the hollow carbon nanotubes are being destroyed after 14 psec, suffering much lower stress.

The BCNT tip deformation was also studied in the work. The change of its atomic framework configuration was observed after 45 psec under the condition of the 10 m/sec speed dynamic loading. Its destruction was found in 51 psec.

Because of all the aforementioned results, and also due to the fact that the BCNTs are ultrathin and heavy-duty nanostructures, one can conclude that they can be used as nanoindenters.

Mathematical model of the infinite bamboo-like carbon nanotubes was developed. Also, the emission properties of the infinite bamboo-like nanotubes with diameter of 2.024 nm and a distance between the bridges 2.811 nm were studied by tight-binding method. It was found that the ionization potential of these nanotubes decreased by 0.1 eV in comparison with the infinite hollow nanotubes. The adsorption of potassium atoms on the surface of an infinity bamboo-likenanotubes was also modeled by tight-binding method. It was revealed that the emission properties of BCNT were improved by inserting the potassium atoms in the atomic structure (with a concentration of 0.59% and above in each repeating cell). In this case, the ionization potential has decreased by 0.35eV. Based on these results, one can conclude that the bamboo-like nanotubes could serve as a more prospective material for the construction of nanoemitters than the hollow nanotubes.

Acknowledgements

We appreciate the financial support The Russian Foundation for Basic Research (12-02-00807 « Experimental and theoretical study of physical basis of field emission of electrons with an abnormally high current density from surface the carbon nanostructured films » (2012-2014)), which is implemented together with our colleagues in the Saratov branch of the Institute of Radio-engineering and Electronics of RAS and the financial support The Federal Target program "Scientific and scientific-pedagogical personnel of innovative Russia" for 2009 - 2013 years, XLI event queue 1.2.1, engineeringthe, number of application in a computerized information system "2012-1.2. 1-12-000-2013-064 ".

References

[1] NANO INDENTERS: http://www.microstartech.com/index/NANOINDENTERS.pdf

[2] Kuznetzov A A, Lee S B, Zhang M, Baughman R H and Zakhidov A A 2010 *Carbon.* **48** 41

[3] Olek M, Ostrander J, Jurga S, Möhwald H, Kotov N, Kempa K and Giersig M 2004 *Nano letters.* **4** 1889

[4] Olek M 2006 *Carbon nanotube composites-mechanical, electrical and optical properties* (Dissertation, Department of Mathematics and Natural Sciences of the Rhein Friedrich Wilhelm University Bonn) p. 132.

[5] Zhao J, Han J and Lu J P 2002 *Phys. Rev. B.* **65.** 193401

[6] Musatov A L , Izraelyants K R, Chirkov E G and Kristinin A V 2011 *Solid State Physics* **53** 1428

[7] Xu S F, Yuan G, Li C, Jia Z J and Mimura H 2010 *Appl. Phys. Lett.* **96** 233111

[8] Cai T, Zhang A, Feng Y P, Zhang C, Teoh H F and Ho G W 2009 *J. Chem. Phys.* **131** 224701

[9] Katar S L , Gonzalez-Berrios A , Jesus J D , Weiner B and Morell G 2008 *J. Nanomaterials* **515890** 1

[10] Glukhova O E and Zhbanov A I 2003 *Solid State Phys.* **45** 189

[11] Glukhova O E and Slepchenkov M M 2011 *Nano and Microsystem Technology* **7** 1

[12] Brenner D W 1990 *Phys. Rev. B.* **42** 9458

[13] Cornell W D, Cieplak P, Bayly C I, Gould I R, Merz K M, Ferguson D M, Spellmeyer D C, Fox T, Caldwell J Wa and Kollman P A 1995 *J. Am. Chem. Soc.* **117** 5179

[14] Kosevich A M 2006 *The Crystal Lattice: Phonons, Solitons, Dislocations, Superlattices.* (WILEY-VCH Verlag GmbH & Co. KGaA , Weinheim)

Dubna-Nano 2012 IOP Publishing
Journal of Physics: Conference Series **393** (2012) 012028 doi:10.1088/1742-6596/393/1/012028

The electronic properties of double wall carbon nanotubes

R. Pincak[1,2] **and M. Pudlak**[1]

[1] Institute of Experimental Physics, Slovak Academy of Sciences, Watsonova 47,043 53 Kosice, Slovak Republic

[2] Joint Institute for Nuclear Research, BLTP, 141980 Dubna, Moscow region, Russia

E-mail: pincak@saske.sk, pudlak@saske.sk

Abstract. The electronic spectra double wall carbon nanotubes using tight binding approximation are investigated. We focus on the double wall armchair nanotubes. The influence of nanotube curvature on the electronic spectra is also treated. The impact of the external magnetic field on the spectral characteristic of double wall nanotubes is computed. We found the big change of the electronic spectra for the double walled carbon nanotubes due to the external magnetic field.

1. Introduction

Carbon based materials has attracted scientific interest in recent years from both an experimental and a theoretical viewpoint. Among them carbon nanotubes are very interesting because of their unique mechanical and electronic properties [1]. A single-wall carbon nanotube can be described as a graphene sheet rolled into a cylindrical shape so that the structure is one-dimensional with axial symmetry and in general exhibiting a spiral conformation called chirality. The electronic states of carbon nanotubes are classified by the chiral vector that assigns the diameter and chirality of the nanotubes [2]. The primary symmetry classification of carbon nanotubes is either achiral or chiral. Achiral carbon nanotubes are defined by a carbon nanotube whose mirror images have an identical structure to the original one. There are only two cases of achiral nanotubes, armchair and zig-zag nanotubes. The names of armchair and zig-zag nanotubes arise from the shape of the cross-section ring at the edge of the nanotubes. Chiral nanotubes exhibit spiral symmetry whose mirror image cannot be superposed onto the original one. There is a variety of geometries in carbon nanotubes where the diameter, chirality and cap structures are different [3]. The electronic structure of carbon nanotubes is derived by a simple tight-binding calculation for the π-electrons of carbon atoms.

2. $(5,5)-(10,10)$ armchair tubules in a static magnetic field

We investigate the armchair nanotubes in a static magnetic field $\vec{B}$ parallel to the nanotube axis. We assume Hamiltonian for an electron in a potential $V(r)$ and in the magnetic field in the form

$$H = \frac{1}{2m}\left(\vec{p} - \frac{e}{c}\vec{A}\right)^2 + V \tag{1}$$

Potential $V(r)$ reflects the structure of the crystal lattice such as the symmetry and periodicity properties. Here this potential describes the structure of armchair. Vector potential $\vec{A}$ in the Landau gauge can be expressed in the form $\vec{A} = (\frac{\Phi}{L}, 0)$, where $\Phi = B\pi r^2$ is the magnetic flux penetrating the cross section of carbon nanotube and $L = 2\pi r$ is a circumference of the nanotube (r-nanotube radius). Here coordinate x is in the circumferential direction and coordinate y denotes the direction parallel to nanotube axis.

$$\psi_\alpha(\vec{k}, \vec{r}) = \frac{1}{\sqrt{M}} \sum_n \exp\left(i\vec{k}(\vec{r_n} + \vec{d}_\alpha) + i\frac{e}{c\hbar}G(\vec{r_n} + \vec{d}_\alpha)\right) |\varphi(\vec{r} - \vec{r}_n - \vec{d}_\alpha)\rangle, \quad (2)$$

where α denotes A or B atoms. Here $\vec{d}_\alpha$ are the coordinate of the α atom in the unit cell and $\vec{r_n}$ is a position of a unit cell, M is the number of the unit cell; $|\varphi(\vec{r})\rangle$ is a π orbital which is generally different for the outer and inner shell. $G(\vec{R})$ is the phase factor associated with the magnetic field and is expressed by

$$G(\vec{R}) = \int_{\vec{R}}^{\vec{r}} \vec{A}(\vec{x}).d\vec{x} = \int_0^1 (\vec{r} - \vec{R}).\vec{A}(\vec{R} + \lambda(\vec{r} - \vec{R}))d\lambda \quad (3)$$

Employing Eq.(2), we get $G(\vec{R}) = \int_0^1 (\vec{r} - \vec{R}).(\frac{\Phi}{L}, 0)d\lambda = (x - X)\frac{\Phi}{L}$. We denote $\epsilon = \langle\varphi^{out}(\vec{r} - \vec{A_i})|H|\varphi^{out}(\vec{r} - \vec{A_i})\rangle = \langle\varphi^{out}(\vec{r} - \vec{B_i})|H|\varphi^{out}(\vec{r} - \vec{B_i})\rangle$,$\widetilde{\epsilon} = \langle\varphi^{in}(\vec{r} - \vec{A_i})|H|\varphi^{in}(\vec{r} - \vec{A_i})\rangle = \langle\varphi^{in}(\vec{r} - \vec{B_i})|H|\varphi^{in}(\vec{r} - \vec{B_i})\rangle$ and the intertube hopping integrals $\langle\varphi^{out}(\vec{r} - \vec{R_i})|H|\varphi^{in}(\vec{r} - \vec{R_j})\rangle = W_{ij}$, ϵ and $\widetilde{\epsilon}$ are Fermi energies of the outer and inner nanotubes; $|\varphi^{out}(\vec{r} - \vec{R_i})\rangle$, $|\varphi^{in}(\vec{r} - \vec{R_j})\rangle$ are π orbitals on site i at the outer and on site j at the inner tubes; $W_{ij} = \frac{\gamma_0}{8}\cos(\theta_{ij})e^{(\xi - d_{ij})/\delta}$, where γ_0 is the hopping integral in the graphene,θ_{ij} is an angle between the i-th atom of the inner shell and the j-th atom of the outer shell, d_{ij} is the interatom distance and ξ is the intertube distance. The characteristic length $\delta = 0.45\mathring{A}$. We can define the vectors connecting the nearest neighbor carbon atoms for armchair nanotubes in the form:$\vec{\tau_1} = a(\frac{1}{\sqrt{3}}; 0)$, $\vec{\tau_2} = a(-\frac{1}{2\sqrt{3}}; -\frac{1}{2})$,$\vec{\tau_3} = a(-\frac{1}{2\sqrt{3}}; \frac{1}{2})$. The distance between atoms in the unit cell is also $|\vec{\tau_i}| = \frac{a}{\sqrt{3}}$. Now we define the

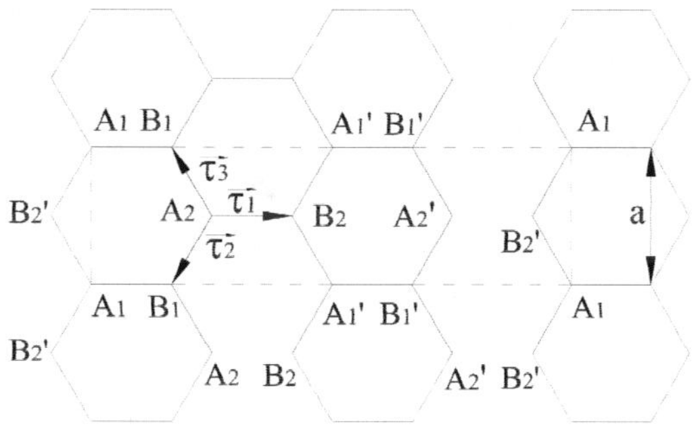

Figure 1. The outer shell part of the unit cell in the case of armchair nanotubes.

intratube hopping integrals $\langle\varphi^{out}(r - A_1)|H|\varphi^{out}(r - B_1)\rangle = \gamma_0\alpha$, $\langle\varphi^{out}(r - A_1)|H|\varphi^{out}(r - B_2^{'})\rangle = \gamma_0\beta$, and $\langle\varphi^{in}(r - A)|H|\varphi^{in}(r - B)\rangle = \gamma_0\widetilde{\alpha}$, $\langle\varphi^{in}(r - A)|H|\varphi^{in}(r - B^{'})\rangle = \gamma_0\widetilde{\beta}$, where $\alpha(\widetilde{\alpha})$, $\beta(\widetilde{\beta})$ are parameters which describe the dependence of hopping integrals on the surface curvature [4]. From Figures 1 and 2 we get the system of equations which describe the electronic properties of the armchair DWCNT [5]. The spectra for some values of $ka/2$ near the Fermi points of double wall armchair nanotubes without and with static magnetic field are depicted in Figure 3,4.

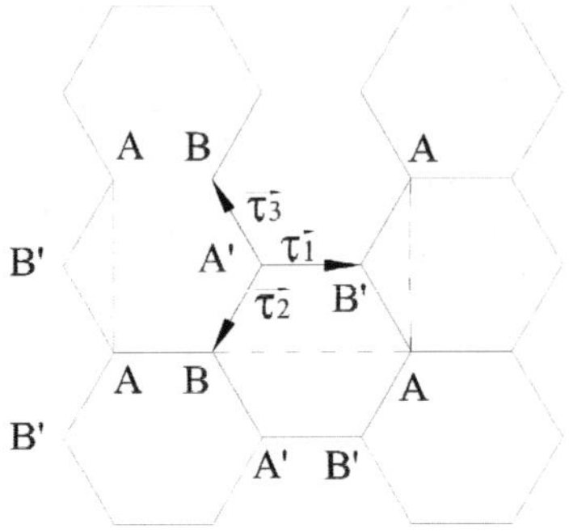

Figure 2. The inner shell part of the unit cell in the case of armchair nanotubes.

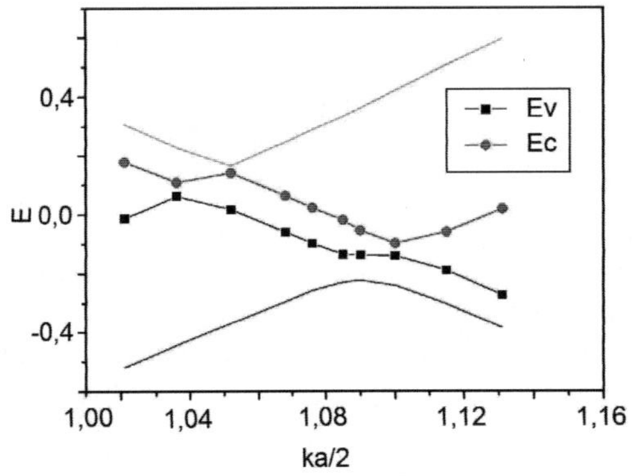

Figure 3. Spectra of armchair DWCNT with the intertube interactions without static magnetic field.

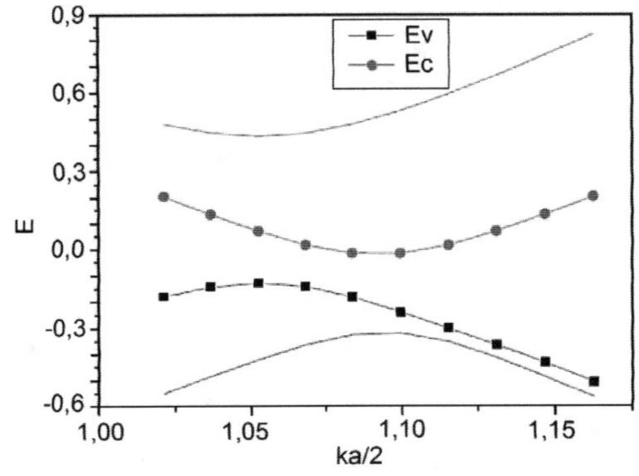

Figure 4. Spectra of armchair DWCNT with the intertube interactions when the static magnetic field $\Phi = 0.5\Phi_0$ is imposed.

3. Conclusion

When the static magnetic field is applied the character of conductivity is changing. For instance we get for armchair carbon nanotube that the conductivity is changed from semimetal to semiconductor.

ACKNOWLEDGEMENTS — The work was supported by the Slovak Academy of Sciences in the framework of CEX NANOFLUID, and by the Science and Technology Assistance Agency under Contract No. APVV 0509-07 , 0171 10, VEGA Grant No. 2/0069/10 and Ministry of Education Agency for Structural Funds of EU in frame of project 26220120021.

References

[1] Meijer F E, Morpurgo A F and Klapwijk T M 2002 *Phys.Rev.* B **66** 033107
[2] Saito R, Dresselhaus G and Dresselhaus M S 1998 *Physical Properties of Carbon Nanotubes* (London:Imperial Colledge Press)
[3] Molnar B ,Peeters F M and Vasilopoulos P 2004 *Phys.Rev.* B **69** 155335
[4] Pudlak M and Pincak R 2009 *Eur.Phys.J.* B **67** 565-76
[5] Pincak R, Pudlak M and Smotlacha J 2012 *Carbon Nanotubes: Synthesis,Properties and Applications* (New York: Nova Science Publisher)

Electronic properties of perturbed cylinder

R. Pincak[1,2], J. Smotlacha[2] and M. Pudlak[1]
[1] Institute of Experimental Physics, Slovak Academy of Sciences, Watsonova 47,043 53 Kosice, Slovak Republic
[2] Joint Institute for Nuclear Research, BLTP, 141980 Dubna, Moscow region, Russia
E-mail: pincak@saske.sk, smota@centrum.cz, pudlak@saske.sk

Abstract. The electronic structure of nanocylinder without and with a small perturbation is investigated with the help of calculation of the local density of states. A continuum gauge field-theory model is used for this purpose. In this model, Dirac equation is solved on a curved surface. The local density of states is calculated from its solution. The case of 2 heptagonal defects is considered. This paper is an extension of our previous work [1] where one heptagonal and one pentagonal defects in hexagonal graphene network were compared. The metallization for the perturbed cylinder structure is found.

1. Introduction

The carbon nanostructures play a key role in constructing nanoscale devices like quantum wires, nonlinear electronic elements, transistors, molecular memory devices or electron field emitters. Their molecules are variously-shaped geometrical forms its surface is composed of disclinated hexagonal carbon lattice. The main structure of this kind is graphene - the carbon lattice plane from which all other kinds are derived. Most often, the heptagons appear in pairs with pentagons in the connecting parts of the folded forms [2].

Because of the applications, the research of the electronic properties of the carbon nanostructures is important. One of the main characteristics is the local density of states ($LDoS$). In the presented model coming from the effective-mass theory, knowledge of the solution of the corresponding Dirac equation is necessary for the calculation [3]. This solution is represented by the wave-function and to find it, we have to know the geometry of the molecular surface. As discovered in [4] for the case of nanocones, the most suitable geometry for the description of the close vicinity of the defects is the hyperboloidal geometry. Very often, for a given geometry, the number of possible defects is limited.

The solutions for spherical, conical and 2-fold-hyperboloidal cases were found in [4, 5, 6]. In [1], we used the presented model for calculation of the electronic properties of the structures with the geometry of the 1-fold hyperboloid. The aim was to describe the electronic properties in the vicinity of the locally negative curvature of an arbitrary nanoparticle. This restriction did not enable us to do the calculations for the case of more than 1 defect. In this paper, we present a model describing the electronic properties of a simple nanocylinder and a curved nanocylinder including 2 heptagons at the opposite sides of the surface. The hyperboloidal geometry is used again. Because a nanocylinder is an opened nanotube, comparison with the case of the capped nanotube could be performed (see e.g. [7] for this purpose).

2. Computational formalism

To research the electronic properties, we have to solve the Dirac equation in (2+1) dimensions. It has the form

$$i\sigma^{\alpha} e_{\alpha}^{\mu}[\partial_{\mu} + \Omega_{\mu} - ia_{\mu} - ia_{\mu}^{W}]\psi = E\psi, \tag{1}$$

where ∂_{μ} means the partial derivation according to the μ parameter, i.e. $\partial_{\mu} = \frac{\partial}{\partial x^{\mu}}$.
In this equation, besides the energy E, the particular constituents have the following sense: $\sigma^{\alpha}, \alpha = 1, 2$, denote the Pauli matrices. The wave function ψ, the so-called bispinor, is composed of two parts:

$$\psi = \begin{pmatrix} \psi_A \\ \psi_B \end{pmatrix}, \tag{2}$$

each corresponding to different sublattices of the curved graphene sheet. The gauge field a_{μ} arises from spin rotation invariance for atoms of different sublattices A and B in the Brillouin zone and the gauge field a_{μ}^{W} is connected with the chiral vector (n, m):

$$a_{\varphi} = N/4, \qquad a_{\varphi}^{W} = -\frac{1}{3}(2m + n). \tag{3}$$

If we write the wave function in the form

$$\begin{pmatrix} \psi_A \\ \psi_B \end{pmatrix} = \frac{1}{\sqrt[4]{g_{\varphi\varphi}}} \begin{pmatrix} u(z)e^{i\varphi j} \\ v(z)e^{i\varphi(j+1)} \end{pmatrix}, \qquad j = 0, \pm 1, ... \tag{4}$$

and substituting (4) into (1) we obtain

$$\frac{\partial_z u}{\sqrt{g_{zz}}} - \frac{\tilde{j}}{\sqrt{g_{\varphi\varphi}}} u = Ev, \qquad -\frac{\partial_z v}{\sqrt{g_{zz}}} - \frac{\tilde{j}}{\sqrt{g_{\varphi\varphi}}} v = Eu, \tag{5}$$

where

$$\tilde{j} = j + 1/2 - a_{\varphi} - a_{\varphi}^{W}. \tag{6}$$

Each of the solutions u, v consists of two linearly independent components such that

$$u(E, z) = C_1(E)u_1(E, z) + C_2(E)u_2(E, z), \tag{7}$$

$$v(E, z) = \frac{C_1}{E}\left(\frac{\partial_z u_1}{\sqrt{g_{zz}}} - \frac{\tilde{j}u_1}{\sqrt{g_{\varphi\varphi}}}\right) + \frac{C_2}{E}\left(\frac{\partial_z u_2}{\sqrt{g_{zz}}} - \frac{\tilde{j}u_2}{\sqrt{g_{\varphi\varphi}}}\right), \tag{8}$$

where for a concrete value of E, the functions $C_1(E), C_2(E)$ stand for satisfying the normalization condition

$$2\pi \int_{-z_{max}}^{z_{max}} (|u(E, z)|^2 + |v(E, z)|^2)\mathrm{d}z = 1. \tag{9}$$

For a given z_0, the $LDoS$ is defined as

$$LDoS(E) = |u(E, z_0)|^2 + |v(E, z_0)|^2. \tag{10}$$

2.1. Perturbed cylinder

In the case of a small perturbation, we have

$$\vec{R}(z,\varphi)=\left(a\sqrt{1+\triangle z^2}\cos\varphi, a\sqrt{1+\triangle z^2}\sin\varphi, z\right), \tag{11}$$

where $\triangle$ is a positive real parameter, $\triangle << 1$. For $\triangle = 0$, we get the defect-free cylinder discussed in previous chapter.

Then

$$g_{zz}=1+\frac{a^2\triangle^2 z^2}{1+\triangle z^2}\sim 1+a^2\triangle^2 z^2, \qquad g_{\varphi\varphi}=a^2(1+\triangle z^2) \tag{12}$$

and

$$g=g_{\varphi\varphi}=a^2(1+\triangle z^2). \tag{13}$$

b For small $\triangle$ and neglecting the second order of $\triangle$, it can be simplified as

$$\partial_z u-\frac{\tilde{j}}{a}\left(1-\frac{1}{2}\triangle z^2\right)u=Ev, \qquad -\partial_z v-\frac{\tilde{j}}{a}\left(1-\frac{1}{2}\triangle z^2\right)v=Eu. \tag{14}$$

The solution is

$$u(z)=C_{\triangle 1}D_{\nu_1}(\xi(z))+C_{\triangle 2}D_{\nu_2}(i\xi(z)), \tag{15}$$

$$v(z)=\frac{C_{\triangle 1}}{E}\left(\partial_z D_{\nu_1}(\xi(z))-\frac{\tilde{j}D_{\nu_1}(\xi(z))}{a}(1-\frac{1}{2}\triangle^2 z^2)\right)+ \tag{16}$$
$$\frac{C_{\triangle 2}}{E}\left(\partial_z D_{\nu_2}(i\xi(z))-\frac{\tilde{j}D_{\nu_2}(i\xi(z))}{a}(1-\frac{1}{2}\triangle^2 z^2)\right),$$

where

$$\nu_1=i\frac{a^2\triangle-4a^2E^2+4ia\sqrt{\triangle}\tilde{j}+4\tilde{j}^2}{8a\sqrt{\triangle}\tilde{j}}, \qquad \nu_2=-i\frac{a^2\triangle-4a^2E^2-4ia\sqrt{\triangle}\tilde{j}+4\tilde{j}^2}{8a\sqrt{\triangle}\tilde{j}}, \tag{17}$$

$$\xi(z)=(-\triangle)^{1/4}\left(\sqrt{\frac{a}{2\tilde{j}}}+\sqrt{\frac{2\tilde{j}}{a}}z\right), \tag{18}$$

$D_\nu(\xi)$ being the parabolic cylinder function. The functions $C_{\triangle 1}=C_{\triangle 1}(E)$, $C_{\triangle 2}=C_{\triangle 2}(E)$ will be calculated in the same way as for the defect-free cylinder, i. e. from the normalization condition

$$\int_{-z_{max}}^{z_{max}}(|u(E,z)|^2+|v(E,z)|^2)\mathrm{d}z\mathrm{d}\varphi=4\pi\int_0^{z_{max}}(|u(E,z)|^2+|v(E,z)|^2)\mathrm{d}z=1. \tag{19}$$

3. Conclusion

In Fig. 1, a comparison of the $2D$ plots of $LDoS$ is made for different values of $\triangle$ and similarly in Fig. 2, where the $3D$ plots are compared. The number of defects $N=2$ and the other values are same as in the defect-free case. On the whole, we can conclude from the comparison that $LDoS$ is slightly increasing for higher $\triangle$. Although this is only the first order approximation, we will see that the real solution of behaves in a similar manner. We found that for the defect-free case, the electronic properties can sometimes correspond to the case of infinitely long nanoribbons. For the case of perturbation, the metallic properties are more manifested.

ACKNOWLEDGEMENTS — The work was supported by the Slovak Academy of Sciences in the framework of CEX NANOFLUID, and by the Science and Technology Assistance Agency under Contract No. APVV 0509-07 , 0171 10, VEGA Grant No. 2/0069/10 and Ministry of Education Agency for Structural Funds of EU in frame of project 26220120021.

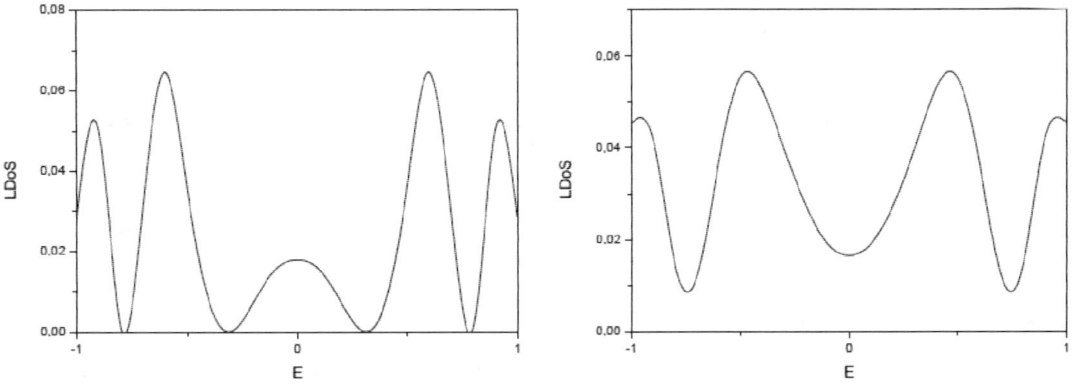

Figure 1. *LDoS* as a function of $E \in (-1, 1)$ and $z = 98$ for perturbed cylinder with $\triangle = 0.05$ (left) and $\triangle = 0.1$ (right)

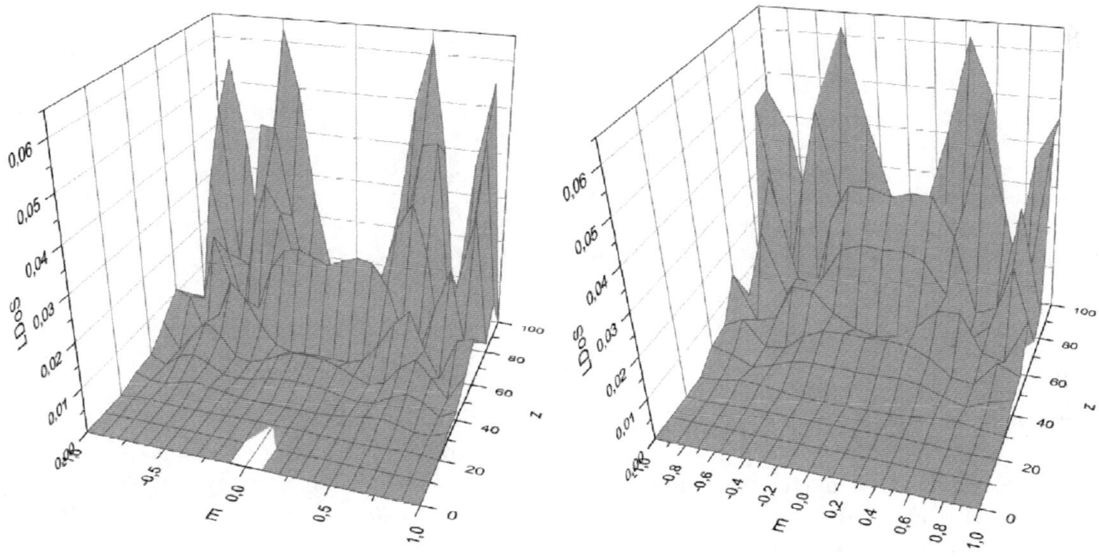

Figure 2. *LDoS* as a function of $E \in (-1, 1)$ and $z \in (0, 100)$ for perturbed cylinder with $\triangle = 0.05$ (left) and $\triangle = 0.1$ (right)

References

[1] Smotlacha J, Pincak R and Pudlak M 2011 *Eur. Phys. J.* B **84** 255-64
[2] Saito R, Dresselhaus G and Dresselhaus M S 1996 *Phys Rev.* B **53** 2044-50
[3] DiVincenzo D P and Mele E J 1984 *Phys. Rev* B **29** 1685-94
[4] Kochetov E A, Osipov V A and Pincak R 2010 *J. Phys.: Condens. Matter* **22** 395502
[5] Pincak R and Pudlak M 2007 *Progress in Fullerene Research* (New York: Nova Science Publishers)
[6] Lammert P E and Crespi V H 2004 *Phys. Rev.* B **69** 035406
[7] Kolesnikov D V and Osipov V A 2007 *EPL* **78** 47002

Bright solitons in cavity-QED arrays containing two-level atoms

E S Sedov[1], A P Alodjants[1], S M Arakelian[1], I-H Chen[2], Y Y Lin[2], Y-C Lai[2] and R-K Lee[2]

[1] Department of Physics and Applied Mathematics, Vladimir State University named after A.G. and N.G. Stoletovs, Vladimir, 600000, Russia
[2] Institute of Photonics Technologies, National Tsing-Hua University, Hsinchu, 300, Taiwan

E-mail: alodjants@vlsu.ru

Abstract. We examine the problem of polariton soliton formation in the array of weakly coupled optical cavities, each containing an ensemble of ultracold non-interacting two-level atoms. An effective complex Ginzburg-Landau equation (GLE) is derived in the continuum limit taking into account the effects of cavity field dissipation and atomic polarization dephasing. Bright solitons are found to be supported under perturbations only in the upper (optical) branch of polaritons, for which the corresponding group velocity is controlled by tuning the atom-light detuning. With the help of perturbation theory for solitons, we also demonstrate that the group velocity of these polariton solitons is suppressed by the diffusion process.

1. Introduction

At present, photonic band-gap micro- and nanostructures, that is photonic crystals and photonic crystal fibres, cavity arrays, etc, are known to be important tools for controlling light propagation in media and for quantum optical information processing, see e.g. [1]. In particular, an effective reduction of the light group velocity is achieved in both hot and cold atomic ensembles [2,3], in resonant photonic band-gap structures, i.e. in coupled resonator optical waveguides (CROW) [4] and also in different types of solid-state multilayer semiconductor systems [5]. The physical reason for observable reduction of the group velocity of light propagation in a medium is determined by the so-called dark and bright polaritons. The objects are bosonic quasi-particles representing a linear superposition of photon states in the external electromagnetic field and the macroscopic (coherent) excitations in the two-level oscillator system.

It is known that quantum optical solitons are a natural candidate for quantum optical information processing purposes, see e. g. [6]. Studies of their transmission, formation, and transformation in coupled cavity-QED arrays represent important steps for achieving this aim. Polariton solitons have some advantages in this case. In particular, as demonstrated in [7], polaritonic nonlinearity can be high enough in comparison with pure optical nonlinearities achieved, e.g. with pure optical cavity solitons in VCSELs [8], and can permit to achieve a soliton regime for essentially smaller particle (polariton) number. More precisely, here we deal with the problem of polariton soliton formation occurring at quantum matter-filed interface in semiconductor quantum wells (QWs) embedded in Fabry-Perot microcavities. Nowadays polariton solitons and relevant superfluid behavior of nonequilibrium exciton-polaritons with

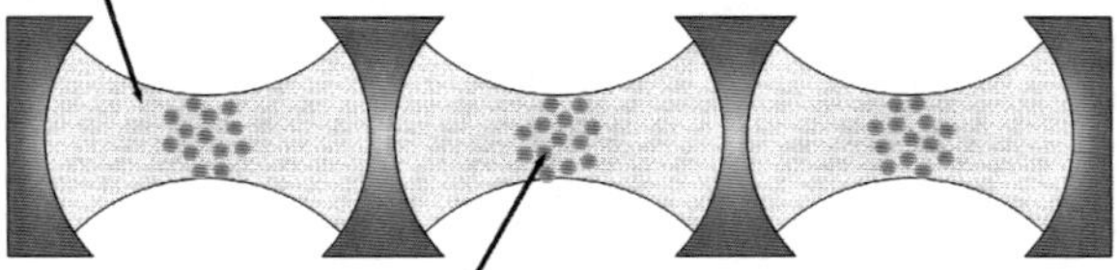

Figure 1. (Color online) Schematic for our proposed 1D cavity QED arrays, in which each cavity contains an ensemble of two-level atoms.

narrow-band (GaAs) semiconductor structures are observed in a few labs, [7, 9]. The main physical features of such solitons are connected with the balance between dissipation effects and external pumping that occurs for parametrical processes allowing to create non-equilibrium exciton-polariton Bose-Einstein condensate (BEC) at lower polariton branch under the temperatures of few kelvins. The solitons in experiments are excited within picoseconds time scale and localized in micrometer-scale sizes.

Recently, in [10] we proposed a 2D polaritonic crystal model that represents cluster material exhibiting high nonlinear properties due to the small but macroscopical number of two-level atoms strongly coupled with optical field in the cavity lattice. Basing on the Holstein-Primakoff approach, we have shown that such nonlinearity can basically be created by means of a saturation effects occurring due to localization of macroscopically small number of atoms at each cavity. Experimentally such a system can be designed by using 2D photonic crystal host with defect cavities [4].

In the paper we consider the problem of soliton formation in 1D weakly coupled cavity arrays containing two-level atoms that interact with single mode optical cavity fields – see Fig. 1. In the continuum limit, we use a multiple-scale envelope function (MSEF) method [11,12] to construct soliton solutions for considering nonlinear interactions among atoms as well as dissipation effects for them. In particular, we derive and examine a complex Ginzburg-Landau equation (GLE) that supports slow soliton propagations in the presence of polariton formation. Being a step in the studies of collective properties of light with interacting media, our results in the proposed coupled-cavity-QED system are shown to be connected with the formation of coupled matter-field states, upper (UB) and lower (LB) branch polaritons, which would be natural carriers for quantum information processing at matter-field interface.

2. The model of 1D cavity-QED array with two-level atoms

As illustrated in Fig. 1 we consider a one-dimensional (1D) array of optical cavities, each containing an ensemble of N_i of non-interacting two-level atoms under the extremely low temperatures, cf. [12–14]. For the configuration of an array of cavities, the atoms are assumed to be confined inside the cavities, while photons can tunnel from one cavity to another. We also assume that the tunneling process for photons is dominated by the hopping between adjacent sites. The value of hopping matrix elements is a function of the distance between cavities. If the distance between each cavity is sufficiently small, a cavity field will only hop to the nearest neighbors. Then the total Hamiltonian of interaction for this cavity-QED array is given as,

$$\hat{H} = \hbar \sum_{i=1}^{M} \left[\Delta_i \hat{a}_i^\dagger \hat{a}_i + \frac{g_i}{\sqrt{N_i}} \left(\hat{a}_i \hat{J}_{+,i} + \hat{a}_i^\dagger \hat{J}_{-,i} \right) - \alpha \left(\hat{a}_i^\dagger \hat{a}_{i+1} + \hat{a}_{i+1}^\dagger \hat{a}_i \right) \right]. \quad (1)$$

where α is a photon-hopping strength from cavity site i to $i+1$, and M counts the number of cavities. Detuning frequency Δ_i has been defined as the difference between atomic transition frequency ω_i and quantized field frequency ω_f, i.e., $\Delta_i = \omega_i - \omega_f$; orbital angular momentum representation, $\hat{J}_{i,j}$ with $j = x, y, z, \pm$, has been chosen to describe the atomic ensemble, while the field is described by its creation (annihilation) operators, $\hat{a}_i^\dagger(\hat{a}_i)$; the collective atom-field coupling are represented by the constants g_i. In the present work, we assume that only one species of atoms is considered and that the number of ensemble atoms is the same in each cavity, i.e., $\omega_i \equiv \omega$, and $N_i \equiv N$. For simplicity, the parameters are taken as real and homogeneous so that the detuning frequency, hopping and coupling strengths be all identical in each cavity site, i.e., $\Delta_i \equiv \Delta$, $g_i \equiv g$, and $\alpha_i \equiv \alpha$, respectively.

If most of atoms stay in the ground state, we can apply Holstein-Primakoff transformation by mapping atomic excitation operator $\hat{J}_{j,i}$ on the Schwinger representation for a two-level system, i.e., $\hat{J}_{+,i} = \sqrt{N}\hat{b}_i^\dagger\left(1 - \hat{b}_i^\dagger\hat{b}_i/N\right)^{1/2}$ and $\hat{J}_{z,i} = \hat{b}_i^\dagger\hat{b}_i - N/2$, along with the commutation relation $[\hat{b}_i, \hat{b}_j^\dagger] = \delta_{ij}$. Here $\hat{b}_i^\dagger\left(\hat{b}_j\right)$ is creation (annihilation) operator of polarization (excitation) of i-th atom. Practically, the number of atoms N is large enough and one can expand the orbital angular momentum representation for the atomic ensemble up to the order of $1/N$, i.e., $[1-(\hat{b}_i^\dagger\hat{b}_i/N)]^{1/2} \approx 1-\hat{b}_i^\dagger\hat{b}_i/2N$. We are working within the so-called low-excitation-density limit when the atoms mostly populate their ground levels. At the same time low-excitation-density limit implies that the average photon number at each cavity should be essentially smaller than the average number of atoms at the excited state, that is $\left\langle\hat{a}_i^\dagger\hat{a}_i\right\rangle \ll \left\langle\hat{b}_i^\dagger\hat{b}_i\right\rangle$ – cf. [15].

Now, since number N is assumed to be large enough, we are able to neglect higher-order terms in the expansion of $\hat{J}_{+,i}$ operator. Then, we can transform the Hamiltonian in Eq. (1) into

$$\begin{aligned}\hat{H}_{\mathrm{HP}} &= \hbar\sum_{i=1}^{M}\{-\alpha\left(\hat{a}_i^\dagger\hat{a}_{i+1} + \hat{a}_{i+1}^\dagger\hat{a}_i\right) + g\left(\hat{a}_i^\dagger\hat{b}_i + \hat{a}_i\hat{b}_i^\dagger\right) \\ &- \frac{g}{2N}\left[\hat{a}_i^\dagger\hat{b}_i^\dagger\hat{b}_i\hat{b}_i + \hat{a}_i\hat{b}_i^\dagger\hat{b}_i^\dagger\hat{b}_i\right] + \Delta\hat{a}_i^\dagger\hat{a}_i\}. \qquad (2)\end{aligned}$$

Based on this Hamiltonian, Heisenberg's equations of motion for the involved two bosonic operators of cavity modes $\hat{a}_i$ and atomic polarization $\hat{b}_i$ are

$$\mathrm{i}\partial_t\hat{a}_i = \Delta\hat{a}_i - \alpha\left(\hat{a}_{i+1} + \hat{a}_{i-1}\right) + g\left[\hat{b}_i - \frac{1}{2N}\left(\hat{b}_i^\dagger\hat{b}_i\hat{b}_i\right)\right], \qquad (3)$$

$$\mathrm{i}\partial_t\hat{b}_i = g\left[\hat{a}_i - \frac{1}{2N}\left(\hat{a}_i^\dagger\hat{b}_i\hat{b}_i + 2\hat{a}_i\hat{b}_i^\dagger\hat{b}_i\right)\right]. \qquad (4)$$

Here, we have neglected quantum fluctuations due to a large number of atoms considered.

To have mean-field solutions, we replace the pair of operators $(\hat{a}_i, \hat{b}_i)$ by the corresponding expectation values (ψ_i, β_i), and approximate this array configuration to a continuous model, i.e., $\psi_{i+1} + \psi_{i-1} \approx 2\psi(x,t) + d^2\frac{\partial^2}{\partial x^2}\psi(x,t)$, with the distance between two adjoint optical cavities denoted by d. By considering the conservation of the total photon number and atomic excitations, we can re-normalize the variables with respect to the number of polaritons N_{pol}, defined as $N_{\mathrm{pol}} = |\psi|^2 + |\beta|^2$. By replacing the cavity field and atomic excitation by $\psi \to \psi/\sqrt{N_{\mathrm{pol}}}$ and $\beta \to \beta/\sqrt{N_{\mathrm{pol}}}$, respectively, now we have a normalization condition for the cavity field and atomic excitation, $|\psi|^2 + |\beta|^2 = 1$. Thus, in this continuum limit, the equations of motion in Eqs. (3,4) become,

$$\mathrm{i}\partial_t\psi = (\Delta - 2\alpha - \mathrm{i}\gamma_{\mathrm{c}})\psi - \alpha d^2\partial_{xx}\psi + g\beta - U_{\mathrm{sat}}|\beta|^2\beta, \qquad (5)$$

$$\mathrm{i}\partial_t\beta = g\psi - U_{\mathrm{sat}}\left[\beta^2\psi^* + 2\psi|\beta|^2\right] - \mathrm{i}\Gamma_{\mathrm{d}}\beta, \qquad (6)$$

where $U_{\mathrm{sat}} \equiv \frac{gn_{\mathrm{pol}}}{2}$; $n_{\mathrm{pol}} = N_{\mathrm{pol}}/N$ is a polariton number density. In Eqs. (5,6) we have a phenomenologically introduced decay rate for photonic field γ_{c} that characterizes the leakage of photons in the cavity and dephasing rate Γ_d for the atomic system. Physically, parameter U_{sat} characterizes a nonlinear saturation effect under the atom-light interaction.

Let us briefly discuss the applicability of our model (Eqs. (5,6)) for real atomic systems. In the paper we examine ultracold two-level rubidium atoms with resonance frequency $\omega_{ab}/2\pi = 382$THz that corresponds to mean weighted rubidium D-lines [16]. The atomic polarization dephasing rate and the minimal value of each cavity field decay rate can be taken as several tens of megahertz's that corresponds to cavity quality factor $Q = \omega_{ab}/2\gamma_c \sim 10^6$. The strength of interaction of a single atom with a quantum optical field is taken as $g_0/2\pi = 89.5$MHz at each cavity with the effective volume of atom-field interaction $V = 5000\mu\text{m}^3$. To achieve a strong atom-field coupling regime, see Eqs. (9) below, one can propose a macroscopically large number of atoms at each cavity, say $N = 5 \times 10^5$. This number is relevant to the density $\rho = 10^{14}\text{cm}^{-3}$ of ultracold atoms that implies a collective atom-filed coupling parameter as $g \equiv g_0\sqrt{N} = 2\pi \times 63.3$GHz at each cavity. Noticing, that the parameter that characterizes atom-atom interaction $\eta = 4\pi\hbar a_{sc}\rho/m$ in the Born approximation can be estimated as $\eta = 2\pi \times 0.73$kHz, where a_{sc} is atomic scattering length that we choose as $a_{sc} = 5$nm. Thus, for atomic QED array we can neglect atom-atom interaction processes in Eqs. (5,6).

The set of coupled nonlinear equations (5,6) is the starting point of our work, and soliton solutions both for the wave-packet envelope of cavity field ψ (order parameter) and for atomic excitations β are considered analytically below.

3. Polariton solitons under perturbations

To construct solitons in this cavity-QED arrays, we seek for wave-packet solutions by the *multiple-scale envelope function* method [11, 12]. With the introduction of different length and time scales, i.e., $x_m = \lambda^m x$ ($\lambda \ll 1, m = 0, 1, 2\ldots$) and $t_m = \lambda^m t$ ($\lambda \ll 1, m = 0, 1, 2\ldots$), we can expand photon field $\psi = \sum_{n=1}\lambda^n\psi^{(n)}$ and atomic excitation $\beta = \sum_{n=1}\lambda^n\beta^{(n)}$. By substituting these expansions in Eqs. (5-6), we can gather all terms that are proportional to the first order of λ. This first-order expansion for the atomic excitation $\beta^{(1)}$ supports a plane wave solution, from which we can find a corresponding dispersion relation by using the solution in the form of $\psi^{(1)} = E^{(1)}e^{\mathrm{i}(kx_0-\omega t_0)}$ and $\beta^{(1)} = \frac{g}{\omega+\mathrm{i}\Gamma_d}E^{(1)}e^{\mathrm{i}(kx_0-\omega t_0)}$. The corresponding carrier frequencies are

$$\omega \equiv \omega_\pm = \frac{1}{2}\left[\delta - \mathrm{i}\left(2\gamma_c - \gamma\right) \pm \left(\left(\delta - \mathrm{i}\gamma\right)^2 + 4g^2\right)^{1/2}\right], \tag{7}$$

where $\gamma = \gamma_c - \Gamma_d$ is an effective atom-field decay rate for the system and $\delta = \Delta - 2\alpha\left(1 - d^2k^2/2\right)$ is a total *momentum dependent* atom-light detuning, k is a wave number. Physically, Eq. (7) reproduces a familiar result for frequencies of two branches of polaritons (cf. [10]), which are denoted as UB (ω_+) and LB (ω_-), respectively.

It is important to note that field ($\psi^{(1)}$) and atomic excitation ($\beta^{(1)}$) variables become exact solutions of Eqs. (5,6) neglecting nonlinear effects, i.e., setting $U_{sat} = 0$. Hence, it is easy to understand that MSEF method developed here is valid if dispersion characteristics of polaritons, see Eq. (7), cannot be essentially modified taking into account nonlinear effects. Strictly speaking, the condition $U_{sat} \ll 2g$ should be fulfilled for coupled atom-light systems described by Eqs. (5,6). In practice this condition implies achieving low-excitation density limit $n_{pol} \ll 1$ discussed above.

Next, for the second order of multiple-scales, i.e., λ^2, we can have a linear wave equation for a wave-packet envelope $\left(\partial_{t_1} + v_\pm\partial_{x_1}\right)E^{(1)} = 0$, from which one can find the corresponding group velocities,

$$v_\pm = \partial_k\omega_\pm = \frac{2\alpha k d^2\Omega_\pm^2}{\Omega_\pm^2 + g^2} \tag{8}$$

Journal of Physics: Conference Series **393** (2012) 012030 doi:10.1088/1742-6596/393/1/012030

that are defined for UB (v_+) and LB (v_-) polariton wave-packets. Below we consider variable $E^{(1)}$ in the moving frames $\xi_\pm = x_1 - v_\pm t_1$. In Eq. (8) we have also introduced characteristic frequencies $\Omega_\pm = \frac{1}{2}\left[\delta - \mathrm{i}\gamma \pm \left((\delta - \mathrm{i}\gamma)^2 + 4g^2\right)^{1/2}\right]$.

It is noted that the group velocities $v_\pm$ defined in Eq. (8) are complex numbers in general. It is known that this leads to the additional deformation of a pulse envelope propagated in the medium [17]. The above MSEF method works only for small decay rates if a strong coupling atom-light condition is valid. In particular, we require that conditions

$$\gamma_{\rm c},\ \Gamma_{\rm d} \ll 2g, \qquad \mathrm{Im}\,(\omega_\pm) \ll \mathrm{Re}\,(\omega_\pm)\,. \tag{9a, b}$$

should be fullfilled. The inequality (9b) means that dissipation effects are not dominant under atom-field interaction in the cavity array.

Now, we do a multiple-scale expansion up to the third order by collecting terms with λ^3, and reach a complex GLE, i.e.,

$$\mathrm{i}\frac{\partial \Psi}{\partial t} + D_\pm \frac{\partial^2 \Psi}{\partial X_\pm^2} + C_\pm \left|\Psi\right|^2 \Psi = 0, \tag{10}$$

where variables $\Psi = \lambda E^{(1)}$, $X_\pm = \xi_\pm/\lambda$, and $t_2 = \lambda^2 t$ have been introduced. The coefficients $D_\pm$ and $C_\pm$, appearing in the second and third terms of Eq. (10) respectively, have the forms,

$$D_\pm = \frac{\alpha d^2 \Omega_\pm^3 + \mathrm{v}_\pm^2 g^2}{\Omega_\pm\left(\Omega_\pm^2 + g^2\right)}, \qquad C_\pm = \frac{n_{\rm pol} g^4 \left[(3\Omega_\pm + \Omega_\pm^*)\right]}{2\left|\Omega_\pm\right|^2\left(\Omega_\pm^2 + g^2\right)}. \tag{11a, b}$$

Coefficients defined in Eqs. (11a,b) are complex and can be evaluated as $D_\pm = D_\pm^{(1)} + \mathrm{i}D_\pm^{(2)}$, $C_\pm = C_\pm^{(1)} + \mathrm{i}C_\pm^{(2)}$. Real part $D_\pm^{(1)}$ of $D_\pm$-coefficient describes diffraction effects occurring with the wave-packet; while the imaginary part $D_\pm^{(2)}$ characterizes diffusion processes. Parameter $C_\pm^{(1)}$ is responsible for a Kerr-like nonlinearity that occurs due to polariton-polariton scattering. At the same time an imaginary part, that is $C_\pm^{(2)}$, is relevant to nonlinear absorption effects.

By taking into account Eq. (7), it is helpful to introduce convenient UB (m_+) and LB (m_-) polariton masses

$$m_\pm = \hbar\left[\partial_k^2 \omega_\pm\Big|_{k=0}\right]^{-1} = \frac{m_{\rm ph}\left(\Omega_\pm^2 + g^2\right)}{\Omega_\pm^2}, \tag{12}$$

where $m_{\rm ph} = \hbar/2\alpha d^2$ is an effective photon mass in the cavity, cf. [15]. Neglecting the kinetic energy of polaritons and supposing that the solutions are taken at the bottom of dispersion curve, one can approximate

$$D_\pm \simeq \frac{\hbar}{2m_\pm}, \qquad C_\pm \simeq \pm\frac{2n_{\rm pol} g Y_\mp^3}{Y_\pm}, \tag{13a, b}$$

where we have introduced Hopfield coefficients such as $Y_\pm = \frac{1}{\sqrt{2}}\left[1 \pm \frac{\delta}{\sqrt{\delta^2 + 4g^2}}\right]^{1/2}$ and for the sake of simplicity we suppose that $g > 0$. The latter parameters determine the contributions from photons and atomic excitations to polaritons. In particular, we can express field variable Ψ as following

$$\Psi = Y_+ \Xi_{\rm UB} - Y_- \Xi_{\rm LB}, \tag{14}$$

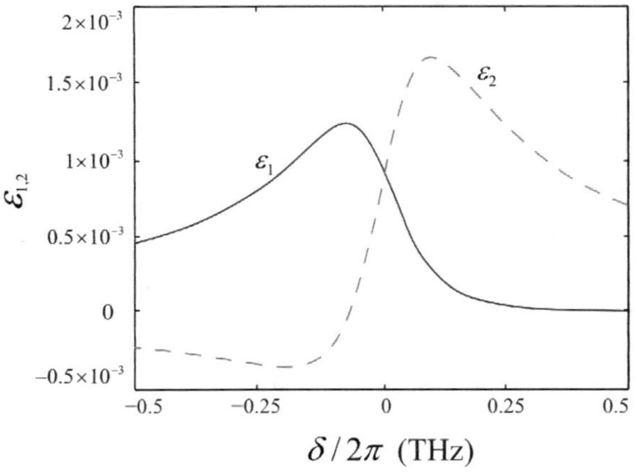

Figure 2. (Color online) Perturbation coefficients $\varepsilon_{1,2}$ versus atom-light detuning δ. The parameters are: $g = 2\pi \times 63.3$GHz, $\alpha = 2\pi \times 10$GHz, $\gamma_c = 2\pi \times 30$MHz, $\Gamma_d = 2\pi \times 6$MHz, $d = 5\mu$m, $n_{pol} = 0.01$.

where Ξ_{UB} and Ξ_{LB} are new variables characterizing UB and LB polaritons respectively. They are defined in a convenient way by using Bogoliubov transformation, cf. [16].

In particular, for a positive and large frequency detuning δ, that is $|\delta| \gg g$ and $\delta > 0$, we have $Y_+ \approx 1$ and $Y_- \approx \frac{g}{|\delta|}$, respectively, which corresponds to *photon-like UB polaritons* ($\Xi_{UB} \simeq \Psi$) with mass $m_+ \approx m_{ph}$ and group velocity $v_+ = \hbar k/m_+$, see Eqs. (12-14). Remarkably, at the same time the group velocity of LB polaritons with mass $m_- \approx m_{ph}\frac{\delta^2}{g^2}$ approaches $v_- = \hbar k/m_- \ll c$, where c is speed of light in vacuum. Now, we have a "slow" (matter-like) soliton formation in the cavity array. For $|\delta| \gg g$ but $\delta < 0$, a physical picture becomes opposite and a propagating optical pulse relevant to *LB polaritons* ($\Xi_{LB} \simeq -\Psi$) with mass $m_- \approx m_{ph}$ and group velocity $v_- = \hbar k/m_- \gg v_+$. In the presence of atom-light resonance condition, $\delta = 0$, UB and LB polaritons equally contribute to the photonic state, that is $Y_\pm = 1/\sqrt{2}$. The masses of polaritons are equal to each other, i.e., $m_{pol} \equiv m_\pm = 2m_{ph}$. Their group velocities are $v_\pm = \hbar k/m_{pol}$.

It is useful to examine Eqs. (10-11) in the limiting (dissipationless) case for the system by setting $\gamma_c = \Gamma_d = 0$. In this limit, both coefficients $D_\pm$ and $C_\pm$ become real and GLE in Eq. (10) is reduced to a standard nolinear Schrödinger equation, that possesses exact *bright* and *dark* soliton solutions depending on the sign of the coefficients $C_\pm = C_\pm^{(1)}$ and $D_\pm = D_\pm^{(1)}$. Here, we suppose that coefficients $D_\pm > 0$. Then, for the case of $C_\pm > 0$, we have a *bright* soliton solution; while for the case of $C_\pm < 0$, *dark* solitons can be found [18]. By using Eq. (11b) it is possible to show that we can only have a bright soliton solution supported in UB; while a dark one in LB.

Now we are going to examine a coupled atom-light system in respect of bright soliton formation; the system being non-equilibrium. In this case we are interested in soliton solutions of Eq. (10) for optical field amplitudes Ψ in the presence of polariton formation. It is useful to bring Eq. (10) to a dimensionless form, rewriting it as follows

$$i\frac{\partial\Psi}{\partial\tau} + \frac{1}{2}\frac{\partial^2\Psi}{\partial x^2} + |\Psi|^2\Psi = -i\varepsilon_1|\Psi|^2\Psi + i\varepsilon_2\frac{\partial^2\Psi}{\partial x^2}, \tag{15}$$

where $x = X_+/d$ and $\tau = \frac{2D_+^{(1)}}{d^2}t$ are new dimensionless spatial and temporal coordinates respectively; $\varepsilon_1 = \frac{C_+^{(2)}}{\left|C_+^{(1)}\right|}$, $\varepsilon_2 = \frac{|D_+^{(2)}|}{2D_+^{(1)}}$ are perturbation coefficients ($\varepsilon_{1,2} > 0$). In Eq. (15) we also use the fact that the characteristic time scale of a dispersion action and the influence of

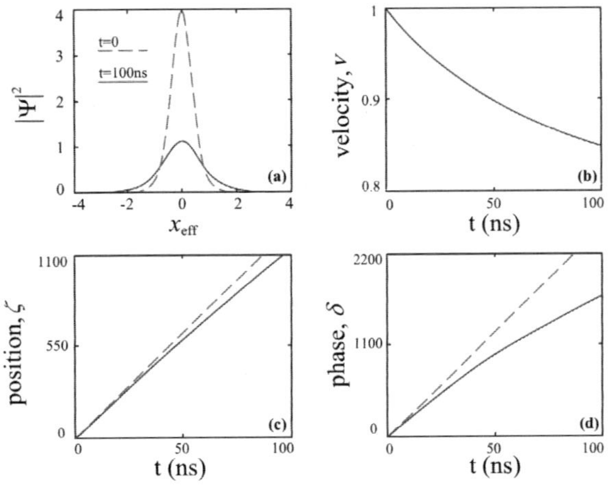

Figure 3. (Color online) (a) Snapshots of bright soliton profile $|\Psi|^2$ as a function of effective coordinate $x_{\rm eff} = x - \zeta(\tau)$ at different time of propagation t; (b) – (d) reflect temporal behavior of soliton parameters. Dashed lines in (c,d) correspond to solution of Eq. (15) for unperturbed solitons with $\varepsilon_{1,2} = 0$. The numerical values of parameters characterizing atom-light system for solid curves are the same as in Fig. 2 under the atom-light resonance condition $\delta = 0$.

nonlinearity on wave packet spreading should be the same [18]. In particular, the condition

$$n_{\rm pol} = \frac{4D_+^{(1)}\left|\Omega_+\right|^2}{d^2 g^4}\left|\mathrm{Re}\left[\frac{3\Omega_+ + \Omega_+^*}{\Omega_+^2 + g^2}\right]\right|^{-1} \tag{16}$$

should be met for polariton number density $n_{\rm pol}$ in this case.

Behavior of perturbation coefficients $\varepsilon_{1,2}$ as a function of atom-light detuning δ for soliton formation is shown in Fig. 2. It is important that parameters $\varepsilon_{1,2} \ll 1$ within wide range of variation of δ for appropriate effective decay rate γ. Noticing that for $|\delta| \gg g$ and $\delta < 0$ the coefficient ε_2 becomes negative. However, in this limit inequality (9b) is violated and we are not able to consider dissipation effects as a small perturbations. Below we examine the GLE in Eq. (15) using perturbation theory for solitons by consulting the reference [19]. In particular, we are looking for the solution of Eq. (15) in the form

$$\Psi(\tau, x) = 2\nu \mathrm{sech}\left[2\nu\left(x - \zeta(\tau)\right)\right] e^{i\varphi(\tau,x)}, \tag{17}$$

where ν, $\zeta(\tau)$, and $\varphi(\tau, x)$ are related soliton amplitude, position, and phase, respectively. In the absence of perturbation ($\varepsilon_{1,2} = 0$), the ansatz used in Eq. (17) represents an exact solution for Eq. (15) with the parameters evolving in time as $\zeta(\tau) = v\tau/2 + \zeta_0$, $\varphi(\tau, x) = v\left(x - \zeta(\tau)\right)/2 + \delta(t)$, $\delta(\tau) = v^2\tau/2 + 2\nu^2\tau + \delta_0$, where parameters ζ_0 and δ_0 represent the initial soliton position and phase respectively. Here, $v \equiv v_+$ is soliton velocity that we can associate with the velocity of UB polaritonic wave-packet by setting $\gamma = \Gamma_{\rm d} = 0$. In the presence of small perturbation ($\varepsilon_{1,2} \neq 0$), we consider the soliton amplitude ν and velocity v to be time dependent. Applying soliton perturbation theory, we arrive at the set of equations for solitons parameters under the adiabatic approximation,

$$\dot{\nu} = -\frac{8}{3}\left(2\varepsilon_1 + \varepsilon_2\right)\nu^3 - \frac{1}{2}\varepsilon_2 v^2 \nu, \quad \dot{v}_+ = -\frac{16}{3}\varepsilon_2 v \nu^2, \quad \dot{\zeta} = \frac{v}{2}, \quad \dot{\delta} = \frac{v^2}{2} + 2\nu^2, \tag{18a, b, c, d}$$

where dot denotes derivative in respect of normalized time, i.e., $d/d\tau$.

In Fig. 3, we represent time dependent solutions of Eqs. (18) (solid lines). In particular, Fig. 3a demonstrates soliton profile taken at different time moments. Based on real parameters of atomic system the initial shape of this soliton is characterized by the dashed (red) curve. From Fig. 3a, it is evident that a soliton amplitude vanishes due to effective nonlinear absorption. At the same time the group velocity of soliton is slightly modified due to diffusion processes characterized by parameter ε_2 shown in Eq. (18a). The characteristic time of disturbing soliton group velocity is tens of nanoseconds which seems to be reasonable taking into account polariton lifetime in atomic structures, cf. [10, 11].

4. Conclusions

In this work, we consider the problem of soliton formation in the array of weakly coupled atomic ensembles interacting with the optical field in a tunnel-coupled cavity array. The effects of cavity field dissipation and atomic dephasing have been taken into account. We use multiple-scale envelope function method to obtain an effective complex Ginzburg-Landau equation, containing nonlinear absorption and diffusion terms. Mean-field soliton solutions for this coupled matter-field states are revealed in the representation of UB and LB polaritons. We have demonstrated that bright solitons are formed for UB polariton wave packets and characterized by slowly varying soliton amplitude, momentum, position, and phase. Notably, the group velocity of soliton decreases in time due to diffusion processes within a nanoseconds domain. Our analytical and numerical results represented here and connected with a cavity-QED array provide a platform for the studies of collective properties of light with interacting media.

5. Acknowledgments

This work was supported by RFBR Grants No. 10-02-13300, 11-02-97513, 12-02-90419, 12-02-97529 and by Russian Ministry of Education and Science under the contracts 14.740.11.0700.

References

[1] Joannopoulos J D, Johnson S G, Winn J N and Meade R D 2008 *Photonic Crystals: Molding the Flow of Light 2nd edn* (Princeton, NJ: Princeton University Press)
[2] Karpa L, Vewinger F and Weitz M 2008 *Phys. Rev. Lett.* **101** 170406
[3] Liu C, Dutton Z, Behroozi C H and Hau L V 2001 *Nature* **409** 490.
[4] Lee R K, Painter O, Kitzke B, Scherer A and Yariv A 2000 *J. Opt. Soc. Am. B* **17** 629; Vuckovic J, Loncar M, Mabuchi H and Scherer A 2001 *Phys. Rev. E* **65** 016608.
[5] Turukhin A V, Sudarshanam V S, Shahriar M S, Musser J A, Ham B S and Hemmer P R 2002 *Phys. Rev. Lett.* **88** 0236021.
[6] Haus H A and Wong W S 1996 *Reviews of Modern Physics* **68** 423.
[7] Sich M *et al.* 2012 *Nature Photonics* **6** 50.
[8] Barland S *et al.* 2002 *Nature* **419** 699.
[9] Amo A *et al.* 2011 *Science* **332** 1167.
[10] Sedov E S, Alodjants A P, Arakelian S M, Lin Y Y and Lee R-K 2011 *Phys. Rev. A* **84** 013813.
[11] de Streke C M and Sipe J E 1988 *Phys. Rev. A* **38** 5149.
[12] Paternostro M, Agrarwal G. S and Kim M S 2009 *New J. Phys.* **11** 013059.
[13] Lei S-C and Lee R-K 2008 *Phys. Rev. A* **77** 033827.
[14] Tomadin A and Fazio 2010 *J. Opt. Soc. Am. B* **27** A130; Greentree A D, Tahan C, Cole J H and Hollenberg L C L. 2006 *Nature Physics* **2** 856.
[15] Alodjants A P, Barinov I O and Arakelian S M 2010 *J. Phys. B* **43** 095502.
[16] Alodjants A P, Chestnov I Yu and Arakelian S M 2011 *Phys. Rev. A* **83** 053802.
[17] Vainstein L A 1976 *Sov. Phys. Usp.* **19** 189; Bukhman N S 2001 *Rus. J. Quant. Electron.* **31** 774.
[18] Agrawal G P 1989 *"Nonlinear Fiber Optics"* (Academic Press: Boston).
[19] Karpman V I, Maslov V I 1977 *Zh. Eksp. Teor. Fiz.* **75** 537 [1977 *Sov. JETP* **46** 281]; Karpman V I, Solov'ev V I 1981*Physica 3D* 487; Kivshar Yu S, Malomed B 1989 *Rev. Mod. Phys.* **61** 763.

Dubna-Nano 2012 IOP Publishing
Journal of Physics: Conference Series **393** (2012) 012031 doi:10.1088/1742-6596/393/1/012031

A model of small-angle scattering from three-phase fractal systems

Alexander Yu Cherny[1], Eugen M Anitas[1,2], Vladimir A Osipov[1] and Alexander I Kuklin[1]

[1]Joint Institute for Nuclear Research, Dubna 141980, Moscow region, Russian Federation
[2]Horia Hulubei National Institute of Physics and Nuclear Engineering, RO-077125 Bucharest, Romania

E-mail: `cherny<at>theor.jinr.ru`

Abstract. In small-angle scattering (SAS) intensities on a double logarithmic scale, one generically observes "knees" arising from changes of the power-law exponents. While the appearance of power-law behavior is associated usually with a fractal structure, the explanation of the knees is a more intricate problem in general. It is shown that the crossover between the exponents can be accounted for by various levels of the structures with different scattering length densities (SLD). Thus, the whole system is a multiphase system, composed of the homogeneous structures embedded into each other. In this way, one can explain the crossover between arbitrary exponents whose values vary from -4 to 0. In order to confirm this conclusion, we calculate the SAS intensity from a set of non-interacting, randomly oriented and uniformly distributed three-phase fractals with controllable Hausdorff dimension. We observe the appearance of crossover (knee), which depends on the values of the SLD of each phase and estimate its position. Random fractals are described as well within the developed model by taking the fractal sizes at random, which leads to polydispersity and, as a consequence, to smearing of the SAS intensity curves.

1. Introduction

Small-angle scattering (SAS) [1] has been proved to be a very useful technique to study structural properties of fractals [2]. In general, the SAS technique enables us to obtain the fractal dimension and the edges of the fractal region [3,4]. As was shown recently, the SAS intensity from *deterministic* fractal systems (DFS) can be used to extract the *additional* fractal parameters: the fractal iteration number, the scaling factor and the number of structural units of which the fractal is composed [5].

For a two-phase system, fractals are composed of homogeneous units, which are embedded into a homogeneous solid matrix. Then the SAS intensity from the system clearly shows a separate contribution of the two structural levels: the fractal itself and its building units [4–7]. Therefore, at low q the intensity is characterized by the Guinier region of the fractal (a plateau on a double logarithmic scale) and the fractal region [$I(q) \propto q^{-D}$, where D is the fractal dimension], while at high q the intensity is characterized by the second Guinier and Porod [$I(q) \propto q^{-4}$] regions of the building units. When the Guinier region of the building units is small enough, the intensity shows a crossover point (knee) at some value of the wave vector q_{c}, where the scattering exponent changes from $-D$ to -4.

However, at the knee position, SAS experimental data often exhibit a transition between *arbitrary* values of the scattering exponents, whose values vary from -4 to 0 [8–11]. In general, the values of the exponent lie within $-3 < D_{\mathrm{m}} < 0$ for a mass fractal, and within $-4 < D_{\mathrm{s}} < -3$ for a surface fractal, where D_{m} and D_{s} are mass and, respectively, surface fractal dimension [3,4]. We argue that the crossover between the exponents can be accounted for by various levels of structures with different

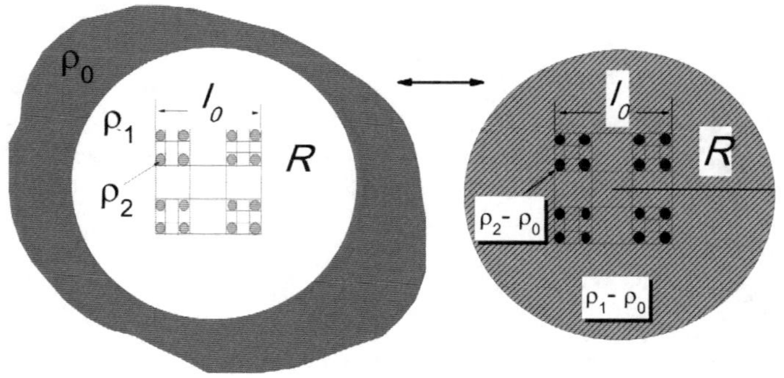

Figure 1. A three-phase fractal system is equivalent to a two-phase one in a vacuum, where the building units of the fractal have SLD $\rho_2 - \rho_0$ and the surrounding ball of radius R has SLD $\rho_1 - \rho_0$.

scattering length densities (SLD). Thus, this kind of SAS intensity is typical for a multiphase system, composed of the homogeneous structures embedded into each other.

In order to check this statement, we suggest a model of three-phase system, where one of the phase is a DFS. In this model, a uniform matrix with scattering length density (SLD) ρ_0 (phase 1) contains balls with SLD ρ_1 (phase 2), in which a DFS with SLD ρ_2 (phase 3) are embedded, see Fig. 1. We model the DSF phase by means of the generalized Cantor fractals (GCF) [7], which is a mass fractal with controllable fractal dimension (Fig. 2). A random fractal can be modeled by this system as well if we consider an ensemble of DSFs, whose sizes are taken at random (polydisperse DFS system) [4–7].

This paper examines both monodisperse and polydisperse SAS intensities from a set of randomly oriented and uniformly distributed three-phase fractals and focuses on studying of the knee on the scattering intensity. We investigate the effects of variation of the SLD values on the contribution of the individual structural levels to the total scattering intensity.

2. The model

The three-phase model is a solid matrix with uniform SLD ρ_0 that includes a macroscopic number of spherical regions with SLD ρ_1, and each of them contains the 3D generalized Cantor fractals (GCF) with SLD ρ_2 (see the left diagram in Fig. 1). The fractal dimension can be controlled by varying its scaling factor β_{s}. The GCF is constructed [7] with the help of a cube by iterations called approximants (Fig. 2). The zero-order iteration (initiator) is a ball of radius $r_0 = l_0/2$, where l_0 is the length of the cube edge. The radius of the ball surrounding the fractal is $R = kl_0$, where $k \geqslant \sqrt{3}/2$ (Fig. 1). It is assumed that the spatial positions and orientations of these constructions in the solid matrix are uncorrelated.

A constant shift of the SLD in the overall sample is important only for small values of wave vector, unattainable with the SAS technique. Therefore, by subtracting the solid matrix density ρ_0, the model can be reduced to the fractals with SLD $\rho_2 - \rho_0$ and the volume $V_2 = 8^m \beta_{\mathrm{s}}^{3m} 4\pi r_0^3/3$, embedded into the surrounding balls with SLD $\rho_1 - \rho_0$ and the volume $V_{\mathrm{b}} = 4\pi R^3/3$ (Fig. 1). Thus, the model can be considered as if the balls were "frozen" in a vacuum and had the density

$$\rho(\boldsymbol{r}) = \begin{cases} \rho_2 - \rho_0 & \text{in } V_2, \\ \rho_1 - \rho_0 & \text{in } V_{\mathrm{b}} \backslash V_2. \end{cases} \tag{1}$$

Effectively, this procedure allows us to reduce the three-phase system to the two-phase system. We will call this construction, shown in Fig. 1, the three-phase fractal.

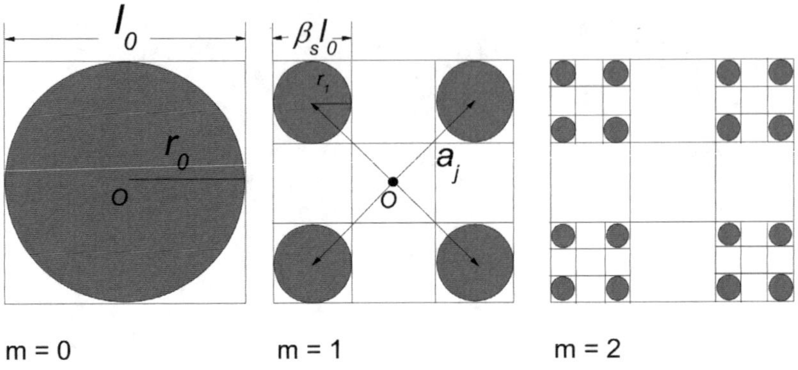

Figure 2. The 2D projection of the 3D generalized Cantor fractal (GCF): initiator ($m = 0$) and first two iterations ($m = 1$ and $m = 2$). Vectors $\mathbf{a}_j$ connect the center of the ball of radius r_0 with the centers of balls of radii r_1 (see Ref. [7] for details).

3. Calculation of the scattering intensity

The scattering intensity (the differential cross section per unit volume of the sample) can be calculated if we take into account the randomness of orientations and spatial positions of the three-phase fractals. Then they scatter the incident beam independently, which leads to [1]

$$I(q) = n\langle|A(\boldsymbol{q})|^2\rangle, \tag{2}$$

$$A(\boldsymbol{q}) \equiv \int_{V_{\rm b}} \rho(\boldsymbol{r}) e^{-i\boldsymbol{q}\cdot\boldsymbol{r}} \mathrm{d}\boldsymbol{r}, \tag{3}$$

where the quantity $A(\boldsymbol{q})$ is the scattering amplitude of the three-phase fractal. Its effective SLD is given by Eq. (1). Here n is the number of fractals per unit volume, the brackets $\langle\cdots\rangle$ stand for the ensemble averaging over all orientations of $\boldsymbol{q}$.

It is convenient to write down the intensity in terms of the normalized scattering amplitudes. The normalized amplitude is defined in general as

$$F(\boldsymbol{q}) \equiv \int_V \rho(\boldsymbol{r}) e^{-i\boldsymbol{q}\cdot\boldsymbol{r}} \mathrm{d}^3 r \Big/ \int_V \rho(\boldsymbol{r}) \mathrm{d}^3 r \,. \tag{4}$$

It follows from the definition that $F(0) = 1$, and for a homogeneous system $A(\boldsymbol{q}) = \rho V F(\boldsymbol{q})$. A straightforward substitution of Eq. (1) into Eq. (3) yields after a little algebra

$$A(\boldsymbol{q}) = V_{\rm b}(\rho_1 - \rho_0)\alpha^{-1}[\alpha F_{\rm b}(qR) + (1-\alpha)F_2^{(m)}(\boldsymbol{q}l_0)], \tag{5}$$

where we introduce by definition the "contrast parameter"

$$\alpha \equiv \left(1 + \frac{\rho_2 - \rho_1}{\rho_1 - \rho_0}\frac{V_2}{V_{\rm b}}\right)^{-1}, \tag{6}$$

and $F_{\rm b}(qR)$ and $F_2^{(m)}(\boldsymbol{q}l_0)$ are the normalized form factors of a homogeneous ball of radius R [1] and of the fractal at the mth iteration [7], respectively. They are calculated analytically in terms of elementary functions. Thus, the final expression for the scattering amplitude takes the form

$$I_m(q) = nV_{\rm b}^2(\rho_1 - \rho_0)^2\alpha^{-2}\left\langle\left|\alpha F_{\rm b}(qR) + (1-\alpha)F_2^{(m)}(\boldsymbol{q}l_0)\right|^2\right\rangle. \tag{7}$$

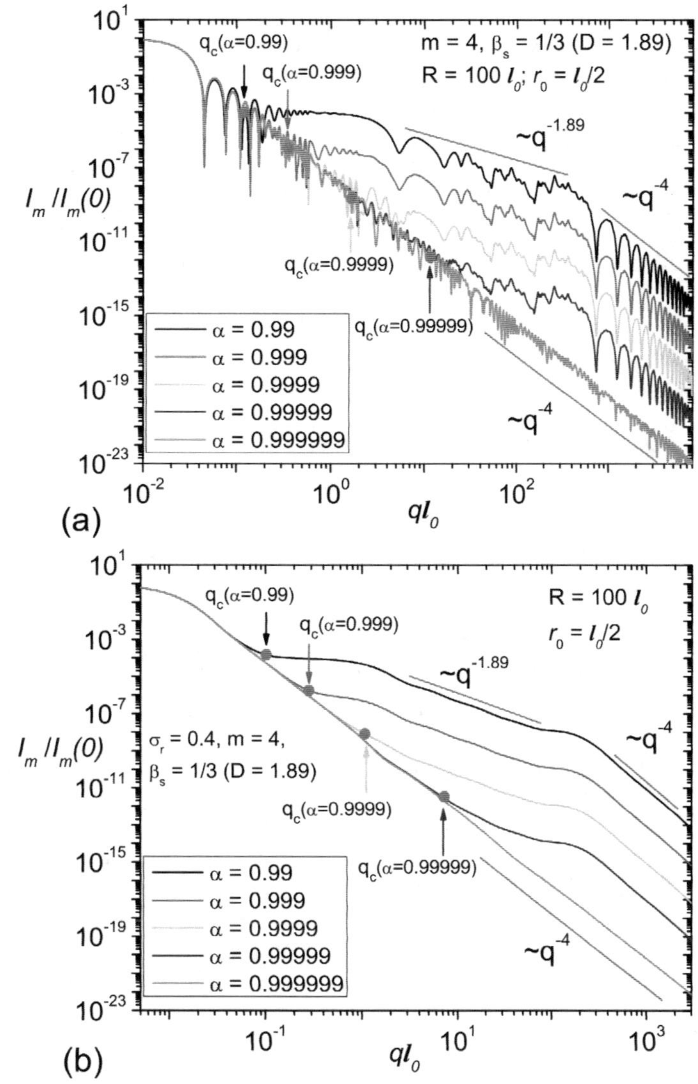

Figure 3. (Color online) The scattering intensities from (a) monodisperse and (b) polydisperse three-phase deterministic model. The model parameters are shown in the figures. The dimensionless parameter σ_{r} is the relative variance of the distribution function [5, 7]. The magenta (light gray) discs represent the crossover's positions (knees) as estimated from Eq. (9).

In a real physical system, scatterers almost always have different sizes. Therefore, a more realistic description should involve size polydispersity. Here we consider an ensemble of the three-phase fractals with different sizes. Specifically, we choose the log-normal distribution of the sizes with the average fractal size l_0 and its relative dispersion σ_{r} (see Refs. [5, 7] for details). Thus, the average in Eq. (7) is taken over the both angles and sizes. Polydispersity obviously smears the intensity curves, and the oscillations disappear at strongly developed polydispersity [5, 7].

4. Results and Discussions

The numerical results for the scattering intensity (7) are represented in Fig. 3 for different values of the contrast parameter α. One can clearly see the appearance of the crossover points (knees on a double logarithmic scale). In order to estimate their positions, let us make simple approximations in the limit of strongly developed polydispersity (see Fig. 3b)

$$\langle |F_{\mathrm{b}}(qR)|^2\rangle \propto \begin{cases} 1, & q \lesssim 1/R, \\ (qR)^{-4}, & q \gtrsim 1/R, \end{cases} \quad \text{and } \langle |F_2^{(m)}(ql_0)|^2\rangle \propto \begin{cases} 1, & q \lesssim 1/l_0, \\ (ql_0)^{-D}, & 1/l_0 \lesssim q \lesssim 1/l_m, \\ (qr_m)^{-4}, & q \gtrsim 1/r_m, \end{cases} \tag{8}$$

where $l_m = \beta_{\mathrm{s}}^m l_0$ and $r_m = \beta_{\mathrm{s}}^m r_0$ are the mean length of cube's edge and the mean radius of the composing balls at the mth iteration. Note that the radius of surrounding sphere R is always bigger than the fractal size l_0.

For a given q, the ball amplitude $F_{\mathrm{b}}(qR)$ and the fractal amplitude $F_2^{(m)}(ql_0)$ in the r.h.s. of Eq. (7) can be of *different orders of magnitude*, provided the *corresponding exponents are not equal*. Then one of the amplitudes dominates. The knee position can be found by equating the corresponding intensities $\alpha^2\langle |F_{\mathrm{b}}(q_{\mathrm{c}}R)|^2\rangle \simeq (1-\alpha)^2\langle |F_2^{(m)}(q_{\mathrm{c}}l_0)|^2\rangle$ and using the estimations (8). This gives us

$$q_{\mathrm{c}} \simeq \begin{cases} \dfrac{1}{R}\sqrt{\dfrac{|\alpha|}{|1-\alpha|}}, & \dfrac{1}{R} \lesssim q_{\mathrm{c}} \lesssim \dfrac{1}{l_0}, \\ \dfrac{1}{l_0}\left(\dfrac{|\alpha|}{|1-\alpha|}\dfrac{l_0^2}{R^2}\right)^{\frac{1}{2-D/2}}, & \dfrac{1}{l_0} \lesssim q_{\mathrm{c}} \lesssim \dfrac{1}{l_m}. \end{cases} \tag{9}$$

If $1/R \lesssim q_{\mathrm{c}} \lesssim 1/l_0$ (say, for $\alpha = 0.999$ in Fig. 3b), one can see all the regions in the intensity curve: the Guinier and Porod regions of the surrounding ball, and the Guinier, fractal, and Porod regions of the fractal. When the contrast $\rho_2 - \rho_1$ between the ball and fractal is suppressed or the fractal volume is small enough, the contrast parameter α gets closer to one. As a consequence, a part of the fractal curve is "absorbed" by the ball curve. For instance, at $\alpha = 0.9999$ one can see the direct transition between the Porod and fractal regions. At sufficiently small values of $|1-\alpha|$, the fractal region can not be observed by the SAS technique at all.

5. Conclusions

We develop a model that describes multi-phase fractal systems, where one phase (GCF) is embedded into another phase (ball) and altogether the combined system is further put into a third phase (matrix). We derive an analytical expression for the SAS intensity from such a system and observe the appearance of various knees between the power-law regimes. The knee positions and the contributions of the different structural levels to the total scattering intensity are controlled by the effective contrast parameter α (6), depending on the relative values of the SLD of each phase and their volumes. The found estimations (9) for the knee positions can be used to answer the question of principle: whether the fractal region is observable at a given contrast by the SAS method or not.

In order to explain the transition between arbitrary scattering exponents, the developed model can be generalized to complex multilevel structures embedded into each other.

References

[1] Feigin L A and Svergun D I, *Structure Analysis by Small-Angle X-Ray and Neutron Scattering* (Plenum Press, 1987, NY).
[2] Mandelbrot B B, *The Fractal Geometry of Nature*, (Freeman, 1982, San Francisco).
[3] Martin J E and Hurd A J, *J. Appl. Cryst.*, **20**, 61 (1987).
[4] Schmidt P W, *J. Appl. Cryst.*, **24**, 414 (1991).
[5] Cherny A Yu, Anitas E M, Osipov V A and Kuklin A I, *Phys. Rev. E*, **84**, 036203 (2011).

[6] Cherny A Yu, Anitas E M, Kuklin A I, Balasoiu M and Osipov V A, *J. Surf. Invest.*, **4**, 903 (2010).
[7] Cherny A Yu, Anitas E M, Kuklin A I, Balasoiu M and Osipov V A, *J. Appl. Cryst.*, **43**, 790 (2010).
[8] Schaefer D W and Justice R S, *Macromolecules*, **40**, 8501 (2007).
[9] Kravchuk K V, Gomza Yu P, Pashkova O V,V'yunov O I, Nesin S D and Belous A G, *J. Non. Crys. Sol.*, **355**, 2557 (2009).
[10] Schneider G J and Goritz D, *J. Chem. Phys.*, **132**, 154903 (2010).
[11] Singh M, Sinha I, Singh A K and Mandal R K, *J. Nanopart. Res.*, **13**, 69 (2011).

Journal of Physics: Conference Series **393** (2012) 012032 doi:10.1088/1742-6596/393/1/012032

Prospects for investigating deterministic fractals: Extracting additional information from small angle scattering data

A Yu Cherny[1], E M Anitas[1,2], V A Osipov[1] and A I Kuklin[1]
[1]Joint Institute for Nuclear Research, Dubna 141980, Russia
[2]Horia Hulubei National Institute of Physics and Nuclear Engineering, RO-077125 Bucharest-Magurele, Romania

E-mail: alexander.yu.cherny<at>gmail.com

Abstract. We examine a general problem of small angle scattering from deterministic fractals: whether one can extract some additional information from the small angle scattering data apart from the fractal dimension and the edges of the fractal region. It is shown that for single-scale mass fractals one can additionally obtain the fractal iteration number, the scaling factor, and the number of structural units composing the fractal, provided the quality of fractal samples is high enough.

Fractals, one of the most beautiful and interesting group of objects, appeared in the scientific literature quite recently [1]. Right after that, the fractals attracted much attention of a broad community of investigators in various fields. Nearly at the same time, the connections were revealed between fractal structures and the intensities of small-angle neutron or X-ray scattering (see, e.g., reviews [2,3] and references therein). In particular, it was shown that within the fractal region of a mass fractal, the SAS intensity decays in the momentum space according to the power law $I(q)\sim 1/q^D$, where D is the fractal dimension. The majority of SAS studies focus on determining the fractal dimension and the edges of the fractal region, within which the object behaves as a fractal. For random (stochastic) fractals, it is rather difficult to extract more information from the SAS data, because the fine structure of spatial correlations of particles, forming the fractal, is usually smeared due to the randomness. Presently, various deterministic fractals can be artificially created due to a rapid progress in nanotechnologies. As was recently shown [4], deterministic fractals, being exact self-similar objects, allow us to obtain more information from the scattering data.

In the papers [5,6], we constructed the generalized 3D Cantor and Vicsek sets (Fig. 1), whose dimensions are controlled by the scaling factor of fractal. The SAS intensities from these deterministic fractals were calculated analytically, and the obtained results exhibit a number of general features, common for ***deterministic mass fractals with a single scale*** [4]. If the fractals, composing a sample, are randomly oriented and placed, then the SAS intensities represent ***minima and maxima superimposed on the power-law decay*** $I(q)\sim 1/q^D$ (the generalized power-law decay). In the reciprocal fractal region, the curve $I(q)q^D$ is approximately ***log-periodic with the period equal to the logarithm of the fractal scaling factor*** (Fig. 2), and this log-periodicity of the scattering curves is a consequence of the self-similarity of the fractal. As was shown [4-6], the minima and maxima amplitudes are

Journal of Physics: Conference Series **393** (2012) 012032 doi:10.1088/1742-6596/393/1/012032

damped with increasing fractal polydispersity, i.e., variance in fractal sizes in the sample. Physical reasons for such a behaviour are quite clear: the fractal dimension dictates the power law $I(q) \sim 1/q^D$ for the intensity only on the average. Polydispersity smears the spatial correlations between the units composing the fractal and, hence, the intensity becomes smoother.

For a mass fractal with a single scale, one can extract a number of parameters from the scattering intensity even in the presence of polydispersity (Fig. 2): *(i)* ***The fractal dimension*** from the generalized power-law decay. *(ii)* ***The fractal scaling parameter*** from the period on the logarithmic scale. *(iii)* ***The number of fractal iteration***, which is equal to the number of periods of function $I(q)q^D$. *(iv)* ***The lower and upper fractal edges*** from this diagram as the beginning and end of the "periodicity region". The edges allow us to estimate the fractal size and the smallest distance between fractal units in real space. *(v)* ***The total number of structural units***, of which the fractal is composed, by the relation $N_m=(1/\beta s)^{m\,D}$, where m is the iteration number.

The analysis of SAS data for deterministic fractals represents an important step in the structural investigations of complex systems. One can expect that the required samples of deterministic fractals of higher quality will be obtained with the help of modern nanotechnologies and investigated by SAS in near future.

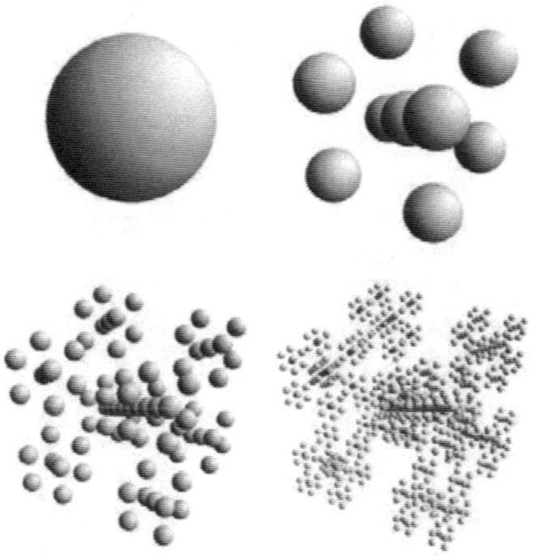

Figure 1. The initiator and first three iterations for the generalized self-similar Vicsek fractal, a deterministic fractal immersed in 3D, at the scaling factor $\beta s = 1/6$.

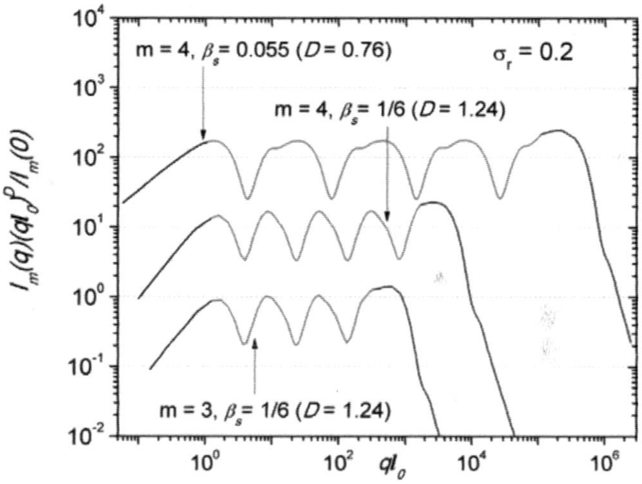

Figure 2. The log-periodicity of the scaled SAS intensity $I(q)q^D$ in the fractal region of the generalized Vicsek sets. The period in the log-scale is equal to $\log_{10}(1/\beta s)$. Here l_0 and □$_r$ are the average fractal size and its relative variance, respectively, and m is the iteration number.

References

[1] Mandelbrot B B 1982 *The Fractal Geometry of Nature* (Freeman: San Francisco)
[2] Teixeira J 1988 *J. Appl. Cryst.* **21** 781
[3] Schmidt P W 1991 *J. Appl. Cryst.* **24** 414
[4] Cherny A Yu, Anitas E M, Osipov V A and Kuklin A I 2011 *Phys Rev* E **84** 036203
[5] Cherny A Yu, Anitas E M, Kuklin A I, Balasoiu M and Osipov V A 2010 *J. Surf. Invest.* **4** 903
[6] Cherny A Yu, Anitas E M, Kuklin A I, Balasoiu M and Osipov V A 2010 *J. Appl. Cryst.* **43** 790

Dubna-Nano 2012 IOP Publishing
Journal of Physics: Conference Series **393** (2012) 012033 doi:10.1088/1742-6596/393/1/012033

Vibron Self–trapped States in Biological Macromolecules: Comparison of Different Theoretical Approaches

D. Čevizović[1,2], S. Galović[1,2], A. Reshetnyak[3], and Z. Ivić[1]

[1] University of Belgrade, "Vinča" Institute of Nuclear Sciences, Laboratory for theoretical and condensed matter physics, Belgrade, Serbia
[2] Joint Institute for Nuclear Research, Bogoliubov Laboratory for Theoretical Physics, Dubna, Russia
[3] Laboratory of Non–Linear Media Physics, Institute of Strength Physics and Material Science SB RAS, Tomsk 634021, Russia

E-mail: reshet@ispms.tsc.ru

Abstract. We present a study of the applicability of the variational treatments based on using of the modified Lang–Firsov unitary transformation (MLF method) in the investigation of the vibron self–trapped states in biological macromolecular chains. Here we compare the values of the ground state energy predicted by MLF methods with the values of the ground state energy predicted by the standard small–polaron theory, for various values of the basic energy parameters of the system. We obtain regions in system parameter space where MLF approach gives better description of the vibron states.

1. Introduction

There exist many unresolved fundamental problems in molecular biology and biochemistry and one from them is the problem of long–distance energy transport in biological macromolecules. It is believed that the energy released by the hydrolysis of adenosine triphosphate (ATP) appears by an universal energy source allowing many biological processes. Mentioned energy released by the ATP hydrolysis may be captured by the protein molecules and excite the high–frequency amide–I vibration of a peptide group. Due to the dipole–dipole coupling between the neighbouring peptide groups the excited amide–I vibration is delocalized causing formation of the vibron state which, in turn, can move along the polypeptide chain. However, it is yet not clearly understood how this energy can be transported along macromolecule at long distances, without being dissipated or dispersed. An earlier explanation of this process is based on the soliton theory [1], [2], [3]. According to those models, a soliton arises due to the self–trapping (ST) of the amide–I quanta.

According to the general theory of ST phenomena [4], ST problem exhibits two asymptotic solutions depending on the values of three basic energy parameters of the system: the vibron bandwidth $2|J|$, the phonon characteristic frequency ω_C and the SP bending energy E_b (which is proportional to the strength of the vibron–phonon coupling). Thus, in the adiabatic limit ($2|J| > \hbar\omega_C$) lattice distortion with large inertia can not follow the exciton motion and it forms an essentially static potential well where that particle may be trapped. Depending on the value of the mutual ratio of E_b and $\hbar\omega_C$, the spatial extent of an entity created in this way (polaron) may vary from being concentrated around one site only ($E_b >> \hbar\omega_C$) i.e. adiabatic SP (ASP), while in the case when $E_b < \hbar\omega_C$ it is spread over the large number of lattice sites and we have adiabatic large polaron (ALP) or soliton. By contrast, in non–adiabatic limit ($2|J| < \hbar\omega_C$) the quantum nature of the phonons plays a crucial role. In such case a vibron is dressed by a virtual cloud of phonons which yields a lattice distortion essentially located on a single site and instantaneously follows the vibron motion. The vibron dressed by the virtual phonon cloud forms a small polaron (SP) whose properties are well described by performing the so–called Lang–Firsov (LF) unitary transformation [5]. As a consequence, conventional SP theories are applicable in the strong–coupling, non–adiabatic limit.

Unfortunately, the values of the basic energy parameters in biological macromolecular chains belong to non–adiabatic limit. For that reason, it has been suggested by some authors [6],[7], [8] that vibron self–trapping in hydrogen–bonded macromolecular chains might result in the formation of a small–polaron, rather than a soliton. From the other side, the values of the energy parameters indicate that the strength of vibron–phonon coupling in hydrogen–bonded macromolecules falls in the weak to intermediate limits [9]. In addition, some recent numerical [10] studies of the ST phenomena in hydrogen–bonded macromolecular chains indicate that its proper theoretical description requires an approach that goes beyond the conventional strong–coupling SP theories. Namely, according to results from [10] it seems that depending on the temperature and the values of the system parameters the abrupt transition from partially dressed (light and mobile) to self–trapped (practically immobile quasi particle) may occur. This situation can not be described in the framework of the standard SP approach (based on using of the Lang–Firsov unitary transformation) and as a consequence its theoretical investigations require the means which involve the concept of partial dressing. This concept is based on modified Lang–Firsov (MLF) unitary transformation and is applicable in a wide part of system parameter space. Such an approach is considered in a close correspondence with the supplementary variational treatment of the problem by means of the Toyozawa ansatz [11]. In our recent paper [12] we used slightly more flexible method in order to investigate the temperature dependence of the SP states character.

In this paper we consider two variational formalisms based on using of the modified Lang–Firsov approach and compare obtained results with ones predicted by the standard SP theory. The variational parameter(s) introduced here characterizes the degree in which the vibron distorts the lattice and the lattice feedback on the vibron, i.e. vibron dressing. The first variational approach (so called "f_q–approach") is based on the model used in [12], while the second one ("δ–approach") exploits the method introduced by Brown and Ivić in [7], [8] in the form used in [13]. The last approach intermediates between non-adiabatic and adiabatic limit with use of only one variational parameter. We calculate and compare the system ground state energies for various basic system parameter values in the case when vibron interacts with both optical and acoustical phonon subsystems. Our attention was restricted to the single–vibron case.

2. Theoretical methods

The system under consideration consists of single vibron excited on n–th structural element of the macromolecule which, in turn, interacts with thermal oscillations of the chain through the linear short–ranged deformation potential interaction. We suppose that vibron excitation can move along the chain from n–th to its nearest–neighbouring structural element. The corresponding Hamiltonian can be written in the following form [14],

$$H = \Delta \sum_n a_n^\dagger a_n - \sum_n J a_n^\dagger (a_{n+1} + a_{n-1}) + \sum_q \hbar\omega_q b_q^\dagger b_q + \frac{1}{\sqrt{N}} \sum_q \sum_n F_q e^{iqnR_0} a_n^\dagger a_n (b_q + b_{-q}^\dagger), \quad (1)$$

where Δ is the vibron excitation energy, $a_n^\dagger(a_n)$ describes the presence (absence) of the vibron quanta on the structural element, which is positioned on n–th lattice site, $b_q^\dagger(b_q)$ creates (annihilates) phonon quanta, and ω_q is the phonon frequency. The inter–site overlap integral J characterize the vibron transfer between neighbouring structural elements in the chain. Finally, $F_q = F_{-q}^*$ is the vibron–phonon coupling parameter which governs the character of ST states. In order to investigate the vibron self–trapping in macromolecular chains, we shall consider the following quasi particle–phonon interaction:

(a) interaction with dispersionless optical phonon modes [14],[15], [16]

$$F_q = \chi\sqrt{\frac{\hbar}{2M\omega_q}},$$

where χ is the vibron–phonon coupling constant, which accounts the influence of the external motion of the vibron on the $(n \pm 1)$–th structural elements and on the modulation of the vibron frequency in the n–th structural element, $\omega_q = \omega_C$ is phonon dispersion law, $\omega_C = 2\sqrt{\kappa/M}$, κ appears by the stiffness of the chain, M is the molecular group mass;

(b) interaction with acoustic phonon modes through the short–ranged deformation potential (acoustic deformation potential, ADP) [15], [16]

$$F_q = 2i\chi\sqrt{\frac{\hbar}{2M\omega_q}}\sin qR_0,$$

with $\omega_q = \omega_C \sin|qR_0/2|$ being by phonon dispersion law, and R_0 is a lattice constant.

In order to examine the degree of the vibron dressing, we pass to the small–polaron picture using unitary transformation operator,

$$U = \exp\left\{-\frac{1}{\sqrt{N}}\sum_q\sum_n f_q e^{-iqnR_0} a_n^\dagger a_n (b_{-q} - b_q^\dagger)\right\}.$$

Here $f_q = f_{-q}^*$ are variational parameters, while operators $a_n^\dagger(a_n)$ represent (partially) dressed vibrons. In the case of the passing to the standard small–polaron picture, parameter f_q has standard Lang–Firsov form: $f_q = \frac{F_q^*}{\hbar\omega_q}$. Operators $a_n^\dagger(a_n)$ that appear in the transformed Hamiltonian, represent the (fully) dressed quasi particle. In all cases the operators $b_q^\dagger(b_q)$ describe the new phonons in the lattice with shifted equilibrium positions of molecular groups.

Partially dressed vibron states represent the dynamically stable eigen-states of the system provided that these variational parameters correspond to the minimum energy state. Thus, we search for their optimized values minimizing the system ground–state energy. According to the method described in [17], optimization procedure will be performed in a mean–field manner, which gives better upper bound of the system free energy and consequently the better upper bound of the system ground state energy. In particular, we define (in wave number representation) an effective, mean–field Hamiltonian $\tilde{H}_0$ in the following way,

$$\tilde{H} = \tilde{H}_0 + \tilde{H}_{int}, \tag{2}$$

where $\tilde{H}_0 = \tilde{H}_v + \tilde{H}_{ph}$, $\tilde{H}_v = \left\langle \tilde{H} - \tilde{H}_{ph}\right\rangle_{ph}$ and $\tilde{H}_{int} = \tilde{H} - \tilde{H}_{ph} - \left\langle \tilde{H} - \tilde{H}_{ph}\right\rangle_{ph}$. The symbol $\langle\,\rangle_{ph}$ denotes the average over the new–phonon ensemble, which is considered in the thermal equilibrium state at the temperature T. In such a way we derive,

$$\tilde{H}_0 = \sum_k E_{SP}(k) a_k^\dagger a_k + \sum_q \hbar\omega_q b_q^\dagger b_q, \tag{3}$$

where $a_k = \frac{1}{\sqrt{N}}\sum_n e^{iknR_0} a_n$, and

$$E_{SP}(k) = \Delta - \frac{1}{N}\sum_q \left\{(f_q + f^*_{-q})F_q - \hbar\omega_q |f_q|^2\right\} - 2Je^{-W(T)}\cos(kR_0) \tag{4}$$

denotes the variational energy of the small–polaron band states, while $W(T) = \frac{1}{N}\sum_q |f_q|^2 (2\bar{\nu}_q + 1)(1 - \cos(qR_0))$ denotes the vibron–band narrowing factor, which characterizes the degree of the reduction of the corresponding overlap integral or equivalently the enhancement of the polaron effective mass. For obvious reasons, they are sometimes called the "'dressing fractions'" or the "'dressing parameters'". At last, the quantity $\bar{\nu}_q$, $\bar{\nu}_q = 1/(e^{\hbar\omega_q/k_BT} - 1)$, denotes the phonon average number.

The polaron–phonon interaction term may be neglected provided that dressed vibrons represent sufficiently well eigen states of system. In such a way it cannot affect SP equilibrium properties substantially and at $T = 0$ K the further procedure simply reduces to minimization of the SP ground state energy. We note the ground–state vector of the effective Hamiltonian $\tilde{H}_0$ reads, $|\psi_{GS}\rangle = a^\dagger_k |0\rangle_{vib} \otimes \prod_q |0\rangle_q$, where $\prod_q |0\rangle_q$ is the phonon vacuum vector so that the ground–state energy corresponds to the lowest level of SP energy, i.e.,$E_{GS} = \left\langle \psi_{GS}\middle|\tilde{H}_0\middle|\psi_{GS}\right\rangle = E_{SP}(k)$. Thus, from the requirement, $\left.\frac{\partial E_{SP}(k)}{\partial f_q}\right|_{k_0} = 0$, we get,

$$f_q = \frac{F^*_q}{\hbar\omega_q + 4|J|e^{-W}(2\bar{\nu}_q + 1)\sin^2\left(\frac{qR_0}{2}\right)}. \tag{5}$$

From the Eq.(5) it follows that for each value of q there is the corresponding value of the variational parameter f_q. For that reason, the optimization procedure for this approach becomes extremely difficult problem, even numerically. Note, however, that only $k_0 = \pi/R_0$ (when $J < 0$), or $k_0 = 0$ (when $J > 0$), should be considered since it will correspond to the polaron ground state only. This case is also the most relevant for the optical spectroscopy since vibron bandwidths in these media are very narrow and its dispersion in all practical examination of optical spectra can be neglected [10], [15], [16] [18], [19].

The expressions for vibron ground state energy (normalized on the characteristic phonon energy $\hbar\omega_C$ and measured from the energy level Δ) and vibron dressing factor are obtained by the substitution of the variational parameters into (4). Performing the summation over the phonon quasi momenta q by virtue of the rule $\frac{1}{N}\sum_q \cdots \to \frac{R_0}{2\pi}\int_{-\pi/R_0}^{\pi/R_0} \cdots dq$ and by introducing two dimensionless parameters (that determine the system parameter space) $S = \frac{E_b}{\hbar\omega_C}$ and $B = \frac{2|J|}{\hbar\omega_C}$, we obtain,

$$\mathcal{E}_{GS} = -W - 3BWe^{-W} - Be^{-W}, \tag{6}$$

and

$$W = \frac{S}{(1 + 2Be^{-W})^{3/2}} \tag{7}$$

in the case when vibron interact with optical phonon mode, and

$$\mathcal{E}_{GS} = -\frac{W}{2}\frac{I_E}{I_W} - Be^{-W}, \qquad W = 8SI_W, \tag{8}$$

when the vibron interacts with acoustic phonon modes. The quantities I_E, I_W that appear in the above relations have the integral representation,

$$I_E = \frac{1}{\pi}\int_0^{\pi/2} \cos^2 x \frac{1 + 4Be^{-W}\sin x}{(1 + 2Be^{-W}\sin x)^2}dx, \text{ and } I_W = \frac{1}{\pi}\int_0^{\pi/2} \frac{\sin x\cos^2 x}{(1 + 2Be^{-W}\sin x)^2}dx.$$

In the case of simple "δ–approach" we choose f_q in the form: $f_q = \frac{\delta F_q^*}{\hbar\omega_q}$, where $0 < \delta < 1$ is variational parameter. In that case, we have more simple variational technique in which we should determine only one variational parameter (δ). On the other hand, this assumption implies the equal dressing for all phonon modes. At the first sight, it looks like a very strong assumption since the whole set of variational parameters (one for each mode) is substituted by a single one. However, here we showed that predictions obtained by this approach are very close with ones, obtained by more rigorous "f_q approach". Repeating above described procedure we obtained condition which determines parameter δ,

$$4J\delta W_{LF} \mathrm{e}^{-\delta^2 W_{LF}} - 2E_b(1-\delta) = 0, \tag{9}$$

where $E_b = \frac{1}{N}\sum_q \frac{|F_q|^2}{\hbar\omega_q}$ is small–polaron binding energy, while $W_{LF} = \frac{1}{N}\sum_q \frac{|F_q|^2}{(\hbar\omega_q)^2}(1-\cos qR_0)$ is SP band narrowing factor calculated in the non–adiabatic, strong coupling limit (standard LF approach). Then, performing summation over phonon quasimomenta, we derive the relation for the ground state energy (in (S, B) parameter space) in the form,

$$\mathcal{E}_{GS} = -\delta(2-\delta)S - B\mathrm{e}^{-\delta^2 S} \tag{10}$$

in the case when vibron interact with optical phonon, and

$$\mathcal{E}_{GS} = -\delta(2-\delta)S - B\mathrm{e}^{-\frac{8\delta^2 S}{3\pi}}, \tag{11}$$

when vibron interact with acoustical phonons.

3. Results

According to the available literature data, the hopping constant in α–helix proteins is typically equal to, $J = 7.8$ cm^{-1} (in the case of the hydrogen bonds between peptide units) and $J = -12.4$ cm^{-1} (between different spines of hydrogen–bonded peptide units, i.e. in the case of the covalent bonds between peptide unit) [19, 20]. The mass of the peptide unit ranges between $M = 2\cdot 10^{-25}$ kg and $M = 5.7\cdot 10^{-25}$ kg [9, 19].

The above mentioned values of the system parameters in macromolecular chains indicate that adiabatic parameter in these structures ranges between $B = 0.01$ and $B = 0.5$. In the same time, coupling constant ranges between $S = 0.01$ and $S = 0.3$. The last fact indicates that these structures belong to the non-adiabatic coupling limit from weak to intermediate one.

Obtained numerical results are presented on the Figs. 1–4. The calculations obtained by "δ–approach" are presented by full lines, the ones obtained by the "f_q–approach" are presented by dashed lines, and the calculations obtained by the standard LF approach are presented by dotted lines. As one can remark, the results obtained by both variational approaches are strongly different compared with ones obtained by standard LF approach, for small values of coupling constant S. This difference disappears on large values of S for all values of B. For small and intermediate values of S, the difference between two variational approaches increases with increasing of B. These results are expectable, since standard LF approach is not applicable in weak coupling limit. On the other side, results obtained by both variational approaches are very similar in the whole range of S, for small values of B (non-adiabatic limit). There are significant difference only for $B > 1$ (adiabatic limit), for small and intermediate values of S.

One should be noted in addition, it is evident that both variational approaches predict a multivalued dependence of $\mathcal{E}_{GS}$ and W on S, in certain region of S, only in the adiabatic limit. According to the variational principle, the only lowest energy value is physically meaningful. Consequently, only lower values of $\mathcal{E}_{GS}$ (and its corresponding values of W) have physical significance and truly variational energy is presented by one line which is not smooth (at certain value of S it have the point of cusp). In the same time, W undergoes an abrupt change. That means that vibron ST state have abrupt transition from weakly dressed to heavy dressed polaron state. Obtained results may look as a consequence of the applied variational approach. However, it seems that these results are in good agreement with some recent numerical investigations [10], where authors interpreted similar results as a coexistence of two types of SP states instead of abrupt transition from one to another quasiparticle type.

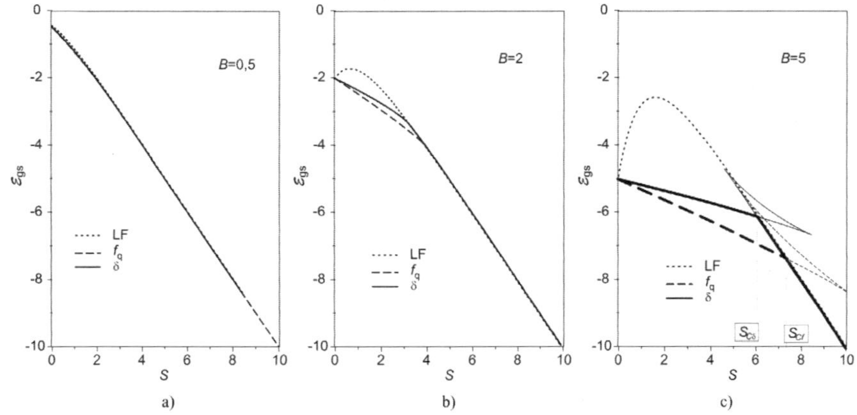

Figure 1. The dependence of the $\mathcal{E}_{GS}$ versus S, for different values of the B. The case of vibron that interacting with optical phonon modes.

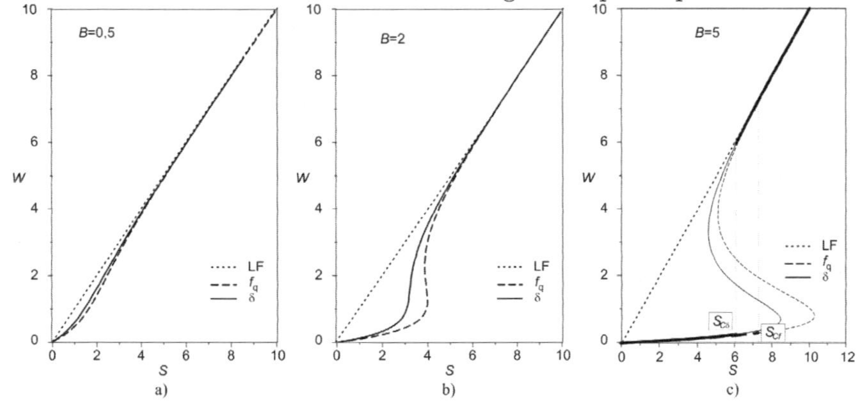

Figure 2. The dependence of the W on the S, for different values of the B. The case of vibron that interaction with optical phonon modes.

4. Conclusions

In accordance to the obtained results, one may note that both variational approaches give very similar qualitative and quantitative predictions, especially in non-adiabatic region. With the rise of the adiabatic parameter, the difference between variational approaches, as well as between variational approaches and standard small–polaron approach vanishes. In the case of the vibron interacting with the optical phonon modes, ground state energies predicted by the "f_q–approach" are something lower than ground state energies predicted by the "δ–approach". But, in the non–adiabatic region of the parameter space this difference may be practically neglected, and "δ–approach" is quite good method for the investigation of the vibron self–trapping states. Similar results are obtained in the case when vibron interacts with acoustic phonon modes.

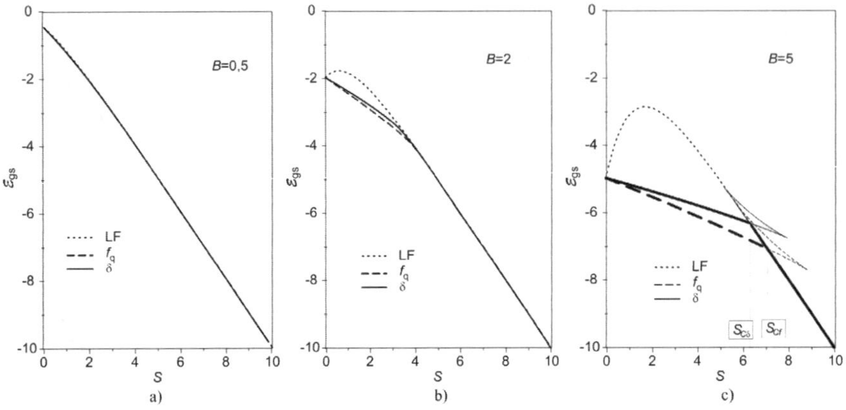

Figure 3. The dependence of the $\mathcal{E}_{GS}$ versus S, for different values of the B. The case of vibron that interacting with acoustic phonon modes.

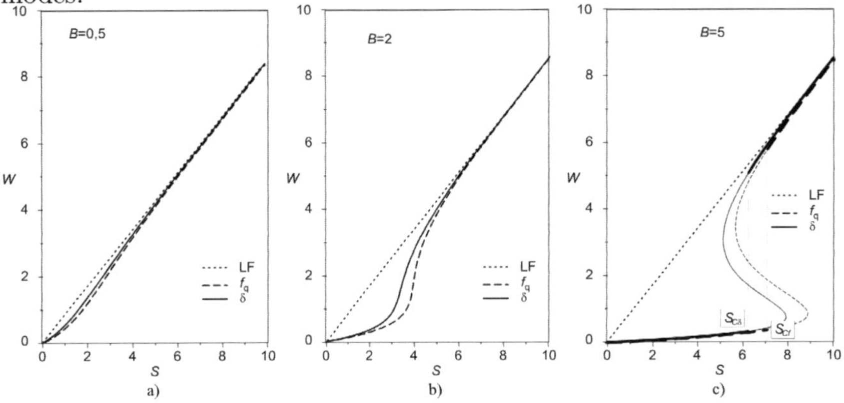

Figure 4. The dependence of the W on the S, for different values of the B. The case of vibron that interaction with acoustic phonon modes.

In the adiabatic region of the system parameter space the difference between two variational approaches is significant.

Due to the fact that practically all biological macromolecular structures belongs to the non-adiabatic, weak or intermediate coupling limit, it seems that simple "δ–approach" may be by the hopeful method for investigations of the self–trapped vibron states in such structures.

Acknowledgments
The work was supported by the Ministry of Education and Science of Republic of Serbia under contract numbers III–45005 and III–45010, and by the project within the Cooperation Agreement between the JINR, Dubna, Russian Federation and Ministry of Education and Science of Republic of Serbia.

References

[1] Davydov A S 1991 *Solitons in Molecular Systems* (Dordrecht: Kluwer Academic Publishers)
[2] Davydov A S 1981 The role of solitons in the energy and electron transfer in one -dimensional molecular systems *Physica* D **3** 1
[3] Davydov A S 1982 *Biology and Quantum Mechanics* (New York: Pergamon Press)

[4] Rashba E I 1982 *Excitons* ed E I Rashba and M D Struge (Amsterdam: North Holland) p 543
[5] Lang I J and Firsov Yu A 1962 Sov. Phys. JETP **16** 1301
[6] Alexander D M and Krumhansl J A 1986 Localized excitations in hydrogen–bonded molecular crystals *Phys. Rev.* B **33** 7172
[7] Brown D W and Ivić Z 1989 Unification of polaron and soliton theories of exciton transport *Phys. Rev.* B **40** 9876
[8] Ivić Z, Kapor D, Skrinjar M and Popović Z 1993 Self–trapping in quasi–one–dimensional electron– and exciton–phonon systems *Phys. Rev.* B **48** 3721
[9] Pouthier V 2010 Vibron phonon in a lattice of H-bonded peptide units: A criterion to discriminate between the weak and the strong coupling limit *J. Chem. Phys.* **132** 035106
[10] Hamm P and Tsironis G 2008 Barrier crossing to the small Holstein polaron regime *Phys. Rev.* B **78** 092301
[11] Toyozawa Y 1961 Self–Trapping of an Electron by the Acoustical Mode of Lattice Vibration I *Progr. Theor. Phys.* **26** 29
[12] Čevizović D, Galović S and Ivić Z 2011 Nature of the vibron self–trapped states in hydrogen–bonded macromolecular chains *Phys. Rev.* E **84** 011920
[13] Ivić Z 1998 The role of solitons in charge and energy transfer in 1D molecular chains *Phys.* D **113** 218
[14] Holstein T 1959 Studies of Polaron Motion *Ann. Phys.* **8** 325
[15] Davydov A S 1976 *Theory of solids (Teoriya tverdogo tela),* (Moscow: Nauka, on Russian)
[16] Mahan G D 1990 *Many-Particle Physics* (New York:Plenum Press)
[17] Yarkony D and Silbey R 1976 Comments on exciton phonon coupling: Temperature dependence *J. of Chem. Phys.* **65** 1042
[18] Emin D 1973 On the existence of free and self–trapped carriers in insulators: an abrupt temperature–dependent conductivity transition *Advan. In. Phys.* **22** 57
[19] Scott A C 1992 Davydov's soliton *Phys. Rep.* **217** 1
[20] Scott A C 1982 Dynamics of Davydov solitons *Phys. Rev.* A **26** 578

Dubna-Nano 2012 IOP Publishing
Journal of Physics: Conference Series **393** (2012) 012034 doi:10.1088/1742-6596/393/1/012034

Momentum and spin dynamics of Dirac particles at effective dimensional reduction

Alexander J. Silenko
Research Institute of Nuclear Problems, Belarusian State University, Minsk 220030, Belarus
E-mail: silenko@inp.bsu.by

Oleg V. Teryaev
Bogoliubov Laboratory of Theoretical Physics, Joint Institute for Nuclear Research, Dubna 141980, Russia
E-mail: teryaev@theor.jinr.ru

Abstract. We consider the dynamics of Dirac particles moving in the curved spaces of variable dimension interpolating smoothly between 3- and 2-dimensional spaces and considered as a toy model for 2-dimensional structures in solid state physics. Performing the Foldy-Wouthuysen (FW) transformation of Dirac equation and passing to the classical limit, we derive the equations of motion of momentum and spin. The spin precesses with the variable angular velocity and may "flick" appearing in the remnant 2-dimensional space twice during the period.

1. Introduction

Modern solid state physics intensively investigates (almost) two-dimensional structures: graphene, fullerene, carbon nanotubes, carbon nanobuds, etc. Properties of particles placed into such structures are usually described by considering quantum theory at two dimensions. However, there is no doubt that real space remains 3-dimensional which may lead to qualitative difference in some observables. This especially concerns the particle (electron) spin properties which are crucially different at two and three dimensions (see, e.g., Ref. [1]). To investigate this problem, we consider the toy model [2] of the curved space of *variable* dimension smoothly changing from two and three. We use the Dirac equation for a correct description of spin-1/2 particle motion in the curved space rather than for taking into account relativistic effects (which are negligible in condensed matter physics) although the latter are included in our approach "for free" allowing its further applications to the processes at Large Hadron Collider in the case [3] of variable (momentum) space dimension. We use the earlier developed method of the FW transformation to obtain the equations of motion for particle momentum and *spin*. In the present work, we focus our attention just on spin properties.

2. Hamiltonian for the metric admitting the effective dimensional reduction

We consider an example of an effective (because we actually remain in the 3-dimensional space) dimensional reduction for a Dirac particle in a curved spacetime using the metric [2]:

$$ds^2 = dt^2 - \rho_1^2(z)d\Phi_1^2 - \rho_2^2(z)d\Phi_2^2 - \rho_3^2(z)dz^2, \tag{1}$$

where $\rho_3(z)^2 = 1 + (\rho_1'(z))^2 + (\rho_2'(z))^2$. Hereinafter, primes denote derivatives with respect to z. The (3+1)-dimensional manifold defining this metric is a hypersurface in a flat pseudo-Euclidean (5+1)-dimensional space.

The transverse part of the metric (if z is assumed to be a longitudinal coordinate) has a structure of the Clifford torus which is the product of two unit circles in the 4-dimensional Euclidean space:

$$y_1^2 + y_2^2 = y_3^2 + y_4^2 = 1. \tag{2}$$

We consider Clifford tori as a toy model of condensed media. Note that they are used for analyzing twisted materials [4] and vesicles [5, 6, 7]. There is also some qualitative similarity to projection of a tube in a 6-dimensional space onto a 3-dimensional space which was used for construction of theory of quasicrystals [8].

Taking the limit $\rho_1(z) \to 0$ or the limit $\rho_2(z) \to 0$ may lead to the reduction [9] of dimension of the physical space from $d = 3$ to $d = 2$.

We apply the Dirac equation for curvilinear coordinates:

$$(i\hbar\gamma^a D_a - mc)\Psi = 0, \qquad a = 0, 1, 2, 3, \tag{3}$$

where the Dirac matrices γ^a are defined in local Lorentz frame defined by tetrad e_a^μ. The spinor covariant derivatives are given by

$$D_a = e_a^\mu D_\mu, \qquad D_\mu = \partial_\mu + \frac{i}{4}\sigma^{ab}\Gamma_{\mu\, ab}. \tag{4}$$

Here $\Gamma_\mu{}^{ab} = -\Gamma_\mu{}^{ba}$ are the Lorentz connection coefficients, Ψ is the four-component wave function, and $\sigma^{ab} = \frac{i}{2}\left(\gamma^a\gamma^b - \gamma^b\gamma^a\right)$.

To find the Hamiltonian form of Dirac equation, one can use the general formulas obtained in Refs. [10, 11] or substitute the given metric into the general equation for the Hermitian Dirac Hamiltonian (Eq. (2.21) in Ref. [12]). The Hermitian Dirac Hamiltonian for given metric (1) was first derived in Ref. [11] and can be presented in the form

$$\mathcal{H}_D = \beta m - \frac{i}{\rho_1}\alpha_1\frac{\partial}{\partial\Phi_1} - \frac{i}{\rho_2}\alpha_2\frac{\partial}{\partial\Phi_2} - \frac{i}{2}\alpha_3\left\{\frac{1}{\rho_3}, \frac{\partial}{\partial z}\right\}, \tag{5}$$

where $\{\ldots, \ldots\}$ is an anticommutator.

We transform it to the FW representation by the method elaborated in Ref. [13] which was earlier applied to non-Minkovskian spacetimes in Refs. [12, 14, 15]. After the exact FW transformation, we get the result

$$\mathcal{H}_{FW} = \beta\sqrt{a + \hbar\mathbf{\Sigma}\cdot\boldsymbol{b}}, \tag{6}$$

where

$$a = m^2 + \frac{p_1^2}{\rho_1^2} + \frac{p_2^2}{\rho_2^2} + \frac{1}{4}\left\{\frac{1}{\rho_3}, p_3\right\}^2, \quad \boldsymbol{b} = \frac{\rho_2'}{\rho_2^2\rho_3}p_2\boldsymbol{e}_1 - \frac{\rho_1'}{\rho_1^2\rho_3}p_1\boldsymbol{e}_2, \tag{7}$$

and $(p_1, p_2, p_3) = \left(-i\hbar\frac{\partial}{\partial\Phi_1}, -i\hbar\frac{\partial}{\partial\Phi_2}, -i\hbar\frac{\partial}{\partial z}\right)$ is the generalized momentum operator. For the given time-independent metric, the operators $\mathcal{H}_{FW}, p_1$, and p_2 are integrals of motion.

If terms of order of $\hbar^2$ and higher orders in $\hbar$ are neglected, Hamiltonian (6) takes the form

$$\mathcal{H}_{FW} = \beta\sqrt{a} + \frac{\hbar\mathbf{\Pi}}{4}\cdot\left\{\boldsymbol{b}, \frac{1}{\sqrt{a}}\right\}, \tag{8}$$

where $\mathbf{\Pi} = \beta\mathbf{\Sigma}$ is the polarization operator.

It can be shown that the classical limit of this Hamiltonian coincides with the purely classical Hamiltonian of a spinning particle obtained from general Eq. (3.17) in Ref. [12] for the given metric.

3. Dynamics of momentum and spin

The effect of the spin on the momentum dynamics is defined by the second term in Eq. (8) which usually is much smaller than the first one. In the classical limit, neglecting this effect leads to the following equation of longitudinal motion of the particle:

$$v_z \equiv \frac{dz}{dt} = \frac{\mathcal{P}_3}{E\rho_3^2} = \pm\frac{\sqrt{E^2 - m^2 - \frac{\mathcal{P}_1^2}{\rho_1^2(z)} - \frac{\mathcal{P}_2^2}{\rho_2^2(z)}}}{E\rho_3(z)}, \tag{9}$$

where E is the energy which is conserved. Different signs correspond to two different directions of the longitudinal particle motion.

When we consider the nontrivial case of $\mathcal{P}_1 \neq 0,\ \mathcal{P}_2 \neq 0$, there exists the z_f point in which $v_z = 0$. It can be shown that the particle acceleration at this point is nonzero, except for a rather specific relation between $\rho_1(z)$ and $\rho_2(z)$. Therefore, the z_f point is a turning point. After reaching this point, the particle moves back.

In the classical limit, the equation of (average) spin motion obtained with Hamiltonian (8) is given by

$$\frac{d\boldsymbol{s}}{dt} = \boldsymbol{\Omega} \times \boldsymbol{s}, \tag{10}$$

where

$$\boldsymbol{\Omega} = \left(\frac{\rho_2'(z)}{\rho_2^2(z)\rho_3(z)E}\mathcal{P}_2,\ -\frac{\rho_1'(z)}{\rho_1^2(z)\rho_3(z)E}\mathcal{P}_1,\ 0\right). \tag{11}$$

Hereinafter, $\boldsymbol{\Omega}$ is the angular velocity of spin rotation, $\boldsymbol{s}$ is the unit spin vector, and $(\mathcal{P}_1, \mathcal{P}_2, \mathcal{P}_3)$ is the classical generalized momentum ($\mathcal{P}_1 = const,\ \mathcal{P}_2 = const$). Dynamics of spin depends only on the particle motion along the longitudinal axis, z.

Because $\frac{d\boldsymbol{s}}{dt} = v_z(z)\frac{d\boldsymbol{s}}{dz}$, Eqs. (10),(11) define a solvable system of first order homogeneous linear differential equations.

As a simplest example, we consider the case when $\rho_1(z) = K/z, \rho_2(z) = const$. In this case the angular velocity

$$\boldsymbol{\Omega} = \left(0,\ \frac{\mathcal{P}_1}{KE\sqrt{1 + K^2/z^4}},\ 0\right) \tag{12}$$

is almost constant if $z \gg \sqrt{K}$. As a result, the spin components change as

$$s_1(t) = s_\perp \cos\left(\frac{\mathcal{P}_1}{KE}t + \phi\right),\quad s_2(t) = s_\parallel,\quad s_3(t) = s_\perp \sin\left(\frac{\mathcal{P}_1}{KE}t + \phi\right),\quad (s_\parallel)^2 + (s_\perp)^2 = 1, \tag{13}$$

where $s_\perp$ and $s_\parallel$ are the spin projections onto the $\boldsymbol{e}_2$ axis and $\boldsymbol{e}_1\boldsymbol{e}_3$ subspace.

When the particle spin crosses the above plane, it becomes observable in the (2+1)-dimensional spacetime. In this spacetime, spin "flickering" takes place so that the imaginable observer residing in the respective plane meets for a short time the pseudovector (violating therefore the parity invariance) whose origin is completely unexplainable in terms of 2-dimensional space.

4. Summary

We considered the Dirac fermion motion in the model curved space of variable dimension. Applying FW transformation to the Dirac equation we concentrated on the spin dynamics in the (semi)classical limit. In the simplest case under consideration the average spin precesses with the almost constant velocity. It spends most of the time out of the 2-dimensional space remaining after the effective dimensional reduction and appears in this space twice during each precession period, to the great surprise of the observers residing in this space.

We hope that the development of this model may provide the way to take into account the spin degrees of freedom of electrons confined to 2-dimensional surfaces in condensed matter physics.

The spin dynamics in spaces of variable dimension may be also important for the description of polarization effects at Large Hadron Collider in the presence of extra dimensions.

5. Acknowledgments

We are indebted to P. Fiziev, V.P. Neznamov, Yu. N. Obukhov and D.V. Shirkov for stimulating discussions. This work was supported in part by the RFBR (grants 11-02-01538 and 11-02-01454) and BRFFR grant Φ10D-001.

References

[1] Yip P 1983 *J. Math. Phys.* **24** 1206 (1983)
[2] Fiziev P P 2010 Riemannian $(\mathbf{1}+\mathbf{d})$-Dim Space-Time Manifolds with Nonstandard Topology which Admit Dimensional Reduction to Any Lower Dimension and Transformation of the Klein-Gordon Equation to the **1**-Dim Schrödinger Like Equation *Preprint* math-ph/1012.3520
[3] Anchordoqui L A, Dai D C, Goldberg H, Landsberg G, Shaughnessy G, Stojkovic D and Weiler T J 2011 *Phys. Rev. D* **83** 114046
[4] Sadoc J F and Charvolin J 2009 *J. Phys. A: Math. Theor.* **42** 465209
[5] Ou-Yang Z-C 1993 *Phys. Rev.* A **41** 4517; 1993 *Phys. Rev.* E **47** 747
[6] Hu J-G and Ou-Yang Z-C 1993 *Phys. Rev.* E **47** 461
[7] Fourcade B, Mutz M and Bensimon D 1992 *Phys. Rev. Lett.* **68** 2551
[8] Kalugin P A, Kitaev A Yu, Levitov L S 1985 *JETP Lett.* **41** 145
[9] Fiziev P P and Shirkov D V 2011 *Theor. Math. Phys.* **167**, 680
[10] Gorbatenko M V and Neznamov V P 2011 *Phys. Rev.* D **83**, 105002
[11] Gorbatenko M V and Neznamov V P 2011 A Modified Method for Deriving Self-Conjugate Dirac Hamiltonians in Arbitrary Gravitational Fields and Its Application to Centrally and Axially Symmetric Gravitational Fields *Preprint* gr-qc/1107.0844
[12] Obukhov Yu N, Silenko A J, and Teryaev O V 2011 *Phys. Rev.* D **84** 024025
[13] Silenko A J 2003 *J. Math. Phys.* **44** 2952
[14] Silenko A J, and Teryaev O V 2005 *Phys. Rev.* D **71** 064016; 2007 *Phys. Rev.* D **76** 061101(R)
[15] Obukhov Yu N, Silenko A J, and Teryaev O V 2009 *Phys. Rev.* D **80** 064044

Dubna-Nano 2012 IOP Publishing
Journal of Physics: Conference Series **393** (2012) 012035 doi:10.1088/1742-6596/393/1/012035

Analytical approximation of the nanoscale dose distribution in an irradiated medium with an embedded nanoparticle

V Chernov[1*], G Chernov[2] and M Barboza-Flores[1]

[1]Departamento de Investigación en Física de la Universidad de Sonora, A.P. 5-088, Hermosillo, Sonora, 83190 México

[2]Departamento de Física de la Universidad de Sonora, Apartado Postal 5-088, Hermosillo, Sonora, 83190 México

[*]chernov@cifus.uson.mx

Abstract. In this work we propose an analytical approach describing the dose distribution around a NP embedded in a medium. The approach describes the following sequence of events: The homogenous and isotropic creation of secondary electrons under incident photon fluence; travel of the created electrons toward the NP surface and their escaping from the NP with different energies and angles; deposition of energy in surrounding medium. The radial dose distribution around the NP was found as the average energy deposited by the escaped electrons in a spherical shell at a distance r from the NP center normalized to its mass. The continuous slowing down approximation and the assumption that created electrons travel in a straight-line path were used. As result, a set of analytical expressions describing the dose distribution was derived. The expressions were applied to the calculation of the dose distribution around spherical gold NPs of different size embedded in water. It was shown that the dose distribution is close to the $1/r^2$ dependence and practically independent of the NP radius.

1. Introduction

Many different types of nanoparticles are currently being studied for applications in nanomedicine. In particular, the use of gold nanoparticles (NPs) in radiotherapy may provide a high dose to the tumor while delivering a reduced dose to the surrounding normal tissue [1]. Numerous theoretical studies have investigated dose localization through the introduction of NPs. Most works have focused on the macroscopic dose enhancement averaging effects over volumes much larger than a single NP. This approach is fundamentally flawed because it ignores the significant dose inhomogeneity in the vicinity of embedded NPs. An alternative view on NP radiosensitisation was presented in [2], where Monte Carlo (MC) simulations were carried out to calculate dose in the nanoscale vicinity of individual NPs. Despite the fact that MC methods provide precise numerical results, it is desirable to have an analytical solution that, while unavoidably simplifying the physical model, would provide valuable insight into the nature of the studied problem and gives a possibility to alter parameters and conditions without having to rerun costly MC simulations.
The goal of this paper is to derive analytical expressions, alternative to Monte Carlo simulations, describing the dose distribution around NPs embedded in a medium.

2. Basic equations

In this part we present basic equations describing the energy deposition of secondary electrons created in an irradiated NP. We do not consider here the primary absorption of photon radiation by NPs and will start from the creation of a secondary electron in the NP volume. Also, we will only consider the additional absorbed dose delivered by electrons emitted by a NP under photon irradiation. The contribution of the secondary particles produced in the surrounding medium to dose is more or less the same in a vicinity of NP and can be found by standard methods for the same medium without NPs.
An incident photon interacts with the NP and creates, at a certain point of the NP, a fast electron of kinetic energy E_0 that travels toward the NP surface. During travel, the electron losses energy and slows down. If E_0 is sufficiently high, electron crosses the NP surface, continues to travel in the medium and eventually stops. In our calculations, we will use the continuous slowing down approximation (CSDA). In this approximation, the rate of energy loss is assumed to be equal to the total stopping power and energy-loss fluctuations are neglected [3]. We will also assume that emitted electrons travel in a straight-line path.
Geometry assumed for estimation of energy deposition by electrons created in NP is shown in figure 1. A spherical NP of radius a_0 is centered at the origin O of a spherical coordinate system. An electron is created at a point A_0 at distance l from the NP surface with the initial energy E_0. At a point A_1 the electron crosses the NP surface and travels at the angle α relative to the normal of the NP surface at the point A_1, continuing to travel in the surrounding medium. At a point A_2 the electron reaches a sphere of radius $_r$ centered on the center of the NP.

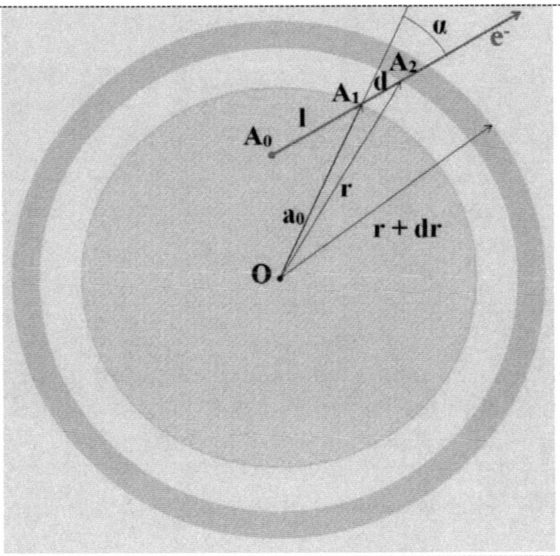

Figure 1. Geometry assumed for the estimation of the energy deposition by electrons created in NP.

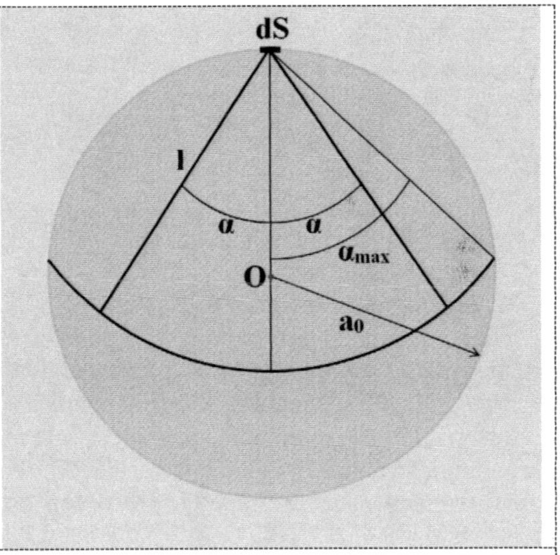

Figure 2. Geometric model for the calculation of $\psi(l,\alpha)$.

The radial dose distribution, $D(r)$, around the NP is defined as the average energy deposited in a spherical shell with the radii between r and r + dr, normalized to its mass:

$$D(r) = -\frac{dE_{av}(r)}{4\pi r^2 \rho_m dr} \qquad (1)$$

where ρ_m is the surrounding medium density, $E_{av}(r)$ is the average energy of escaped electrons deposited at the distance r from the NP center. The minus sign comes from the fact that $E_{av}(r)$ always decreases when r increases.

In the case of CSDA, the rest electron energy at the NP surface E_1 and $E_2(r)$ is the electron energy at distance r from the NP center are determined by its initial energy E_0 and distances l and d that an electron passes in the NP and in the medium (the lengths A_0A_1 and A_1A_2 in figure 1), respectively. The energies can be found from the equations:

$$R_p(E_1) = R_p(E_0) - l,\ R_m[E_2(r)] = R_m(E_1) - d \tag{2}$$

where the functions $R_p(E)$ and $R_m(E)$ are the electron range - energy relationships for NP and medium materials that relate a residual range and an electron energy. The distance d is equal to

$$d = \sqrt{r^2 - a_0^2 \sin^2\alpha} - a_0 \cos\alpha \tag{3}$$

where α is the angle at which an electron escapes the NP (relative to the normal of the NP surface). Inversion the first part of (2) with respect to E_1 and the second one with respect to E_2 gives

$$E_1 = R_p^{-1}[R_p(E_0) - l],\ E_2(r) = R_m^{-1}\left[R_m(E_1) - \sqrt{r^2 - a_0^2 \sin^2\alpha} + a_0 \cos\alpha\right] \tag{4}$$

where the inverse functions $R_p^{-1}(R)$ and $R_m^{-1}(R)$ are the energies that has electron with given range R. As follows from (4), the energy E_2 depends on both l and α, therefore a two-dimensional distribution function $\psi(l,\alpha)$ should be used to find $E_{av}(r)$:

$$E_{av}(r) = \iint_{E_2(r)>0} \psi(l,\alpha) E_2(r)\, dl\, d\alpha \tag{5}$$

where $\psi(l,\alpha)dld\alpha$ is the probability that the created electron has passed the distance between l and l + dl in the NP and left the NP at an angle between α and α + dα relative to the normal of the NP surface at the point of escaping. The integral (5) is taken over those l and α, for which $E_2(r) > 0$.
The set of equations (1) – (5) allows to find the radial dose distribution around a NP per one created electron with energy E_0. If photon irradiation creates electrons of various energies, the equation (1) or (5) should be averaged over a spectrum of the created electrons.

3. The distribution function $\psi(l,\alpha)$

In CSDA the residual electron energy is a single-valued function of l, therefore the distribution function $\psi(l,\alpha)$ represents, in fact, the energy and angle distribution of the electrons escaping from an irradiated NP. The function $\psi(l,\alpha)$ depends on the distribution of the created electrons over the NP volume and their directions of travel. In the case of the homogenous and isotropic creation of secondary electrons, $\psi(l,\alpha)$ can be found from a geometrical consideration, similar to that used in the MIRD formalism to estimate the absorbed dose at the cellular level from intracellularly localized radionuclides [4].
To find $\psi(l,\alpha)$ it is convenient to introduce the cumulative distribution function (CDF) $F(l,\alpha) = \Pr(L \le l, \mathrm{A} \le \alpha)$ representing the probability that the random distance L and random angle A take on a value less than or equal to l and α, respectively. Figure 2 shows an area dS situated at a surface point of a NP. The bold arc indicates the NP points, from which the distance to dS is equal to l. The bold radii indicate a given angle α to the normal, at which the electron leaves the NP.
The CDF $F(l,\alpha)$ corresponds to the fraction of the NP that lies inside the spherical cone limited by the bold lines in figure 2:

$$F(l,\alpha) = \frac{V_{con}}{V_{Np}} = \frac{(1-\cos\alpha)l^3}{2a_0^3} \tag{6}$$

where V_{con} is the volume of the spherical cone of radius l and angle α, V_{Np} is the NP volume.
The distribution function $d\psi(l,\alpha)$ for the area dS can be written as

$$d\psi(l,\alpha) = \frac{\partial F(l,\alpha)}{\partial l \partial \alpha} \frac{dS \cos\alpha}{4\pi l^2} = \frac{3\sin\alpha l^2}{2a_0^3} \frac{dS\cos\alpha}{4\pi l^2} \tag{7}$$

The second fraction in (7) presents the solid angle at the direction α, on which the surface element dS is seen from the point of the electron creation.Integration over the NP surface gives $\psi(l,\alpha)$ for the sphere. In the case of homogenous and isotropic creation of a secondary electron all the points on the NP surface are equivalent, therefore to integrate over the surface we should only replace dS with the NP area:

$$\psi(l,\alpha) dl d\alpha = \frac{3\sin\alpha\cos\alpha}{2a_0} dl d\alpha \tag{8}$$

The length l is changed from 0 to the maximal value $2a_0$. The angle α is changed from 0 to the maximal value α_{max}, which depends on l and is determined by the condition (see figure 2)

$$0 \le l \le 2a_0, 0 \le \alpha \le \alpha_{\max} = \arccos(l/2a_0) \tag{9}$$

4. The linear relationship between energy and range of a secondary electron

To apply the equations (1 - 5) to any specific NP embedded in a medium we should select the approximation formulas describing the electron range - energy relationships for NP and medium material. In this work we will use the simplest supposition that the residual range of electron is proportional to its energy:

$$R_p(E) = \frac{E}{E_0} R_{p0}, R_m(E) = \frac{E}{E_0} R_{m0} \tag{10}$$

where $R_{po} = R_p(E_0)$ and $R_{mo} = R_m(E_0)$ are the ranges of an electron with initial energy E_0 in the NP and medium materials, respectively. It is easy to show that (4) and (10) lead to:

$$E_2(r) = E_0\left(1 - l/R_{p0} - d/R_{m0}\right) \tag{11}$$

where the distance d is given by (3). It is convenient to introduce dimensionless variables $x = l/a_0$, $y = r/a_0$, $f_m = a_0/R_{m0}$, $f_p = a_0/R_{p0}$. With these variables (11) can be rewritten as

$$E_2(r)/E_0 = 1 - f_p x + f_m \mu - f_m \sqrt{y^2 - 1 + \mu^2} \tag{12}$$

where $\mu = cos\alpha$. Substitution (12) into (5) together with (7) gives

$$\frac{E_{av}(r)}{E_0} = \frac{3}{2} \int_0^{x_{\max}} \int_{\mu_{\min}}^{1} \mu\left(1 - f_p x + f_m \mu - f_m \sqrt{y^2 - 1 + \mu^2}\right) d\mu dx \tag{13}$$

where x_{max} and μ_{min} are the integration limits that should be found from the conditions $E_2(r) > 0$ and (9). The straightforward integration of (13) is possible, for instance, with Maple computer algebra system, but final expression are too large to be presented here.

5. The dose distribution around a gold NP in water

As an example of using the derived equations we present here the calculation of the dose distribution around a gold NP in water. This system was selected because in the recent years, application of gold NP in diagnosis and treatment of cancer has been the issue of extensive research [5]. Comparison of the mass energy-transfer coefficients of gold and water shows that the highest achievable enhancement factor with gold NPs could be realized with the photon energy between 10 and 100 keV. The photoelectric effect is the predominant process for photons with these energies. The kinetic energy of created photoelectron can be found by subtracting the K-edge electron binding energy of gold (80.7 keV) from the energy of the incident photon.

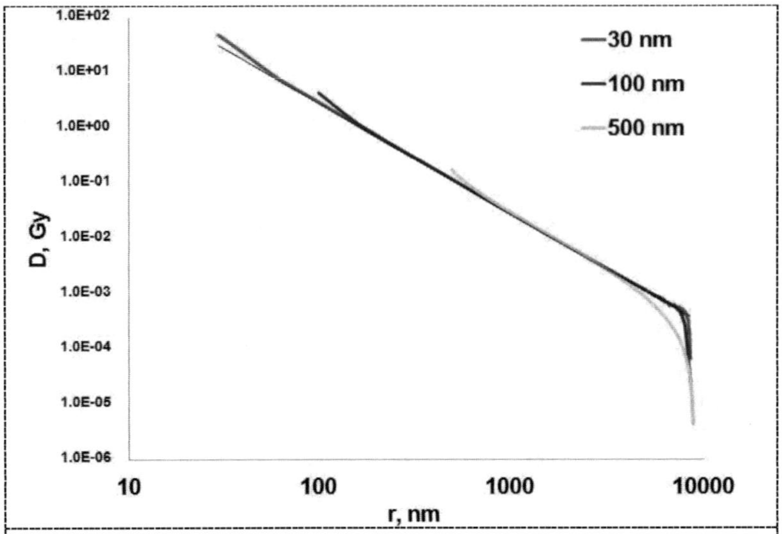

Figure 3. The dose distribution in water around embedded gold NPs. The bold lines corresponds to the additional dose delivered by one 19.7 keV electron created in the NPs of indicated radii. The thin line is the power fit with $D = 2.9\times10^4/r^2$ Gy, where r is the distance from the NP centre in nm.

Figure 3 presents the dose distributions in water around gold NPs irradiated with 100 keV photons per one created electron. These photons create secondary electrons with initial kinetic energy of 19.3 keV having the ranges of about 1200 and 8600 nm in gold and water, respectively [3]. The calculations were carried out with equations (13) and (1) for the NP radii of 30, 100 and 500 nm. All the distributions are similar and closed to the C/r^2 dependence with the constant C independent of the NP radius (thin line in figure 3). The effect of the NP size is important at the distances closed to the NP surface and at the large distances.

6. Conclusion

The analytical expressions describing the dose distribution around a NP embedded in a medium have been derived. From this simple model it was inferred that the dose distribution is close to the 1/r2 dependence and practically independent of the NP radius. The proposed approach is an alternative to Monte Carlo simulations and can be used for fast and inexpensive estimation of the dose enhancement factor in a tumor filled with NPs. In future work we plan to apply the derived expressions to a complex multi-NP geometry.

References

[1] Mesbahi A 2010 Reports of practical oncology and radiotherapy **15** 176
[2] McMahon S J et al. 2011 Scientific Reports **1** 18
[3] National Institute of Standards and Technology 2012 *Stopping-power and range tables for electrons*, available from: http://physics.nist.gov/PhysRefData/Star/Text/ESTAR.html
[4] Goddu S M, Howell R W and Rao D V 1994 J. Nucl. Med. 35 303
[5] Lechtman E, Chattopadhyay N, Cai Z, Mashouf S, Reilly R and Pignol J P 2011 Phys. Med. Biol. **56** 4631

AUTHOR INDEX

Institute of Physics
Dirac House, Temple Back
Bristol BS1 6BE UK

CURRAN ASSOCIATES INC
proceedings.com